AUTOMATIC CONTROL SYSTEMS

BASIC ANALYSIS AND DESIGN

AUTOMATIC CONTROL SYSTEMS

SYSTEMS

BASIC ANALYSIS AND DESIGN

William A. Wolovich, Ph.D.
Brown University

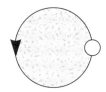

SAUNDERS COLLEGE PUBLISHING
Harcourt Brace College Publishers

Fort Worth • Philadelphia • San Diego • New York • Orlando • Austin

San Antonio • Toronto • Montreal • London • Sydney • Tokyo

Text Typeface: Times Roman
Compositor: ETP Services
Acquisitions Editor: Emily Barrosse
Managing Editor: Carol Field
Project Editor: Martha Brown, Sarah Fitz-Hugh
Copy Editor: Charlotte Nelson
Manager of Art and Design: Carol Bleistine
Art Director: Anne Muldrow, Caroline McGowan
Text Designer: Nanci Kappel
Cover Designer: Lawrence R. Didona
Text Artwork: Grafacon
Director of EDP: Tim Frelick
Production Manager: Joanne Cassetti
Marketing Manager: Marjorie Waldron

Printed in the United States of America

ISBN 0-03-023773-4

Library of Congress Catalog Card Number: 93-086060

3456 042 987654321

This book was printed on paper made from waste paper, containing 10% post-consumer waste and 40% pre-consumer waste, measured as a percentage of total fiber weight content.

To Laura

PREFACE

This text is intended for use in an introductory undergraduate course on automatic control system analysis and design. The student is assumed to have had a course on linear systems or circuits, where Laplace transforms are used to solve ordinary, linear, differential equations. Some of this material is reviewed in Appendix B. Knowledge of matrix algebra is useful, but not necessary, as the material in Appendix A can be introduced as required.

As an introductory text in automatic control systems, this book focuses on the design of linear controllers for single-input/single-output (SISO) systems, the relatively large class of systems that have a *single* reference *input* and a *single* measured *output*. Generally speaking, an effective SISO control system must simultaneously attain several design goals that usually conflict with one another. Therefore, flexible design procedures are necessary that allow relatively simple trade-offs among these conflicting goals. In particular, a controlled system generally is expected to robustly respond to a command input, while continuously rejecting undesirable features that disrupt its ability to do so. These undesirable features can include "reasonable" plant parameter variations and uncertainty, noisy measurements, external disturbances, and limited signal ranges that, if exceeded, can cause unpredictable nonlinear performance.

ORGANIZATION AND SPECIAL FEATURES

1. The text has three distinct parts, with all of the relevant control topics placed in a logical pedagogical sequence, beginning with equivalent system descriptions (Part I), followed by performance goals and tests (Part II), concluding with an updated discussion of classical design methods and an inclusive presentation of modern design methodology, which introduces a number of innovative techniques for achieving a variety of design goals simultaneously, independent of any particular system description (Part III).

The first part, **Dynamic System Representations,** concentrates on the *analysis* of control systems, and outlines the primary ways that dynamic system performance is described, as well as various procedures for transferring from one "equivalent" representation to another. A major point made is that transfer functions, Bode diagrams, differential equations, and state-space representations all present essentially the same information relative to the dynamic behavior of a controllable and observable system, so that it doesn't matter which description one begins with.

Because the first part of the text covers only system representations and not controller design, students avoid making any restrictive associations between specific system descriptions and particular types of compensation. Too often, certain forms of compensation have been introduced along with specific system descriptions, for example lag/lead compensation with Bode diagrams, or linear state feedback with state-space models. In this text, students are encouraged to use any compensator with any system description, such as a state feedback design for a system described by a transfer function, or a PID compensator for a system described in state-space form.

The second major part of the text, **Performance Goals and Tests,** represents a transitional one that both illustrates analysis and motivates the need for design. It explains when and why control systems are required, as well as the fundamental distinction between "loop goals" and "response goals," an important point that is often overlooked or understated. Classical tests for nominal closed-loop stability, such as the Routh-Hurwitz criterion, the root locus, and the Nyquist stability criterion, are introduced as methods for evaluating system performance before their use in design. The various *loop performance goals* that an appropriate controller should achieve, such as robust stability, sensitivity reduction, and disturbance and noise attenuation are then discussed. The loop gain and the sensitivity function are shown to represent important measures of loop performance, which generalize the classical concepts of gain margin and phase margin. Some relatively new $\mathbf{H_\infty}$ results are employed to quantify the presentation.

The simultaneous requirement of "good output tracking," as specified by the *response performance goals* of a small steady-state error, a fast output transient response with minimal overshoot, and a non-saturating plant input, is then covered. One of the main points made is that there are many different reasons why system performance may be unsatisfactory, and that one must consider all possible loop and response performance goals when designing an effective control system. A *two degree-of-freedom* (2 DOF) procedure is introduced that enables a designer to maintain the same (desired) response performance while varying the loop performance of the system.

Parts I and II provide the background material required for the most important third part of the text, **Compensation.** By the time they begin Part III, students will have learned when and why control is necessary and what performance goals require improvement. The design techniques presented in Part III increase in complexity depending on the amount of improvement desired and the control difficulty. For example, a stable, minimum phase system is shown to be far easier to control than an unstable system with nonminimum phase zeros. A number of new and somewhat innovative 2 DOF design techniques are introduced, such as *low loop gain* (LLG) and *loop transfer recovery* (LTR), for simultaneously obtaining desired loop and response performance goals in the more difficult to control cases.

2. The text thoroughly integrates classical and modern methods.

Most classical techniques have a modern "interpretation," and vice versa, and control system designers should have a good understanding of these interrelationships in order to employ the best possible controller, regardless of the way the system initially is defined. Each part of the text integrates relevant classical and modern methods with this overall goal in mind. Part I, for example, shows that the fundamental concepts of controllability and observability transcend state-space representations, and illustrates how any particular system representation can be transformed to any other equivalent representation rather directly. Part II employs the internal model principle to generalize the classical notion of system *type* using the modern explanation of a system lacking complete state observability. In Part III, the student not only learns that classical PID and lag/lead compensators are closely related, but also sees why they both represent a "restrictive form" of modern linear state feedback compensation. Also, observer state feedback is shown to directly imply the general 2 DOF compensator introduced in Part II, and virtually all other linear compensators are shown to be equivalent to some form (possibly restrictive) of 2 DOF compensation.

3. The approach taken separates loop goals from response goals, thereby providing significant motivation for the 2 DOF designs that are presented. A major distinguishing feature of this text is that it provides students the opportunity to attain desired loop and response goals somewhat independently through the particular controller designs that are employed.

These 2 DOF designs, which represent the most general possible linear compensators, are shown to imply much better closed-loop performance in many cases than the more conventional, but restrictive, 1 DOF designs. This text teaches students how to design 2 DOF controllers to obtain robust closed-loop stability with respect to both parameter uncertainty and unmodeled dynamics, disturbance elimination, noise attenuation, and robust zero error tracking. In summary, *the text emphasizes the design and employment of the most general possible type of linear controllers for SISO systems to simultaneously achieve a variety of desired response and loop performance goals, irrespective of any particular system description.*

USE OF THE COMPUTER

Many of the homework problems presented at the end of each chapter require that the student verify and extend particular text examples through the use of appropriate computer-aided design packages, such as the Control System Toolbox on MATLAB. Many of the numerical values associated with these examples were obtained using MATLAB. Also, an asterisk (*) placed next to a problem number indicates that a computer-aided design package, such as MATLAB, would be particularly useful in minimizing the numerical computations required to solve the problem. Students are encouraged throughout the text to complement their analytical investigations by using such routines, or their equivalent. Several other computer-aided control system design packages also can be employed, such as MATRIX$_x$ from Integrated Systems, and CTRL-C, a product of Systems Control Technology.

USE OF THIS TEXTBOOK

This text contains more than enough material to support a one-semester (14-week) introductory course. The sections and subsections marked in the table of contents with a $\diamond$ contain material that enhances the development, but is not as fundamental as that presented in the remainder of the text. This material can be covered in a less rigorous manner in class, assigned to the students to read outside of the classroom, or omitted entirely, at the discretion of the instructor. More emphasis then can be placed on the control-specific analysis and design methods presented in Parts II and III.

ACKNOWLEDGMENTS

Many individuals have contributed to this text. I wish to thank the many reviewers from across the country for their useful comments and suggestions. These include Professors Roy Colby (North Carolina State University), Robert Egbert (Wichita State University), Frederick M. Ham (Florida Institute of Technology), Walter Higgins (Arizona State University), Y. P. Kakad (University of North Carolina at Charlotte), Richard W. Longman (Columbia University), Charles P. Newman (Carnegie Mellon University), Zvi Roth (Florida Atlantic University), Thordur Runolfsson (Johns Hopkins University), Michael Sain (University of Notre Dame), and Baxter Womack (University of Texas at Austin). A special thanks goes to Professors Steve Morse (Yale University) and Allan Pearson (Brown University) who offered many constructive suggestions during the earlier stages of text preparation, as well as to my student George Mihalacoupoulos, who carefully read several versions of the manuscript. Also, it is hoped that the numerous comments made by my students in Engineering 166 (*Automatic Control Systems*) have made this text more readable than it otherwise would have been.

I am grateful to all of the staff at Saunders College Publishing who contributed their time and energies to this effort. Special recognition is given to Emily Barrosse, Martha Brown, Sarah Fitz-Hugh, Laura Shur, Sara Tenney, and Monica Wilson. Finally, much needed moral support and encouragement were provided on numerous occasions by my wife, Laura, to whom this text is gratefully dedicated.

William A. Wolovich
Providence, Rhode Island
September 1993

CONTENTS

xiii

INTRODUCTION AND HISTORICAL PERSPECTIVE

1.1 INTRODUCTION

Automatic control systems abound in our modern technologic societies, from simple, switch-controlled thermostats to highly advanced autopilots for supersonic aircraft. Other control systems enable elevators to move precisely from one floor to another in a smooth manner; cruise controllers maintain the desired speed of an automobile despite abrupt terrain changes; and multiple links of computer-controlled robots rotate in a coordinated manner in response to a variety of preprogrammed tasks, such as welding, spray painting, and parts assembly. Overall, automatically controlled systems play a very important role in today's world, continuously regulating many of the motors, machines, vehicles, and devices required to increase productivity and maintain our quality of life. However, human controllers are still employed in many situations, especially when safety is involved, although they sometimes act in supervisory and monitoring roles with "override" capabilities, as in a commercial aircraft or a nuclear power plant.

Recently, significant advances in both control theory and the means to implement it with modern high-speed computers have made it possible to automatically control not only systems that had previously been manually controlled but also more advanced systems that could not be controlled satisfactorily by humans alone, such as the Apollo lunar vehicle. In the future, an increased understanding of the dynamics of biologic, economic,

1

and political systems will imply an increased ability to effectively control these systems as well.

Controllable Processes

Although control systems are truly interdisciplinary, they do share certain common characteristics that identify them. In particular, a primary function of virtually all automatic control systems is to regulate the behavior of one or more variables, or **outputs**, in a dynamic process or **plant** in some desired manner, for example, the speed of a motor or a motorized vehicle, its heading, its position, or its altitude or depth; the temperature of a room or that of an entire building; the level or flow rate of a liquid in a chemical process; the number and the relative positions of vehicles on a highway or aircraft in a confined airspace.

The process or plant to be controlled is generally assumed to have one or more **inputs** that cause its behavior to change. Therefore, a cause-effect or an **input-output relationship**, as depicted by the **block diagram** of Figure 1.1, is characteristic of virtually all controllable processes. Typical plant inputs include fuel flow to an engine, control surface deflections on an aircraft, and voltage changes in an electric motor.

Mathematical models are usually employed in control system analysis and design, as we will illustrate, and these models are often imprecise and subject to **parameter variations** that mimic reality. For example, the actual internal parameters that define the dynamic behavior of the plant, such as the weight of the passengers in an elevator or the altitude of an aircraft, will often vary. Furthermore, unpredictable disturbances, such as wind gusts, as well as temperature and terrain changes, can also affect the controlled variables. All of these factors complicate control system analysis and design.

In general, the desired behavior of the variable(s) is either constant, such as the speed of an automobile, or time varying, such as a trajectory that the end-effector of a robot must follow to spray paint some surface. In both cases, however, the desired behavior is usually represented as a **reference input** signal, often computer generated, that the plant output is expected to follow.

Feedback

In Figure 1.2, the reference input signal is directly "processed" by the **controller** in order to produce the plant input. However, this form of

● FIGURE 1.1
A Controllable Process

• **FIGURE 1.2**
An Open-Loop Control System

open-loop control is effective only in relatively simple situations, in which plant parameter variations and disturbances do not cause the actual output to deviate significantly from that specified by the reference input. Driving blindfolded is an example of unacceptable open-loop (manual) control, with only the past memory of road variations available to make steering adjustments. On the other hand, an automatic washing machine, with preset wash times, represents an open-loop controlled system that does function effectively.

The most identifiable characteristic associated with almost all controlled systems is *feedback* from the controlled plant output to the reference input. Without feedback there is no means of comparing the actual behavior of the plant or process with its desired behavior to automatically correct or *control* its performance. **Feedback control** can be used to effectively counteract the potentially debilitating effects of plant parameter variations and disturbances.

The presence of a feedback signal implies the need for a physical measurement; hence a **sensor**, such as a potentiometer, a thermometer, or an altimeter, is needed to continuously monitor the output variable. In many cases, the actual measurements obtained from such physical devices are corrupted by noise, which also complicates the design process. Figure 1.3 depicts the block diagram of a "typical" feedback or **closed-loop control system**. Such systems will be depicted in blue throughout the text. In general, the various signals displayed can be vector (multiple) quantities, which would be the case in a multivariable or multi-input/multi-output (MIMO) system, such as a multiple link robot.

The plant depicted in Figure 1.3 might be a large, motor-driven antenna, whose actual output is its angular pointing position. Environmental factors, such as temperature changes, can alter some of the internal parameters that characterize its behavior, while wind gust disturbances can momentarily affect its pointing accuracy. A sensor, which employs a potentiometer, might be used to obtain a measurement of the actual pointing position, which then would be fed back for comparison with the desired reference-input position in the controller. The controller subsequently would produce a corrective controller output/plant input voltage signal to the motor to continuously "drive" the antenna to the desired position, which may be constant or time varying.

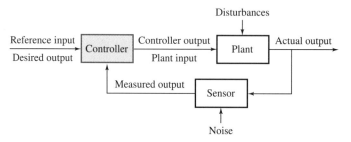

● **FIGURE 1.3**
A Closed-Loop Control System

The design of effective control systems can be rather involved, especially when the plant is difficult to control or there are numerous performance goals that must be obtained. A thorough understanding of the factors that make control difficult, as well as the various performance characteristics that might be required of a controlled system, should be obtained in the simpler cases before more complex systems are considered. Therefore, the main purpose of this textbook is to present a variety of techniques that can be used to design appropriate controllers, primarily in the single-input/single-output (SISO) case, namely, when there is a single controlled output and, as a consequence, a single reference input. In light of Figure 1.3, an appropriate SISO controller therefore would be a dynamic system with two inputs, namely, the reference input and the measured (often noisy) plant output, and a single output, which would correspond to the actual input to the plant.

In many cases, the difference, or **error**, between the reference input and the measured output is used as a single input to the controller, since a minimization of such an error signal often is a primary function of a controller. As we will show, however, greater design flexibility can be obtained when this is not the case, and both of the controller inputs are treated independently.

1.2 AN HISTORICAL PERSPECTIVE

Mayr [48] has written an interesting historical account of feedback control systems. As he notes, the documented use of feedback to automatically control a physical process dates back to a float valve–regulated water clock used in ancient Greece over 2000 years ago. Float valve regulators reappeared in the 18th century, seemingly independent of their ancient heritage, as a means of water-level control in house reservoirs and steam boilers. Ivan Polzunov, an engineer at a coal mine in Barnaul, Altai (Siberia), from 1763 to 1766, constructed a boiler-level regulator that employed a float-actuated control valve, similar to that depicted in Figure 1.4. During

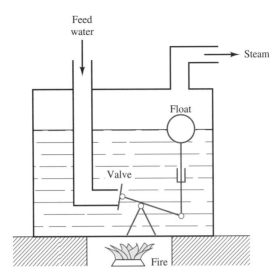

- **FIGURE 1.4**
  Ivan Polzunov's Boiler Level Regulator

the same period of time, float valve regulators also were being used in England to control the level in water tanks, such as those associated with flush toilets.

Perhaps the most important early use of feedback control was to control the speed of a rotating shaft, as found in wind-powered millstones and steam engines. Numerous diverse inventions, based on the concept of feedback regulation, can be attributed to the British millwrights of the 18th century. Although they were an enterprising lot, their designs failed to generate much general interest beyond their own applications. As noted by Mayr,

> Both the virtues and the shortcomings of the millwrights are reflected in their regulating devices. A new idea was grasped with enthusiasm and imagination, but it was not always cultivated to the stage of maturity. It was only in another field, the steam engine, that the idea of feedback control became historically effective.

James Watt's centrifugal governor for controlling the speed of steam engines, which was introduced in 1788, was the first feedback device to attract the attention of the entire engineering community and to eventually become internationally recognized and applauded. Figures 1.5 and 1.6 depict Watt's earlier steam engine and his centrifugal governor, whose main purpose was to maintain the rotational speed of the engine at a desired, preset value despite changes in the steam pressure and the load applied to the engine. The centrifugal governor performs this control

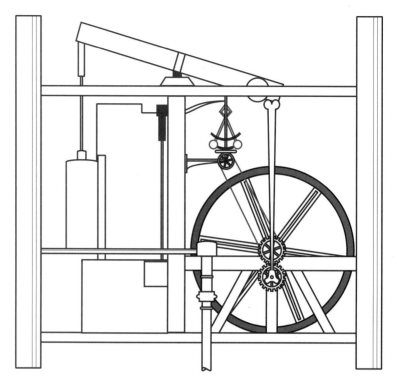

● **FIGURE 1.5**
Watt's Steam Engine with the Centrifugal Governor

task by measuring the actual engine speed by means of its own rotational motion, which is coupled to that of the engine. The centrifugal force generated by its two flyweights continuously varies the position of a sleeve that slides along a shaft coincident with its rotational axis. Subsequent changes in the sleeve position then alter a valve that automatically changes the flow of steam to the engine to maintain the desired speed.

Watt adopted his centrifugal governor from similar devices that were employed earlier by millwrights. Like the millwrights before him, Watt was concerned primarily with solving a specific practical problem, and he probably did not consider or appreciate the generality of the feedback principle on which his design was based. He undoubtedly was completely unaware of the mathematical methods that would eventually be employed to both explain and design feedback control systems. However, the application of a centrifugal governor to control the speed of a steam engine was an innovative breakthrough that solved an important problem of that time, and James Watt well deserved the credit he received. To this day, many individuals believe that Watt's centrifugal governor repre-

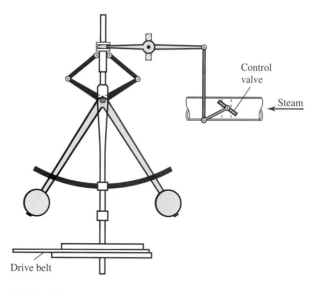

Control valve

Steam

Drive belt

● **FIGURE 1.6**
The Centrifugal Governor

sents the earliest use of feedback for automatically controlling an industrial process.

The theoretical investigations that would eventually manifest themselves in the field of **control systems engineering** have been traced by Fuller [27]. He credits Airy [3] as being the first to employ differential equations to study instability in a feedback control system. In his investigations, Airy discovered that a system controlled by a centrifugal governor could become unstable. However, Maxwell [47] generally is credited with the first comprehensive mathematical treatment of the stability of feedback control systems.

By the mid-19th century, mathematicians understood that the stability of a linear dynamic system was dependent on all roots of the characteristic polynomial having negative real parts. Maxwell developed appropriate necessary and sufficient conditions for this to be true in the case of second- and third-degree polynomials. However, Routh won the Adams Prize of 1877 by solving the problem in general [55], which eventually led to his (and Hurwitz's) now famous stability criterion.

Many other mathematical problems, which were motivated by the stability of feedback control systems, occupied the attention of investigators at the turn of this century, most notably the Russian mathematician A. M. Lyapunov. His fundamental work utilized the state-space (modern) approach to control system analysis and design. However, the stability techniques he developed were not applied to control problems in this country until the late 1950s.

Classical and Modern Control

Before World War II, physical control systems consisted primarily of special mechanical devices that were developed for specific applications, without any extensive theoretical investigations.[1] This was true because general control theories did not exist, and the means of implementing arbitrarily formulated dynamic controllers, the programmable digital computer, had not yet been devised. However, in the communications area, Black's development of the feedback amplifier at Bell Laboratories in the late 1920s, whose performance was enhanced by Nyquist's stability criterion [51] and the graphic design procedures of Bode, not only were making long distance communications more practical but also were laying the foundation for what would soon become the field of control systems engineering.

The technology required for control systems design and implementation received a major impetus during World War II, with engineers, scientists, and mathematicians working together to devise more accurate and reliable weapons systems. During this time, radar appeared, and along with its development at the Radiation Laboratory at the Massachusetts Institute of Technology, a technical group concerned with the control of error-driven systems (servomechanisms) was formed. Collaborative investigations applied frequency response methods from the communications area, such as Bode diagrams and Nyquist plots, to the analysis and design of feedback control systems [36], thereby developing what is now referred to as the **classical approach**.

Immediately after World War II, control systems research continued along these classical lines; and by the early 1950s, generally applicable analysis and design techniques based on frequency-response methods and transfer-function representations, including "loopshaping" PID and lag-lead compensation, as well as Evan's root locus, had been fully developed. Also, increasing emphasis was being placed on the Z transform theory required for sampled data systems, undoubtedly motivated by the parallel development of the programmable digital computers that could implement diverse controller designs.

The need for controlling more complex MIMO systems accelerated during the 1950s. Since classical control theories had been developed primarily for SISO analysis and design, R. E. Kalman and others argued persuasively that state-space methods, based on differential equation models, were required for these more advanced control system investigations. Kalman also believed that the classical methods often obscured the phys-

---

[1]The gyro-stabilized autopilot, developed and built by the Sperry Gyroscope Company of New York in 1910, under the direction of E. A. Sperry, is a prominent example of such a mechanical device [10].

ical, internal nature of dynamic systems and that internal (state-space) descriptions were required to define and explain such fundamental concepts as controllability and observability, as well as to "optimize" controlled system performance. Numerous researchers agreed with Kalman, thereby initiating the so-called **modern approach** to control system analysis and design.

More recently, several investigators have drifted away from the traditional state-space methods and have proposed entirely new analysis and design techniques, often borrowing ideas from both the classical and the modern schools, thereby defining what has been termed **eclectic control** [23]. Still others, who were never convinced of the superiority of the modern approach, have continued to exploit classical frequency-response ideas to produce designs that they believe best handle plant uncertainty, disturbances, and noise. This body of work has been termed **neoclassical control** [26].

Most of the newer methods proposed for linear systems require a level of mathematical sophistication and design complexity that is difficult to introduce at the undergraduate level. The most notable of these include the $\mathbf{H}_\infty$ techniques generally credited to Zames [66], as detailed in references [19], [25], and [50]; the **quantitative feedback theories** (QFT) introduced by Horowitz [32], as outlined in references [33] and [16]; and a variety of MIMO techniques, such as the **inverse Nyquist array** and **multivariable root loci**, which have been developed by several investigators, primarily in the United Kingdom [44].

There still is considerable disagreement today as to which control techniques are appropriate to employ, even in the linear, SISO case. Most basic control systems textbooks present both the classical and the state-space methods, often as distinct topics in different sections or chapters. The majority of these textbooks favor the simpler classical methods, since they are well proven and appear to offer more insight in the SISO case, which basic control systems engineering emphasizes. However, some authors argue that state-space methods are better suited to handle MIMO systems and, therefore, they should be stressed in an introductory course. The eclectic/neoclassical practitioners further complicate the situation by proposing entirely different and often complex methods to control even the simplest systems, although certain of these methods can prove useful when there is "excessive" uncertainty in the plant and the disturbances acting on it [33].

1.3   AN OVERVIEW OF THE TEXT

The primary intent of this overview is to place the particular development employed in this text in proper perspective relative to the current status of linear, SISO control system analysis and design. As additional knowledge and insight are acquired by the student, this discussion will be more illuminating and useful.

The decision to write this text was based, in part, on a perceived need for a comprehensive treatment of linear, SISO control system analysis and design. In particular, although a great deal has been written on this subject, no one has completely integrated classical and modern methods in order to present a single encompassing *eclectic* theory that combines the "best" of both approaches. This will be the primary goal of this text.

The presentation also will dispel the misconception that a *one-degree-of-freedom* (1 DOF) unity feedback design[2] is necessary to ensure a robust, zero, steady-state error, that is, a robust $e_{ss}(t) = 0$. This general misconception has been perpetuated in a variety of ways, originating at an earlier time when programmable computers were not widely used for control purposes. In particular, the American Institute of Electrical Engineers (AIEE) offers the following definition of feedback control [2]:

> A feedback control system is a control system which tends to maintain a prescribed relationship of one system variable to another by means of comparing functions of these variables and *using the difference* as a means of control.

Furthermore, the classical **servomechanism**, a term introduced by Hazen [31], has been defined as follows [8]:

> The device controls some physical quantity by comparing its actual value C with its desired value R and *uses the difference (or error) R − C* to drive C into correspondence with R.

Even one of the most popular of today's control texts [23] contains the following statement relative to the robust $e_{ss}(t) = 0$ that is obtainable using a closed-loop controller characterized by the appropriate *system type:*

> This robustness is the major reason for preferring a *unity feedback* system over any other.... Nonunity feedback renders the integrator ineffective so far as achieving zero steady-state error to a constant input.

Designs based on the (classical) Nichols chart [36] also assume the restrictive correspondence between loop performance and response performance that is characteristic of all 1 DOF compensators, whether unity feedback or not.

It should be noted, however, that a 1 DOF/unity feedback configuration does simplify control system design, and such classical designs very

---

[2]Such a design results when the difference, or error, between the reference input and the measured output in Figure 1.3 is used as a *single* input to the controller.

often produce acceptable closed-loop performance, especially when the given system is stable and minimum phase. Therefore, they represent an important class of controllers that should be covered in any introductory text. However, the student should be aware of their shortcomings, and the alternatives, especially in the more difficult to control situations, for example, when the plant is open-loop unstable and has nonminimum phase zeros.

As we will illustrate, *a 2 DOF design represents the most general, linear, time-invariant compensator possible.* This form of compensation, which results when both of the controller inputs in Figure 1.3 are treated independently, arises naturally when an observer is employed to implement a linear state feedback control law. In particular, the classical transfer function interpretation of 2 DOF control presented here depends on the modern state-space notions of controllability and observer state feedback. Moreover, 2 DOF controllers can be used to obtain a robust $e_{ss}(t) = 0$. As illustrated here, this design goal is achieved via the internal model principle (IMP), which generalizes the classical notion of system type using the modern explanation of a system lacking complete state observability.

In general, the classical frequency response and transfer function methods still provide the best insight regarding the need for control, as well as relatively simple and time-proven 1 DOF methods for achieving basic control objectives. However, the modern state-space methods cannot be ignored, since they provide important additional insight into controlled system performance. These two approaches need not be separated, however, because most classical techniques have a modern "interpretation" (and vice versa), and a control system designer should have a thorough understanding of these interrelationships to develop the best possible controller design for a system, regardless of the way it is defined. The text continuously integrates relevant classical and modern methods with this overall goal in mind.

The particular development employed here has been separated into three distinct parts, with the relevant control topics placed in a logical pedagogic sequence. In particular, Part I introduces Dynamic System Representations, which include ordinary differential equation (or operator) and state-space descriptions in Chapter 2, transfer functions/matrices in Chapter 3, and dynamic response models in Chapter 4. State-space representations are used to introduce modal analysis, which is subsequently employed to define the fundamental concepts of controllability and observability in a natural and state-independent way. Differential operator equations are shown to represent a "bridge" between the state-space and transfer function representations, which can be used to establish a notion of equivalence among the various representations.

Part II of the text deals with performance goals and tests. Here we learn when and why control systems are required, as well as the fundamental distinction between "loop goals" and "response goals," an important point that is often overlooked or understated. Chapter 5 introduces the standard classical tests for nominal stability, such as the Routh-Hurwitz criterion, the root locus, and the Nyquist stability criterion, as methods for evaluating system performance before their use in design. Chapter 6 then outlines the various loop performance goals that an appropriate controller should achieve, such as robust stability, sensitivity reduction, and disturbance and noise attenuation. The loop gain and the sensitivity function are shown to represent important measures of loop performance. Some relatively new $\mathbf{H}_\infty$ results are employed to quantify the presentation.

Chapter 7 covers the simultaneous requirements of "good output tracking," namely, the response performance goals of a small steady-state error, a fast output transient response with minimal overshoot, a nonsaturating plant input, and (ultimately) an input independent $e_{ss}(t) = 0$. One of the main points is that there are many different reasons why system performance may be unsatisfactory and that *all* possible loop and response performance goals must be considered when an effective control system is designed. A 2 DOF procedure is introduced that enables a designer to maintain the same desired output response while varying the loop performance of the system. This part of the text might be thought of as a transitional one that both illustrates analysis and motivates the need for design.

Everything culminates in Part III of the text, which is concerned with compensation. By now, we should know when and why control is necessary and, if so, what performance goals require improvement. The design techniques presented here increase in complexity depending on the amount of improvement that is desired, as well as the control difficulty; that is, a stable, minimum phase system is shown to be far easier to control than an unstable system with nonminimum phase zeros. Chapter 8 concentrates on loop goals that can be achieved by the well-known, classical methods, although these 1 DOF designs are presented with additional insight. In particular, the student not only learns that PID and lag-lead are closely related series loop compensators but also sees how they both represent a "limited form" of linear state feedback compensation. A final example illustrates how the response performance attained by a 1 DOF controller can be improved using a 2 DOF design.

Chapter 9 presents the most general, possible, linear, time-invariant designs, namely, 2 DOF compensators, which are useful and often necessary in certain control situations, for example, when a system is nonminimum phase or unstable and both robust stability and zero error tracking are desired. The presentation illustrates how the loop goals of robust stability, with respect to both plant parameter variations and unmodeled dynamics, disturbance rejection, and noise attenuation, as well as the response goals of low (often zero) error, steady-state tracking with minimal transients but without plant input saturation, can all be considered in an appropriate control system design.

For the most part, the design procedures presented here are based on 2 DOF implementations of root-square locus solutions to the linear quadratic regulator (LQR) problem. This approach represents a classical transfer function interpretation of modern linear state feedback as implemented by an observer. The decision to employ design procedures based on LQR optimal control theory was due to the simplicity and the flexibility afforded by the LQR root-square locus approach when compared with the other available methods, especially in the SISO case.

In particular, although general LQR state-space methods do not deal directly with such performance goals as robust stability and disturbance elimination, a minimization of the magnitude of the sensitivity function is inherent in an appropriately formulated 2 DOF LQR design, thus implying improved stability margins for "reasonable" plant parameter variations, and the IMP can be employed to handle known disturbances, in addition to ensuring a robust $e_{ss}(t) = 0$. Moreover, the variation of a single weighting factor can usually be used to obtain relatively low-order compensators that permit acceptable loop and response performance trade-offs, for example, loop bandwidth adjustments can be made to ensure adequate noise attenuation while retaining a fixed, desired, output response. Chapter 9 contains a number of new and somewhat innovative 2 DOF design techniques, such as low loop gain (LLG) and loop transfer recovery (LTR), for simultaneously obtaining a variety of loop and response performance goals in the cases that are more difficult to control.

In keeping with the introductory nature of this text, many pertinent control topics have been omitted, such as nonlinear, time-varying, digital, and multivariable control. However, it is believed that the material presented is fundamental and the most relevant to linear, SISO control system analysis and design, the cornerstone of automatic control systems. A thorough and comprehensive understanding of the performance goals and design methods available to control this class of systems should be acquired before the other topics are studied. The primary goal of this text is to provide such an understanding in the most natural and direct manner.

PART I

DYNAMIC SYSTEM REPRESENTATIONS

Courtesy of NASA

2

DIFFERENTIAL OPERATOR AND STATE-SPACE MODELS

2.1 INTRODUCTION

In general, control system design usually begins with "some knowledge" of the physical system or plant to be controlled. For example, motor characteristics are usually specified in the case of motor-driven dynamic systems. In certain situations, a motor "transfer function" might be given. Such knowledge generally can be enhanced through experimentation; that is, by making various measurements on the physical plant, from simple weights and lengths to more complex, timed and forced, output response tests. As the result of this knowledge, a mathematical model is developed and compared with the plant. This process may involve simulation studies that compare the response performance of the mathematical model with that of the plant. Mismatches between the plant and the model can then be corrected as needed by appropriately altering the model. To be useful, the eventual mathematical model should not be so detailed that the essential features of the actual system are obscured. On the other hand, the model should not be so simple that important features of the physical plant are omitted.

The dynamic behavior of the systems of primary concern to us will be described by one or more differential equations. A complete describing set of such equations will define the most common form of mathematical models that we will employ. In general, actual dynamic systems are most

17

accurately defined in a mathematical sense by nonlinear differential equations. However, in many cases, linear approximations can be obtained that accurately represent the behavior of such systems around some nominal operating point or along some nominal trajectory. Such **linear mathematical models** are very useful because they quantify the control system design problem in a relatively simple but precise mathematical manner, thus permitting the direct application of a variety of analytic and design algorithms that have been developed in the field of automatic control systems.

For the most part, this text will not deal with modeling,[1] and we will assume that we already have a representative mathematical model of a given dynamic system that requires some form of automatic control to "improve its performance," for example, to balance (stabilize) the inverted pendulum depicted in Figure 2.7. A variety of control system analysis and design techniques based on such knowledge will then be outlined. The controllers that we develop will be mathematical models of systems that are generally placed in series with the system to be controlled, as depicted in Figure 1.3.

The majority of the results that will be presented in this text are directed at the design of mathematical controllers that acceptably alter the closed-loop performance of given physical systems whose dynamic behavior is described by linear, differential equation models. To obtain a corresponding physical system that appropriately controls the actual plant, a mathematical model of the controller must be physically constructed and tested. Specialized digital computers are frequently employed to implement these mathematical controllers, whose outputs subsequently actuate physical devices that are consistent with the system to be controlled. For example, an electric motor-driven system will generally be completely electrical. A gas-pressure control system will often be implemented by a pneumatic motor or valve, and large hydraulic systems, such as those that power machine tools, industrial robots, and aircraft control surfaces, are often activated by hydraulic servomotors. The dynamic behavior of these actuators will usually be considered to be part of the plant. Once an appropriate physical control system has been determined, it should be tested on the actual plant to ensure acceptable performance.

Figure 2.1 displays a **block diagram** that depicts the major steps required to determine an appropriate controller for a given physical plant. It should be noted that this text will deal primarily with the two *mathematical blocks* that are depicted. These blocks emphasize both the analysis of the mathematical models that describe dynamic system behavior and the design of mathematical control systems that improve the closed-loop performance of such systems.

---

[1] The subject of mathematically modeling dynamic systems is covered in a variety of texts, such as [53].

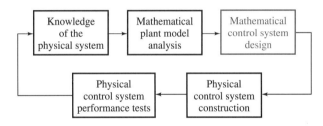

● **FIGURE 2.1**
Control System Analysis and Design

2.2   SOME SIMPLE LINEAR SYSTEMS

In this section we present some relatively simple dynamic systems that can be mathematically modeled using linear differential equations. We assume that the reader is familiar with the basic physical laws that define the dynamic behavior of such systems.

A Series *RLC* Electrical Circuit

We begin with the electrical circuit depicted in Figure 2.2, which consists of a series resistance R, an inductance L, and a capacitance C. The **dynamic response** of the circuit is due to the application of an input voltage $v_i(t)$. If we define $i(t)$ as the loop current and subsequently equate $v_i(t)$ to the sum of the voltage drops around the loop, in accordance with **Kirchhoff's voltage law**, we obtain the relationship:

$$v_i(t) = \underbrace{Ri(t)}_{v_R(t)} + \underbrace{L\frac{di(t)}{dt}}_{v_L(t)} + \underbrace{\frac{1}{C}\int i(t)dt}_{v_C(t)} \tag{2.2.1}$$

When differentiated with respect to time, Eq. (2.2.1) yields the following second-order, linear differential equation mathematical model of the system:

$$R\frac{di(t)}{dt} + L\frac{d^2i(t)}{dt^2} + \frac{i(t)}{C} = \frac{dv_i(t)}{dt} \tag{2.2.2}$$

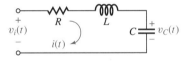

● **FIGURE 2.2**
A Series *RLC* Electrical Circuit

- **FIGURE 2.3**
  An Automobile Suspension System

A Simplified Automobile Suspension System

Next we consider a simplified representation of the suspension system for each wheel of an automobile, as depicted by the spring-mass-damper system of Figure 2.3. The mass M supported by each wheel is acted on by the external gravitational force Mg, as well as a damping force

$$F_B(t) = B\frac{dy(t)}{dt}$$

due to the shock absorbers, and a spring force

$$F_K(t) = Ky(t)$$

so that $y(t) = 0$ defines the relaxed spring position.

If we now employ **Newton's second law**, we obtain the relationship:

$$\Sigma F(t) = -Mg - F_K(t) - F_B(t) = M\frac{d^2y(t)}{dt^2}$$

which implies the following second-order, linear differential equation mathematical model of the system:

$$M\frac{d^2y(t)}{dt^2} + B\frac{dy(t)}{dt} + Ky(t) = -Mg \qquad (2.2.3)$$

Two Coupled Masses

Suppose now that we have two coupled masses that roll on a flat surface in response to an applied force $F_A(t)$, as depicted in Figure 2.4. If (for convenience) we assume point masses and a relaxed spring of zero length, so that the spring force is zero when $x_1(t) = x_2(t)$, and apply Newton's second law to the mass M_1, we obtain the relationship:

$$M_1\ddot{x}_1(t) = K[x_2(t) - x_1(t)] + B[\dot{x}_2(t) - \dot{x}_1(t)] \qquad (2.2.4)$$

and if we apply Newton's second law to the mass M_2, we obtain the relationship:

$$M_2\ddot{x}_2(t) = F_A(t) - K[x_2(t) - x_1(t)] - B[\dot{x}_2(t) - \dot{x}_1(t)] \qquad (2.2.5)$$

Together, these two second-order, linear differential equations mathematically model the dynamic system.

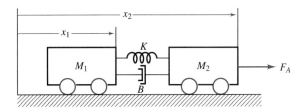

• **FIGURE 2.4**
Two Coupled Masses

A Liquid Level System

We consider next the two-tank, liquid level dynamic system depicted in Figure 2.5. If we assume that linearized flow rates $Q_1(t)$ and $Q_2(t)$ are proportional to corresponding liquid heights, it follows that

$$Q_1(t) = \frac{H_1(t) - H_2(t)}{R_1} \quad \text{and} \quad Q_2(t) = \frac{H_2(t)}{R_2}$$

where R_1 and R_2 are constant flow resistance factors that depend on the positions of the adjustable control valves C_1 and C_2, respectively.

If $Q_i(t)$ represents an external input flow rate, it follows that the net volume-change/unit-time in each tank is given by

$$\frac{d\,[V_1(t) = A_1 H_1(t)]}{dt} = Q_i(t) - Q_1(t)$$

and

$$\frac{d\,[V_2(t) = A_2 H_2(t)]}{dt} = Q_1(t) - Q_2(t)$$

respectively, so that

$$A_1 \frac{d H_1(t)}{dt} = Q_i(t) - \left[\frac{H_1(t) - H_2(t)}{R_1}\right] \tag{2.2.6}$$

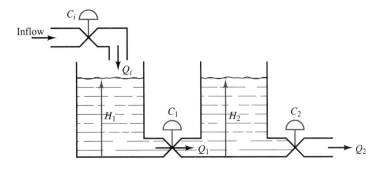

• **FIGURE 2.5**
A Two-Tank Liquid Level System

and

$$A_2 \frac{d H_2(t)}{dt} = \frac{H_1(t) - H_2(t)}{R_1} - \frac{H_2(t)}{R_2} \tag{2.2.7}$$

where A_1 and A_2 represent the (uniform) surface areas in the two tanks. The two first-order, linear time-invariant differential equations defined by Eqs. (2.2.6) and (2.2.7) mathematically model the dynamic behavior of this liquid level system.

An Armature-Controlled DC Servomotor

Finally we consider the armature-controlled DC servomotor depicted in Figure 2.6. If the field current i_f is held constant and a voltage $v_a(t)$ is applied to the armature, which has a resistance R_a and an inductance L_a, the resulting armature current $i_a(t)$ will cause the motor to rotate with a torque $T(t)$ that is proportional to $i_a(t)$, that is,

$$T(t) = K_m i_a(t)$$

for some motor-torque constant K_m. Furthermore, the torque $T(t)$ produced by the motor causes the inertia load J, which is also characterized by a viscous friction constant D, to rotate in accordance with Newton's dynamic relationship:

$$T(t) = J \frac{d^2\theta(t)}{dt^2} + D \frac{d\theta(t)}{dt} = K_m i_a(t) \tag{2.2.8}$$

We next observe that the rotational motion produces a back-*emf* voltage $v_b(t)$ that is proportional to the angular velocity $\omega(t)$ of the motor, so that

$$v_b(t) = K_b \omega(t) = K_b \frac{d\theta(t)}{dt} \tag{2.2.9}$$

for some back-*emf* constant K_b. By equating $v_a(t)$ to the sum of the

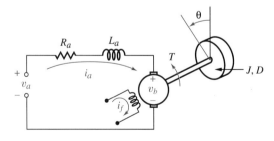

● FIGURE 2.6
An Armature-Controlled DC Servomotor

voltage drops around the armature loop, in accordance with Kirchhoff's voltage law, we obtain the relationship:

$$v_a(t) = R_a i_a(t) + L_a \frac{di_a(t)}{dt} + v_b(t) \qquad (2.2.10)$$

which, together with Eqs. (2.2.8) and (2.2.9), mathematically models the dynamic behavior of this servomotor.

2.3   NONLINEAR SYSTEMS

In this section we present two relatively simple dynamic systems that can be mathematically modeled using nonlinear differential equations. We assume that the reader is familiar with **Lagrange's result** from mechanics [30], the primary physical law that will be employed to derive the defining equations.

An Inverted Pendulum on a Moving Cart

Consider the dynamic system depicted in Figure 2.7. This system consists of an inverted pendulum of point mass M_2, that is free to rotate with an angle $\theta(t)$ around an axis positioned d units away on the top of a cart of mass M_1. The cart rolls on a flat horizontal surface in response to an externally applied force $F(t)$.

The kinetic energy $K_1(t)$ of the cart is due to its translational velocity $v_1(t) = \dot{x}(t)$ alone, so that

$$K_1(t) = \frac{1}{2}M_1 v_1^2(t) = \frac{1}{2}M_1 \dot{x}^2(t) \qquad (2.3.1)$$

We next note that the horizontal and vertical positions of the point mass M_2 are defined by $p_h(t) = x(t) + d\sin\theta(t)$ and $p_v(t) = d\cos\theta(t)$,

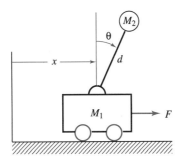

● FIGURE 2.7
An Inverted Pendulum on a Moving Cart

respectively. Therefore, the magnitude squared of its velocity is given by

$$v_2^2(t) = [\dot{p}_h(t)]^2 + [\dot{p}_v(t)]^2 = [\dot{x}(t) + d\dot{\theta}(t) \cos \theta(t)]^2 + [-d\dot{\theta}(t) \sin \theta(t)]^2$$

$$= \dot{x}^2(t) + 2d\dot{x}(t)\dot{\theta}(t) \cos \theta(t) + d^2\dot{\theta}^2(t) \underbrace{[\cos^2 \theta(t) + \sin^2 \theta(t)]}_{1}$$

so that its kinetic energy

$$K_2(t) = \frac{1}{2} M_2 v_2^2(t) = \frac{1}{2} M_2 \left[\dot{x}^2(t) + 2d\dot{x}(t)\dot{\theta}(t) \cos \theta(t) + d^2\dot{\theta}^2(t)\right]$$

(2.3.2)

We now note that the **total kinetic energy** $K(t)$ of the dynamic system consisting of the inverted pendulum and the cart is simply equal to the sum of their respective kinetic energies, as given by Eqs. (2.3.1) and (2.3.2), or that

$$K(t) = \frac{1}{2} M_1 \dot{x}^2(t) + \frac{1}{2} M_2 \left[\dot{x}^2(t) + 2d\dot{x}(t)\dot{\theta}(t) \cos \theta(t) + d^2\dot{\theta}^2(t)\right]$$

(2.3.3)

The mathematical equations that model this dynamic system can now be determined using Lagrange's result, which in this case implies that

$$\frac{d}{dt}\left[\frac{\partial K(t)}{\partial \dot{x}(t)}\right] - \frac{\partial K(t)}{\partial x(t)} = T_1(t) \text{ and } \frac{d}{dt}\left[\frac{\partial K(t)}{\partial \dot{\theta}(t)}\right] - \frac{\partial K(t)}{\partial \theta(t)} = T_2(t) \quad (2.3.4)$$

where

$$T_1(t) = F(t) \text{ and } T_2(t) = M_2 g d \sin \theta(t) \quad (2.3.5)$$

represent the **generalized torques** that act on the cart and the pendulum, respectively, in the direction of possible motion. Since

$$\frac{\partial K(t)}{\partial \dot{x}(t)} = M_1 \dot{x}(t) + M_2 \dot{x}(t) + M_2 d\dot{\theta}(t) \cos \theta(t) \text{ and } \frac{\partial K(t)}{\partial x(t)} = 0$$

in light of Eq. (2.3.3), Eqs. (2.3.4) and (2.3.5) imply that

$$\frac{d}{dt}\left[\frac{\partial K(t)}{\partial \dot{x}(t)}\right] = [M_1 + M_2]\ddot{x}(t) + M_2 d\ddot{\theta}(t) \cos \theta(t) - M_2 d\dot{\theta}^2(t) \sin \theta(t)$$

$$= F(t), \quad (2.3.6)$$

which represents one of the two nonlinear differential equations that define the dynamic behavior of the system.

The second equation can be found in an analogous manner. Since

$$\frac{\partial K(t)}{\partial \dot{\theta}(t)} = M_2 d\dot{x}(t) \cos \theta(t) + M_2 d^2 \dot{\theta}(t)$$

and

$$\frac{\partial K(t)}{\partial \theta(t)} = -M_2 d\dot{x}(t)\dot{\theta}(t) \sin \theta(t)$$

in light of Eq. (2.3.3), Eqs. (2.3.4) and (2.3.5) imply that

$$\frac{d}{dt}\left[\frac{\partial K(t)}{\partial \dot\theta(t)}\right] - \frac{\partial K(t)}{\partial \theta(t)} = M_2 d\ddot{x}(t)\,\cos\theta(t) + M_2 d^2\ddot\theta(t) = M_2 gd\,\sin\theta(t)$$

or, dividing through by $M_2 d$, that

$$\ddot{x}(t)\,\cos\theta(t) + d\ddot\theta(t) = g\,\sin\theta(t) \qquad (2.3.7)$$

which represents the remaining equation. It might be noted that these two defining equations, Eqs. (2.3.6) and (2.3.7), are nonlinear because of the trigonometric multiplier terms $\cos\theta(t)$ and $\sin\theta(t)$, as well as the product term $\dot\theta^2(t)$.

An Orbiting Satellite

We now consider a point mass M satellite rotating in an inverse-square-law force field with a tangential velocity $v_t(t) = r(t)\dot\alpha(t)$ and an orthogonal radial velocity $v_r(t) = \dot{r}(t)$, as depicted in Figure 2.8, so that its total kinetic energy

$$K(t) = \frac{1}{2}Mv^2(t) = \frac{1}{2}M\left[v_t^2(t) + v_r^2(t)\right] = \frac{1}{2}M\left[r^2(t)\dot\alpha^2(t) + \dot{r}^2(t)\right]$$
$$(2.3.8)$$

We will assume that the satellite is capable of thrusting in both the tangential and the radial directions, with force magnitudes $F_t(t)$ and $F_r(t)$, respectively.

The mathematical equations that model the orbiting satellite also can be determined using Lagrange's result, which in this case implies that

$$\frac{d}{dt}\left[\frac{\partial K(t)}{\partial \dot{r}(t)}\right] - \frac{\partial K(t)}{\partial r(t)} = F_r(t) - \frac{kM}{r^2(t)} \qquad (2.3.9a)$$

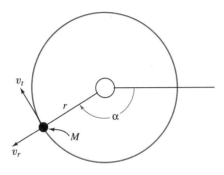

• **FIGURE 2.8**
An Orbiting Satellite

and

$$\frac{d}{dt}\left[\frac{\partial K(t)}{\partial \dot{\alpha}(t)}\right] - \frac{\partial K(t)}{\partial \alpha(t)} = F_t(t) \qquad (2.3.9b)$$

where k is an appropriate gravitational constant. Since

$$\frac{\partial K(t)}{\partial \dot{r}(t)} = M\dot{r}(t) \qquad \text{and} \qquad \frac{\partial K(t)}{\partial r(t)} = Mr(t)\dot{\alpha}^2(t)$$

in light of Eq. (2.3.8), Eq. (2.3.9a) implies that

$$\frac{d}{dt}\left[\frac{\partial K(t)}{\partial \dot{r}(t)}\right] - \frac{\partial K(t)}{\partial r(t)} = M\ddot{r}(t) - Mr(t)\dot{\alpha}^2(t) = F_r(t) - \frac{kM}{r^2(t)}$$

or that

$$\ddot{r}(t) = r(t)\dot{\alpha}^2(t) - \frac{k}{r^2(t)} + \frac{F_r(t)}{M} \qquad (2.3.10)$$

which represents one of the two second-order, nonlinear differential equations that defines the dynamic behavior of the satellite.

The second equation can be found in an analogous manner. Since

$$\frac{\partial K(t)}{\partial \dot{\alpha}(t)} = Mr^2(t)\dot{\alpha}(t) \text{ and } \frac{\partial K(t)}{\partial \alpha(t)} = 0$$

in light of Eq. (2.3.8), Eq. (2.3.9b) implies that

$$\frac{d}{dt}\left[\frac{\partial K(t)}{\partial \dot{\alpha}(t)}\right] = 2Mr(t)\dot{r}(t)\dot{\alpha}(t) + Mr^2(t)\ddot{\alpha}(t) = F_t(t)$$

or, dividing through by $Mr^2(t)$, that

$$\ddot{\alpha}(t) = -\frac{2\dot{r}(t)\dot{\alpha}(t)}{r(t)} + \frac{F_t(t)}{Mr^2(t)} \qquad (2.3.11)$$

which represents the remaining equation.

2.4   STATE-SPACE REPRESENTATIONS

The state-space approach for mathematically modeling the dynamic behavior of physical systems is based on knowledge of a set of n first-order differential equations involving the so-called **state variables**, which are usually denoted as $x_1(t)$, $x_2(t)$, ..., $x_n(t)$. More specifically, the **state** of a dynamic system is a set of n variables whose time evolution completely describes the *internal behavior* of the system. To obtain a "solution" to a set of n state-space equations, knowledge of the entire state at some initial time t_0, namely, $x_1(t_0)$, $x_2(t_0)$, ..., $x_n(t_0)$, as well as the n first-order differential equations that define the system, is required. It may be noted that the specific physical quantities that define the state of a dynamic system are not unique, although their number (n) is unique. This dimensional integer n is called the **order** of the state-space representation.

Although state-space analysis and design is often referred to as the "modern approach" in control system investigations, it is well over 100 years old. Mathematicians have long termed state-variable descriptions of ordinary differential equations the *normal form* of such equations. Moreover, the notion of the state of a dynamic system underlies Newtonian dynamics. In physics, the two-dimensional position versus velocity "plane" of a point mass or rigid body is often termed the **phase-plane**. The **state-space** of a more complex physical process represents a generalization of this basic idea beyond the two-dimensional case.

Nonlinear State-Space Representations

In the general case, the first derivative (with respect to time) of each state-variable

$$\frac{dx_i(t)}{dt} = \dot{x}_i(t)$$

is a nonlinear function of all of the state-variables and an external *control* or *input vector* $\mathbf{u}(t) \stackrel{\text{def}}{=} [u_1(t), u_2(t), \ldots, u_m(t)]^{\text{T}}$, as well as the time t itself, if the system is **time-varying**; that is, in general, a **state-space representation** of a dynamic system is defined by a set of n nonlinear time-varying differential equations in the vector form:

$$\mathbf{\dot{x}}(t) \stackrel{\text{def}}{=} \begin{bmatrix} \dot{x}_1(t) \\ \dot{x}_2(t) \\ \vdots \\ \dot{x}_n(t) \end{bmatrix} = \begin{bmatrix} f_1(\mathbf{x}(t), \mathbf{u}(t), t) \\ f_2(\mathbf{x}(t), \mathbf{u}(t), t) \\ \vdots \\ f_n(\mathbf{x}(t), \mathbf{u}(t), t) \end{bmatrix} \stackrel{\text{def}}{=} \mathbf{f}(\mathbf{x}(t), \mathbf{u}(t), t) \quad (2.4.1)$$

. . . .

EXAMPLE 2.4.2 To illustrate a general state-space representation, recall the orbiting satellite of the previous section, whose dynamic behavior is defined by Eqs. (2.3.10) and (2.3.11). If we define $x_1(t) = r(t)$, $x_2(t) = \dot{r}(t)$, $x_3(t) = \alpha(t)$, $x_4(t) = \dot{\alpha}(t)$, $F_r(t) = u_1(t)$, and $F_t(t) = u_2(t)$, then in light of Eqs. (2.3.10) and (2.3.11), it follows that

$$\dot{x}_1(t) = x_2(t), \qquad \dot{x}_2(t) = x_1(t)x_4^2(t) - \frac{k}{x_1^2(t)} + \frac{u_1(t)}{M}, \qquad \dot{x}_3(t) = x_4(t)$$

and

$$\dot{x}_4(t) = -\frac{2x_2(t)x_4(t)}{x_1(t)} + \frac{u_2(t)}{Mx_1^2(t)}$$

This set of four first-order differential equations can now be written as a single nonlinear, state-space representation of order 4, as in Eq. (2.4.1),

$$
\begin{bmatrix} \dot{x}_1(t) \\ \dot{x}_2(t) \\ \dot{x}_3(t) \\ \dot{x}_4(t) \end{bmatrix} = \begin{bmatrix} f_1(\mathbf{x}(t),\ \mathbf{u}(t),\ t) \\ f_2(\mathbf{x}(t),\ \mathbf{u}(t),\ t) \\ f_3(\mathbf{x}(t),\ \mathbf{u}(t),\ t) \\ f_4(\mathbf{x}(t),\ \mathbf{u}(t),\ t) \end{bmatrix} = \begin{bmatrix} x_2(t) \\ x_1(t)x_4^2(t) - \dfrac{k}{x_1^2(t)} + \dfrac{u_1(t)}{M} \\ x_4(t) \\ -\dfrac{2x_2(t)x_4(t)}{x_1(t)} + \dfrac{u_2(t)}{Mx_1^2(t)} \end{bmatrix}
$$

Linearization

Although the dynamic behavior of most physical systems is nonlinear, many of these systems behave "almost linearly" at and near nominal operating points or along nominal trajectories. In particular, suppose we can find a **nominal solution** to Eq. (2.4.1), that is, some nominal state $\mathbf{x}_0(t)$ that results from the application of some nominal input $\mathbf{u}_0(t)$, so that

$$
\dot{\mathbf{x}}_0(t) = \mathbf{f}(\mathbf{x}_0(t),\ \mathbf{u}_0(t),\ t) \overset{\text{def}}{=} \mathbf{f}|_{\text{nom}} \tag{2.4.3}
$$

If we then expand each nonlinear function $f_i(\mathbf{x}(t),\ \mathbf{u}(t),\ t)$ in Eq. (2.4.1) by means of a Taylor series around the nominal solution, ignoring the higher-order terms, we find[2] that for $i = 1, 2, \ldots, n$,

$$
\dot{x}_i = \underbrace{f_i(\mathbf{x}_0,\ \mathbf{u}_0,\ t)}_{\overset{\text{def}}{=}\ f_i|_{\text{nom}}} + \sum_{j=1}^{n} \left.\frac{\partial f_i}{\partial x_j}\right|_{\text{nom}} \delta x_j + \sum_{k=1}^{m} \left.\frac{\partial f_i}{\partial u_k}\right|_{\text{nom}} \delta u_k \tag{2.4.4}
$$

where $\delta x_i = x_i - x_{0i}$ and $\delta u_k = u_k - u_{0k}$.

Since $\dot{x}_i = \delta \dot{x}_i + \dot{x}_{0i}$ and $\dot{x}_{0i} = f_i|_{\text{nom}}$, Eq. (2.4.4) now implies that for $i = 1, 2, \ldots, n$,

$$
\delta \dot{x}_i = \sum_{j=1}^{n} \left.\frac{\partial f_i}{\partial x_j}\right|_{\text{nom}} \delta x_j + \sum_{k=1}^{m} \left.\frac{\partial f_i}{\partial u_k}\right|_{\text{nom}} \delta u_k \tag{2.4.5}
$$

which can be written in the following expanded, *linear*, vector/matrix

---

[2]For notational simplicity, we will omit the explicit reference to functional time-dependence.

form:

$$
\underbrace{\begin{bmatrix} \delta\dot{x}_1(t) \\ \delta\dot{x}_2(t) \\ \vdots \\ \delta\dot{x}_n(t) \end{bmatrix}}_{\delta\dot{\mathbf{x}}} = \underbrace{\begin{bmatrix} \dfrac{\partial f_1}{\partial x_1} & \dfrac{\partial f_1}{\partial x_2} & \cdots & \dfrac{\partial f_1}{\partial x_n} \\ \dfrac{\partial f_2}{\partial x_1} & \dfrac{\partial f_2}{\partial x_2} & \cdots & \dfrac{\partial f_2}{\partial x_n} \\ \vdots & \vdots & \ddots & \vdots \\ \dfrac{\partial f_n}{\partial x_1} & \dfrac{\partial f_n}{\partial x_2} & \cdots & \dfrac{\partial f_n}{\partial x_n} \end{bmatrix}}_{A_{\text{nom}}} \underbrace{\begin{bmatrix} \delta x_1(t) \\ \delta x_2(t) \\ \vdots \\ \delta x_n(t) \end{bmatrix}}_{\delta\mathbf{x}}
$$

$$
+ \underbrace{\begin{bmatrix} \dfrac{\partial f_1}{\partial u_1} & \cdots & \dfrac{\partial f_1}{\partial u_m} \\ \dfrac{\partial f_2}{\partial u_1} & \cdots & \dfrac{\partial f_2}{\partial u_m} \\ \vdots & \ddots & \vdots \\ \dfrac{\partial f_n}{\partial u_1} & \cdots & \dfrac{\partial f_n}{\partial u_m} \end{bmatrix}}_{B_{\text{nom}}} \underbrace{\begin{bmatrix} \delta u_1(t) \\ \delta u_2(t) \\ \vdots \\ \delta u_m(t) \end{bmatrix}}_{\delta\mathbf{u}}
$$

or simply as

$$
\delta\dot{\mathbf{x}} = A_{\text{nom}}\delta\mathbf{x} + B_{\text{nom}}\delta\mathbf{u} \tag{2.4.6}
$$

where all of the partial derivative terms in A_{nom} and B_{nom} are evaluated at the nominal solution. The **linearized state-space representation** defined by Eq. (2.4.6) approximates the dynamic behavior of the actual, nonlinear system at and "near" the nominal solution.

. . . .

EXAMPLE 2.4.7 To illustrate the process of linearization, consider the nonlinear state-space equations of the orbiting satellite, as given earlier in Example 2.4.2, with the satellite mass M normalized to 1, that is,

$$
\begin{bmatrix} \dot{x}_1(t) \\ \dot{x}_2(t) \\ \dot{x}_3(t) \\ \dot{x}_4(t) \end{bmatrix} = \begin{bmatrix} x_2(t) \\ x_1(t)x_4^2(t) - \dfrac{k}{x_1^2(t)} + u_1(t) \\ x_4(t) \\ -\dfrac{2x_2(t)x_4(t)}{x_1(t)} + \dfrac{u_2(t)}{x_1^2(t)} \end{bmatrix}
$$

It now can be verified that the rotation of the satellite at a constant radial distance $d = r(t)$, at a constant angular velocity ω (so that $\alpha(t) = \omega t$) represents a particular nominal solution that we can linearize around.

In particular, the substitution of d for $x_{01}(t)$, 0 for $\dot{x}_{01}(t) = x_{02}(t)$, ωt for $x_{03}(t)$, and ω for $\dot{x}_{03}(t) = x_{04}(t)$ in these four nonlinear state-space equations implies the single requirement (from the $\dot{x}_{02}(t)$ equation) that

$$\dot{x}_{02} = 0 = d\omega^2 - \frac{k}{d^2} \quad \text{or} \quad \omega^2 = \frac{k}{d^3}$$

which we will assume is true.

We next determine that the nominal partial derivatives

$$\frac{\partial f_2}{\partial x_1} = \omega^2 + 2\frac{k}{d^3} = 3\omega^2, \qquad \frac{\partial f_2}{\partial x_4} = 2d\omega, \qquad \frac{\partial f_4}{\partial x_2} = -\frac{2\omega}{d}, \qquad \text{and} \qquad \frac{\partial f_4}{\partial u_2} = \frac{1}{d^2}$$

with

$$\frac{\partial f_1}{\partial x_2} = \frac{\partial f_3}{\partial x_4} = \frac{\partial f_2}{\partial u_1} = 1$$

Since all of the other nominal partial derivative terms are equal to 0, it follows that the linearized state-space representation

$$
\begin{bmatrix} \delta\dot{x}_1(t) \\ \delta\dot{x}_2(t) \\ \delta\dot{x}_3(t) \\ \delta\dot{x}_4(t) \end{bmatrix}
=
\begin{bmatrix} 0 & 1 & 0 & 0 \\ 3\omega^2 & 0 & 0 & 2d\omega \\ 0 & 0 & 0 & 1 \\ 0 & -\frac{2\omega}{d} & 0 & 0 \end{bmatrix}
\begin{bmatrix} \delta x_1(t) \\ \delta x_2(t) \\ \delta x_3(t) \\ \delta x_4(t) \end{bmatrix}
+
\begin{bmatrix} 0 & 0 \\ 1 & 0 \\ 0 & 0 \\ 0 & \frac{1}{d^2} \end{bmatrix}
\begin{bmatrix} \delta u_1(t) \\ \delta u_2(t) \end{bmatrix}
$$

approximates the dynamic behavior of the orbiting satellite at and near a nominal trajectory. It might be noted that the state $\delta\mathbf{x}(t)$ represents deviations from the nominal. In this particular case, $\delta x_1(t) = r(t) - d$, $\delta x_2(t) = \dot{r}(t)$, $\delta x_3(t) = \alpha(t) - \omega t$, and $\delta x_4(t) = \dot{\alpha}(t) - \omega$.

··

The Linear Time-Invariant Case

In general, A_{nom} and B_{nom} in Eq. (2.4.6) will be time varying, although this will not be the case if the nominal solution is a fixed position. If we further assume time-invariance, then a general state-space representation will take the following special *linear time-invariant* vector/matrix form:

$$
\begin{bmatrix} \dot{x}_1(t) \\ \dot{x}_2(t) \\ \vdots \\ \dot{x}_n(t) \end{bmatrix}
=
\underbrace{\begin{bmatrix} A_{11} & A_{12} & \cdots & A_{1n} \\ A_{21} & A_{22} & \cdots & A_{2n} \\ \vdots & \vdots & \ddots & \vdots \\ A_{n1} & A_{n2} & \cdots & A_{nn} \end{bmatrix}}_{A}
\begin{bmatrix} x_1(t) \\ x_2(t) \\ \vdots \\ x_n(t) \end{bmatrix}
$$

$$
+
\underbrace{\begin{bmatrix} B_{11} & \cdots & B_{1m} \\ B_{21} & \cdots & B_{2m} \\ \vdots & \ddots & \vdots \\ B_{n1} & \cdots & B_{nm} \end{bmatrix}}_{B}
\begin{bmatrix} u_1(t) \\ u_2(t) \\ \vdots \\ u_m(t) \end{bmatrix}
$$

which can be written more concisely as

$$\dot{\mathbf{x}}(t) = A\mathbf{x}(t) + B\mathbf{u}(t) \tag{2.4.8a}$$

with all entries of both the $(n \times n)$ *state matrix* A and the $(n \times m)$ *input matrix* B constant and real. It may be noted that A will always be square while B will almost always have fewer columns than rows, since the number of control inputs (m) will rarely equal or exceed the number of states (n). Indeed, in many of the situations we will encounter, there will be only a single control input $u(t)$. In such single-input (SI) cases, the $(n \times m)$ input matrix B will be an $(n \times 1)$ column vector.

It is also important to note that although the entire n-dimensional state vector $\mathbf{x}(t)$ characterizes the complete internal behavior of a state-space system, it is usually impossible to directly measure or *observe* all n components of $\mathbf{x}(t)$. In most cases, only p functions of the state (and possibly the input) can be measured or directly observed. In the linear time-invariant case, this p-dimensional, external output vector $\mathbf{y}(t)$ can be expressed in the following vector/matrix form:

$$\mathbf{y}(t) \overset{\text{def}}{=} \begin{bmatrix} y_1(t) \\ y_2(t) \\ \vdots \\ y_p(t) \end{bmatrix} = \underbrace{\begin{bmatrix} C_{11} & \cdots & C_{1n} \\ C_{21} & \cdots & C_{2n} \\ \vdots & \ddots & \vdots \\ C_{p1} & \cdots & C_{pn} \end{bmatrix}}_{C} \begin{bmatrix} x_1(t) \\ x_2(t) \\ \vdots \\ x_n(t) \end{bmatrix}$$

$$+ \underbrace{\begin{bmatrix} E_{11} & \cdots & E_{1m} \\ E_{21} & \cdots & E_{2m} \\ \vdots & \ddots & \vdots \\ E_{p1} & \cdots & E_{pm} \end{bmatrix}}_{E} \begin{bmatrix} u_1(t) \\ u_2(t) \\ \vdots \\ u_m(t) \end{bmatrix}$$

which also can be written more concisely as

$$\mathbf{y}(t) = C\mathbf{x}(t) + E\mathbf{u}(t) \tag{2.4.8b}$$

with all entries of both the $(p \times n)$ *output matrix* C and the $(p \times m)$ matrix E constant and real.

In general, a **state-space representation of a linear, time-invariant system** is defined as Eq. (2.4.8a) *together* with Eq. (2.4.8b), or as

$$\dot{\mathbf{x}}(t) = A\mathbf{x}(t) + B\mathbf{u}(t); \qquad \mathbf{y}(t) = C\mathbf{x}(t) + E\mathbf{u}(t) \tag{2.4.8}$$

We often find it convenient to "abbreviate" Eq. (2.4.8) by the matrix quadruple $\{A, B, C, E\}$; that is, $\{A, B, C, E\} \Longleftrightarrow$ Eq. (2.4.8). Also, as in the single-input case, in many physical situations there will be only a single observed output $y(t)$. In such single-output (SO) cases, the $(p \times n)$ matrix C will be a $(1 \times n)$ row vector.

We next observe that the presence of a nonzero E in Eq. (2.4.8) implies a direct and immediate influence of the input $\mathbf{u}(t)$ on the output $\mathbf{y}(t)$.

In the vast majority of physical systems this will not be the case, since most physical systems have some form of "inertia" that must be overcome before output changes (due to applied inputs) can occur. Therefore, when Eq. (2.4.8) is used to model physical systems, E will usually be zero.

However, Eq. (2.4.8) will also be used to model a variety of "controller" or "compensator" systems that involve a direct electrical transmittance from the controller input to the controller output. In such cases, a nonzero gain matrix E will be present. Moreover, in certain situations, time derivatives of the controller input may also directly affect the system output. When this occurs, E must be "modified appropriately" to account for such derivative terms, as we will show later.

EXAMPLE 2.4.9 To illustrate a linear time-invariant, state-space representation, recall the automobile suspension system of Section 2.2, whose dynamic behavior is mathematically modeled by Eq. (2.2.3). If we define $x_1(t) = y(t)$ and $x_2(t) = \dot{y}(t)$, we can then express the dynamic behavior of this system with two first-order, linear differential equations,

$$\dot{x}_1(t) = x_2(t) \qquad \text{and} \qquad \dot{x}_2(t) = -\frac{K}{M}x_1(t) - \frac{B}{M}x_2(t) - g$$

which then can be written as a second-order state-space representation defined by Eq. (2.4.8),

$$\begin{bmatrix} \dot{x}_1(t) \\ \dot{x}_2(t) \end{bmatrix} = \begin{bmatrix} 0 & 1 \\ -\dfrac{K}{M} & -\dfrac{B}{M} \end{bmatrix} \begin{bmatrix} x_1(t) \\ x_2(t) \end{bmatrix} + \begin{bmatrix} 0 \\ -1 \end{bmatrix} g$$

EXAMPLE 2.4.10 Another linear time-invariant, state-space representation can be obtained for the armature-controlled DC servomotor of Section 2.2, whose dynamic behavior is defined by Eqs. (2.2.8), (2.2.9), and (2.2.10). If we define $x_1(t) = i_a(t)$, $x_2(t) = \theta(t)$, and $x_3(t) = \dot{\theta}(t)$, so that $\dot{x}_2(t) = x_3(t)$, the substitution of Eq. (2.2.9) into Eq. (2.2.10) implies that

$$L_a \dot{x}_1(t) = -R_a x_1(t) - K_b x_3(t) + v_a(t)$$

Since Eq. (2.2.8) implies that

$$J \dot{x}_3(t) = K_m x_1(t) - D x_3(t)$$

it follows that the dynamic behavior of the DC servomotor can be described by the following third-order state-space representation:

$$\underbrace{\begin{bmatrix} \dot{x}_1(t) \\ \dot{x}_2(t) \\ \dot{x}_3(t) \end{bmatrix}}_{\dot{\mathbf{x}}(t)} = \underbrace{\begin{bmatrix} -\dfrac{R_a}{L_a} & 0 & -\dfrac{K_b}{L_a} \\ 0 & 0 & 1 \\ \dfrac{K_m}{J} & 0 & -\dfrac{D}{J} \end{bmatrix}}_{A} \underbrace{\begin{bmatrix} x_1(t) \\ x_2(t) \\ x_3(t) \end{bmatrix}}_{\mathbf{x}(t)} + \underbrace{\begin{bmatrix} \dfrac{1}{L_a} \\ 0 \\ 0 \end{bmatrix}}_{B} \underbrace{v_a(t)}_{u(t)};$$

$$y(t) = \underbrace{[0 \quad 1 \quad 0]}_{C} \underbrace{\begin{bmatrix} x_1(t) \\ x_2(t) \\ x_3(t) \end{bmatrix}}_{\mathbf{x}(t)} = \theta(t)$$

Analog Implementations

There are many reasons why state-space representations are useful to control system engineers. One of these is the relative simplicity with which they can be implemented or simulated through the use of three basic analog elements. If we employ electrical analog elements, an **integrator** is a device that integrates (with respect to time) an input voltage $v(t)$, a **summer** is a device that adds (with both $+$ and $-$ signs) two or more input voltages, and a (scalar) **multiplier** is a device that multiplies an input voltage by a real ($+$ or $-$) gain k. Typical representations of these three "building blocks" are shown in Figure 2.9. These three electrical analog elements can be constructed physically using only resistors, capacitors, and high-gain operational amplifiers [34].

We now consider a general second-order ($n = 2$) linear, time-invariant state-space system defined by the equations

$$\begin{bmatrix} \dot{x}_1(t) \\ \dot{x}_2(t) \end{bmatrix} = \begin{bmatrix} A_{11} & A_{12} \\ A_{21} & A_{22} \end{bmatrix} \begin{bmatrix} x_1(t) \\ x_2(t) \end{bmatrix} + \begin{bmatrix} B_{11} \\ B_{21} \end{bmatrix} u(t);$$

$$\begin{bmatrix} y_1(t) \\ y_2(t) \end{bmatrix} = \begin{bmatrix} C_{11} & C_{12} \\ C_{21} & C_{22} \end{bmatrix} \begin{bmatrix} x_1(t) \\ x_2(t) \end{bmatrix} \tag{2.4.11}$$

This system has one control input ($m = 1$) and two observed outputs ($p = 2$) and can be implemented by the block diagram of Figure 2.10 using the three basic analog elements. Moreover, if the individual inputs and outputs in Figure 2.9 are used to denote vector quantities, it then follows that any linear time-invariant state-space system defined by Eq. (2.4.8) can be represented with the block diagram in Figure 2.11.

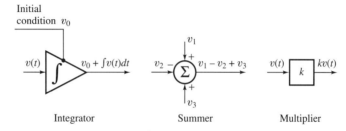

Integrator   Summer   Multiplier

● **FIGURE 2.9**
Three Basic Analog Elements

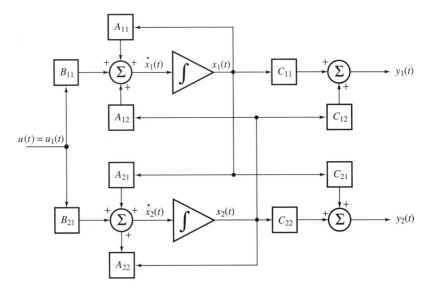

• **FIGURE 2.10**
Analog Implementation of a Second-Order State-Space System

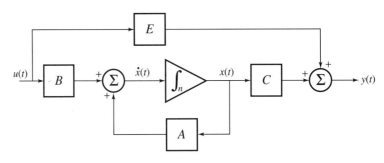

• **FIGURE 2.11**
Analog Implementation of an nth-Order State-Space System

It is important to note that the mathematical equations that define the systems depicted in Figures 2.10 and 2.11 are determined by "reading the signal flow backward" relative to the directions indicated by the arrows. For example, $\dot{x}(t) = B\mathbf{u}(t) + A\mathbf{x}(t) = A\mathbf{x}(t) + B\mathbf{u}(t)$ in Figure 2.11, rather than $\mathbf{u}(t)B + \mathbf{x}(t)A = \dot{\mathbf{x}}(t)$, which is incorrect.

2.5   THE STATE TRANSITION MATRIX

As noted in the previous section, one of the reasons that state-space methods have become widely accepted is the relative simplicity with which they can be simulated using basic analog elements. Another reason, which will form the focus of this section, is the relative ease with which we can express complete solutions to Eq. (2.4.8) in terms of the so-called state transition matrix.

We begin by observing that the solution to the **homogeneous** (or **unforced**, so that $\mathbf{u}(t) = 0$) linear time-invariant state-space equation

$$\dot{\mathbf{x}}(t) = A\mathbf{x}(t) \tag{2.5.1}$$

with initial state $\mathbf{x}_0$ at time t_0; that is, $\mathbf{x}(t_0) = \mathbf{x}_0$, is given by the matrix exponential

$$\mathbf{x}(t) = e^{A(t-t_0)}\mathbf{x}_0 \tag{2.5.2}$$

in which

$$e^{A(t-t_0)} \stackrel{\text{def}}{=} I + A(t - t_0) + \frac{A^2(t - t_0)^2}{2!} + \frac{A^3(t - t_0)^3}{3!} + \ldots \tag{2.5.3}$$

The reader can readily verify that the $\mathbf{x}(t)$ given by Eq. (2.5.2) does indeed satisfy Eq. (2.5.1) because

$$\dot{\mathbf{x}}(t) = \frac{d}{dt} e^{A(t-t_0)}\mathbf{x}_0 = Ae^{A(t-t_0)}\mathbf{x}_0 = A\mathbf{x}(t) \tag{2.5.4}$$

and at $t = t_0$,

$$e^{A(t-t_0)} = e^{A(t_0-t_0)} = I \tag{2.5.5}$$

so that $\mathbf{x}(t = t_0) = \mathbf{x}_0$.

The particular $(n \times n)$ matrix valued function of time, $e^{A(t-t_0)}$, which solves the homogeneous Eq. (2.5.1) is called the **state transition matrix** because it enables us to determine the continuous time transition of the entire n-dimensional state $\mathbf{x}(t)$ through the n-dimensional state-space from any initial state $\mathbf{x}_0$ at any time t_0 to any final state $\mathbf{x}_f$ at any time $t_f > t_0$.

To illustrate this observation, consider any state-space system of order $n = 2$. The time transition of the two-dimensional state, defined by $x_1(t)$ and $x_2(t)$, from any known but arbitrary $x_1(t_0) = x_{10}$ and $x_2(t_0) = x_{20}$ can be depicted on a three-dimensional plot by defining the time t as the third of three mutually orthogonal axes, namely x_1, x_2 and t, as shown in Figure 2.12. The **state transition** can then be depicted as a continuous plot of all points from $\begin{bmatrix} x_{10} \\ x_{20} \\ t_0 \end{bmatrix}$ to $\begin{bmatrix} x_{1f} \\ x_{2f} \\ t_f \end{bmatrix}$ that are defined by the three-vector $\begin{bmatrix} e^{A(t-t_0)}\mathbf{x}_0 \\ t \end{bmatrix}$. Note further that all points along the continuous path from $\mathbf{x}_0$ to $\mathbf{x}_f$ are two-dimensional vectors in the $\{x_1, x_2\}$ **state-space** (the plane perpendicular to the time axis) at all values of time t between t_0 and t_f.

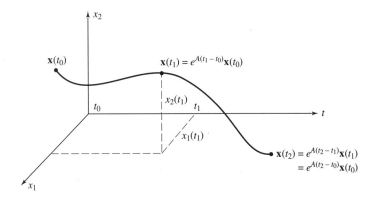

● **FIGURE 2.12**
A Pictorial Representation of the State Transition Matrix

The Zero-Input and Zero-State Responses

We next note that the particular $\mathbf{x}(t)$ given by Eq. (2.5.2) is called the **zero-input response** because it is dependent only on the initial state $\mathbf{x}_0$, when input $\mathbf{u}(t) = 0$.

Once the solution to the homogeneous state equation is determined, it is a relatively simple matter to obtain the **zero-state response**. The initial zero-state solution to the **nonhomogeneous** (or *forced*, so that $\mathbf{u}(t) \neq 0$) state-space Eq. (2.4.8a), namely, $\dot{\mathbf{x}}(t) = A\mathbf{x}(t) + B\mathbf{u}(t)$, is the "convolution" of $B\mathbf{u}(t)$ with the state transition matrix e^{At}. More specifically, if $\mathbf{x}(t_0) = \mathbf{x}_0 = 0$, then the zero-state response

$$\mathbf{x}(t) = \int_{t_0}^{t} e^{A(t-\tau)} B\mathbf{u}(\tau)\, d\tau \tag{2.5.6}$$

as we can now verify by recalling **Leibnitz's rule**, which states that for any "smooth" function $f(t, \tau)$ of two variables,

$$\frac{d}{dt} \int_{t_0}^{t} f(t, \tau)\, d\tau = f(t, t) + \int_{t_0}^{t} \frac{\partial}{\partial t} f(t, \tau)\, d\tau \tag{2.5.7}$$

Looking at Eq. (2.5.7), it follows that if $\mathbf{x}(t)$ is given by Eq. (2.5.6), then

$$\dot{\mathbf{x}}(t) = \frac{\partial}{\partial t} \int_{t_0}^{t} e^{A(t-\tau)} B\mathbf{u}(\tau)\, d\tau =$$

$$B\mathbf{u}(t) + \int_{t_0}^{t} A e^{A(t-\tau)} B\mathbf{u}(\tau)\, d\tau = A\mathbf{x}(t) + B\mathbf{u}(t) \tag{2.5.8}$$

or that Eq. (2.4.8a) holds for this particular choice of $\mathbf{x}(t)$.

The Complete Response

The **complete response** or the **solution** to nonhomogeneous state-space Eq. (2.4.8a), when both $\mathbf{x}_0 \neq 0$ and $\mathbf{u}(t) \neq 0$, is now given as the linear combination of the zero-input response Eq. (2.5.2) and the zero-state response Eq. (2.5.6).

$$\mathbf{x}(t) = \underbrace{e^{A(t-t_0)}\mathbf{x}(t_0)}_{\text{zero-input response}} + \underbrace{\int_{t_0}^{t} e^{A(t-\tau)} B\mathbf{u}(\tau)\, d\tau}_{\text{zero-state response}} \qquad (2.5.9)$$

We note finally that the complete output response $\mathbf{y}(t) = C\mathbf{x}(t) + E\mathbf{u}(t)$ can also be expressed as the sum of a zero-input response and a zero-state response.

$$\mathbf{y}(t) = \underbrace{Ce^{A(t-t_0)}\mathbf{x}_0}_{\text{zero-input response}} + \underbrace{C\int_{t_0}^{t} e^{A(t-\tau)} B\mathbf{u}(\tau)\, d\tau + E\mathbf{u}(t)}_{\text{zero-state response}} \qquad (2.5.10)$$

Modal Analysis

Although Eqs. (2.5.9) and (2.5.10) represent complete mathematical solutions to Eq. (2.4.8), the only expression we have for $e^{A(t-t_0)}$ is the infinite series expression Eq. (2.5.3), which is not particularly practical to evaluate. To remedy this situation, we recall (see Appendix A) that

$$|\lambda I - A| = \lambda^n + a_{n-1}\lambda^{n-1} + \cdots + a_1\lambda + a_0 \qquad (2.5.11)$$

is the **characteristic polynomial of** A, whose n zeros are the **eigenvalues** of A. Therefore, if A has *distinct* eigenvalues $\lambda_1, \lambda_2, \ldots, \lambda_n$,

$$AV = V\Lambda \qquad (2.5.12)$$

where Λ is an $(n \times n)$ diagonal, or **Jordan form matrix**, comprising the eigenvalues of A, that is,

$$\Lambda = \text{diag}[\lambda_i] = \begin{bmatrix} \lambda_1 & 0 & \cdots & 0 \\ 0 & \lambda_2 & \cdots & 0 \\ \vdots & & \ddots & \vdots \\ 0 & 0 & \cdots & \lambda_n \end{bmatrix} \qquad (2.5.13)$$

and

$$V = [V_1, V_2, \ldots, V_n] \qquad (2.5.14)$$

is a matrix whose columns V_i consist of n corresponding **eigenvectors** of A.

In general, both V and Λ will contain complex entries that, for physical system representations, will always appear together with their conjugates, although all of the entries of A are real. We further note that since the eigenvectors V_i are linearly independent, V will be nonsingular. As a consequence, A and Λ will be *similar matrices* in the sense that

$$A = V\Lambda V^{-1} \qquad (2.5.15)$$

Now recall Eq. (2.5.3). If $V\Lambda V^{-1}$ is substituted for A in the expression for the state transition matrix, it follows that

$$e^{A(t-t_0)} = VV^{-1} + V\Lambda V^{-1}(t-t_0) + \frac{V\Lambda^2 V^{-1}(t-t_0)^2}{2!} + \cdots$$

or that

$$e^{A(t-t_0)} = V(e^{\Lambda(t-t_0)})V^{-1} \qquad (2.5.16)$$

Since

$$e^{\Lambda(t-t_0)} = \begin{bmatrix} e^{\lambda_1(t-t_0)} & 0 & \cdots & 0 \\ 0 & e^{\lambda_2(t-t_0)} & \cdots & 0 \\ \vdots & & \ddots & \vdots \\ 0 & 0 & \cdots & e^{\lambda_n(t-t_0)} \end{bmatrix} \qquad (2.5.17)$$

and all of the entries of V and its inverse are constant, it follows that the zero-input response of a state-space system, $e^{A(t-t_0)}\mathbf{x}(t_0) = V(e^{\Lambda(t-t_0)})V^{-1}\mathbf{x}(t_0)$, consists of a linear combination of terms involving $e^{\lambda_i(t-t_0)}$, in which the λ_i are the distinct[3] eigenvalues of the state matrix A.

In light of this observation, the n exponential terms $e^{\lambda_i t}$, for $i = 1, 2, \ldots, n$, will be called the natural or unforced system **modes**, and both the state matrix Λ of Eq. (2.5.13) and the corresponding state transition matrix of Eq. (2.5.17) will be in a modal or Jordan form [28]. Equation (2.5.16) represents a way of determining a closed-form expression for the state transition matrix of a state-space system as a function of its modes.

....

EXAMPLE 2.5.18 Consider an unforced state-space system, $\dot{\mathbf{x}}(t) = A\mathbf{x}(t)$, with a bottom-row **companion form** (see Appendix A) state matrix

$$A = \begin{bmatrix} 0 & 1 & 0 \\ 0 & 0 & 1 \\ 6 & -1 & -4 \end{bmatrix}$$

Since

$$|\lambda I - A| = \lambda^3 + 4\lambda^2 + \lambda - 6 = (\lambda - 1)(\lambda + 2)(\lambda + 3)$$

the eigenvalues of A lie at $+1$, -2, and -3 in the complex plane.

---

[3]The state matrices of the vast majority of dynamic systems will have distinct eigenvalues. However, when any of the eigenvalues λ_i is repeated (k times), "modal terms" of the form $t^j e^{\lambda_i(t-t_0)}$, for $j = 1, 2, \ldots, k$, may also appear in the system response (see Problem 2-7).

If we now define $\lambda_1 = +1$, $\lambda_2 = -2$, and $\lambda_3 = -3$, it follows that the nonsingular **Vandermonde matrix**

$$V = \begin{bmatrix} 1 & 1 & 1 \\ \lambda_1 & \lambda_2 & \lambda_3 \\ \lambda_1^2 & \lambda_2^2 & \lambda_3^2 \end{bmatrix} = \begin{bmatrix} 1 & 1 & 1 \\ 1 & -2 & -3 \\ 1 & 4 & 9 \end{bmatrix} \qquad (2.5.19)$$

represents an appropriate eigenvector matrix that diagonalizes A. In particular, the reader can verify that

$$V^{-1}AV = \Lambda = \begin{bmatrix} 1 & 0 & 0 \\ 0 & -2 & 0 \\ 0 & 0 & -3 \end{bmatrix}$$

so that (setting $t_0 = 0$ for convenience)

$$e^{\Lambda(t-t_0)} = e^{\Lambda t} = \begin{bmatrix} e^t & 0 & 0 \\ 0 & e^{-2t} & 0 \\ 0 & 0 & e^{-3t} \end{bmatrix}$$

As a consequence of Eq. (2.5.16), we obtain a closed-form expression for the state transition matrix,

$$e^{At} = [Ve^{\Lambda t}]V^{-1} = \begin{bmatrix} e^t & e^{-2t} & e^{-3t} \\ e^t & -2e^{-2t} & -3e^{-3t} \\ e^t & 4e^{-2t} & 9e^{-3t} \end{bmatrix} \begin{bmatrix} 0.5 & 0.417 & 0.083 \\ 1 & -0.667 & -0.333 \\ -0.5 & 0.25 & 0.25 \end{bmatrix}$$

$$= \begin{bmatrix} 0.5e^t + e^{-2t} - 0.5e^{-3t}, & 0.417e^t - 0.667e^{-2t} + 0.25e^{-3t}, & 0.083e^t - 0.333e^{-2t} + 0.25e^{-3t} \\ 0.5e^t - 2e^{-2t} + 1.5e^{-3t}, & 0.417e^t + 1.333e^{-2t} - 0.75e^{-3t}, & 0.083e^t + 0.667e^{-2t} - 0.75e^{-3t} \\ 0.5e^t + 4e^{-2t} - 4.5e^{-3t}, & 0.417e^t - 2.667e^{-2t} + 2.25e^{-3t}, & 0.083e^t - 1.333e^{-2t} + 2.25e^{-3t} \end{bmatrix}$$

It may be noted that the direct substitution of $(t - t_0)$ for t in the above matrix eliminates the restriction that $t_0 = 0$, thereby yielding $e^{A(t-t_0)}$.

Equivalent State-Space Systems

It is of interest to note that the diagonalization of the state matrix A by an eigenvector matrix V represents a special case of a more general "transformation of state." In particular, we have already observed at the beginning of Section 2.4 that "the n specific physical quantities that define the state of a dynamic system are not unique, although their number (n) is unique." We now formalize this observation mathematically with the following discussion.

Any two state-space systems $\{A, B, C, E\}$ and $\{\hat{A}, \hat{B}, \hat{C}, \hat{E}\}$ will be called **equivalent** if one can be obtained from the other by a **state transformation,** as defined by the relationship:

$$\hat{\mathbf{x}}(t) = Q\mathbf{x}(t) \qquad (2.5.20)$$

where Q is any $(n \times n)$ nonsingular, **state transformation matrix,** so that

$$\mathbf{x}(t) = Q^{-1}\hat{\mathbf{x}}(t) \qquad (2.5.21)$$

Therefore, for any given state-space representation Eq. (2.4.8), or

$$\dot{\mathbf{x}}(t) = A\mathbf{x}(t) + B\mathbf{u}(t); \qquad \mathbf{y}(t) = C\mathbf{x}(t) + E\mathbf{u}(t)$$

the substitution of $Q^{-1}\hat{\mathbf{x}}(t)$ for $\mathbf{x}(t)$ results in an **equivalent state-space representation.**

In particular, since $\dot{\mathbf{x}}(t) = Q^{-1}\dot{\hat{\mathbf{x}}}(t)$, we have $Q^{-1}\dot{\hat{\mathbf{x}}}(t) = AQ^{-1}\hat{\mathbf{x}}(t) + B\mathbf{u}(t)$ and $\mathbf{y}(t) = CQ^{-1}\hat{\mathbf{x}}(t) + E\mathbf{u}(t)$, or

$$\dot{\hat{\mathbf{x}}}(t) = \underbrace{QAQ^{-1}}_{\hat{A}}\hat{\mathbf{x}}(t) + \underbrace{QB}_{\hat{B}}\mathbf{u}(t); \qquad \mathbf{y}(t) = \underbrace{CQ^{-1}}_{\hat{C}}\hat{\mathbf{x}}(t) + \underbrace{E}_{\hat{E}}\mathbf{u}(t)$$

$$(2.5.22)$$

with $\{\hat{A}, \hat{B}, \hat{C}, \hat{E}\} = \{QAQ^{-1}, QB, CQ^{-1}, E\}$ equivalent to $\{A, B, C, E\}$, a relationship we express as

$$\{A, B, C, E\} \overset{\text{equiv}}{\Longleftrightarrow} \{\hat{A}, \hat{B}, \hat{C}, \hat{E}\} = \{QAQ^{-1}, QB, CQ^{-1}, E\}$$

$$(2.5.23)$$

A "sequential" block diagram interpretation of an equivalence state transformation is depicted in Figure 2.13.

Note that a similarity transformation of the state results in a new but equivalent state-space representation that preserves all of the "essential features" of the original representation. In particular, we have already shown by Eq. (2.5.16) that *the state transition matrices of equivalent state-space systems are similar.* We now further observe that *the modes of equivalent systems are identical,* because similar matrices (such as A and $\hat{A} = QAQ^{-1}$) have identical eigenvalues.

2.6   STATE-SPACE CONTROLLABILITY AND OBSERVABILITY

The ultimate objective of this textbook is to design control systems that improve and often "optimize" the performance of a given dynamic system. Therefore, an obvious question that we will address is *how* do we design an appropriate controller? Before we can resolve this question, however, we first will consider whether or not an appropriate controller exists, that is, *can* we design a satisfactory controller?

Most physical systems are designed so that the control input affects the "complete system" and, as a consequence, an appropriate controller exists. However, this is not always the case. Moreover, in the multi-input and multi-output (MIMO) cases, certain control inputs may affect only "part" of the dynamic behavior. For example, the steering wheel of an automobile does not affect its speed, nor does the accelerator affect its heading; that is, the speed of an automobile is "uncontrollable" by the steering wheel and the heading is "uncontrollable" by the accelerator.

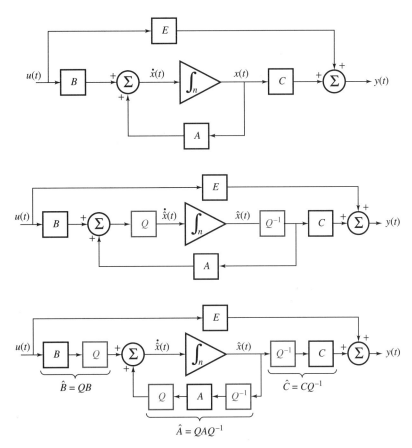

● **FIGURE 2.13**
Equivalent State-Space Systems

In general, it is often of interest to determine whether complete system control is possible if one or more of the inputs (actuators) or outputs (sensors) fails to perform as expected.

The primary purpose of the next two sections is to introduce two fundamental concepts associated with dynamic systems, "controllability" and "observability," which will enable us to resolve the "can" question in the linear time-invariant case. These "dual" concepts, which were first defined by R. E. Kalman [39] using state-space representations, are by no means restricted to systems described in state-space form. Indeed, problems associated with analyzing and controlling systems that were either uncontrollable or unobservable were encountered long before the state-space approach to control system analysis and design was popularized in the early 1960s. This section will deal with the controllability and observability properties of systems described by state-space representations.

The Modal Canonical Form

Consider a *single-input/single-output* (SISO) state-space system defined by Eq. (2.4.8), whose state matrix A has n distinct eigenvalues, $\lambda_1, \lambda_2, \ldots, \lambda_n$. As we noted in the previous section, this A can be diagonalized by any one of its eigenvector matrices V. More specifically, using Eq. (2.5.15), there exists a state transformation matrix $Q = V^{-1}$ that diagonalizes A, so that if

$$\hat{\mathbf{x}}(t) = Q\mathbf{x}(t) = V^{-1}\mathbf{x}(t) \tag{2.6.1}$$

then the dynamic behavior of the equivalent system in **modal canonical form** is defined by the state-space representation:

$$\dot{\hat{\mathbf{x}}}(t) = \begin{bmatrix} \dot{\hat{x}}_1(t) \\ \dot{\hat{x}}_2(t) \\ \vdots \\ \dot{\hat{x}}_n(t) \end{bmatrix} = \underbrace{\begin{bmatrix} \lambda_1 & 0 & 0 & \cdots \\ 0 & \lambda_2 & 0 & \cdots \\ \vdots & & \ddots & \\ 0 & 0 & \cdots & \lambda_n \end{bmatrix}}_{QAQ^{-1} = \hat{A}} \begin{bmatrix} \hat{x}_1(t) \\ \hat{x}_2(t) \\ \vdots \\ \hat{x}_n(t) \end{bmatrix} + \underbrace{\begin{bmatrix} \hat{B}_{11} \\ \hat{B}_{21} \\ \vdots \\ \hat{B}_{n1} \end{bmatrix}}_{QB = \hat{B}} u(t);$$

$$y(t) = \underbrace{[\hat{C}_{11}, \hat{C}_{12}, \ldots, \hat{C}_{1n}]}_{CQ^{-1} = \hat{C}} \begin{bmatrix} \hat{x}_1(t) \\ \hat{x}_2(t) \\ \vdots \\ \hat{x}_n(t) \end{bmatrix} + Eu(t) \tag{2.6.2}$$

as depicted in Figure 2.14.

Controllability

We now observe that if $\hat{B}_{k1} = 0$ for any $k = 1, 2, \ldots, n$, the state $\hat{x}_k(t)$ is **uncontrollable** by the input $u(t) = u_1(t)$, since its time behavior is characterized by the mode $e^{\lambda_k t}$, independent of $u(t)$; that is,

$$\hat{x}_k(t) = e^{\lambda_k(t-t_0)}\hat{x}_k(t_0) \tag{2.6.3}$$

The lack of controllability of the state $\hat{x}_k(t)$ (or the mode $e^{\lambda_k t}$) by $u(t)$ is reflected by a completely zero kth row of the so-called **controllability matrix** of the system, namely, the $(n \times n)$ matrix

$$\hat{C} \stackrel{\text{def}}{=} [\hat{B}, \hat{A}\hat{B}, \ldots, \hat{A}^{n-1}\hat{B}] = \begin{bmatrix} \hat{B}_{11} & \lambda_1\hat{B}_{11} & \cdots & \lambda_1^{n-1}\hat{B}_{11} \\ \hat{B}_{21} & \lambda_2\hat{B}_{21} & \cdots & \lambda_2^{n-1}\hat{B}_{21} \\ \vdots & \vdots & \ddots & \vdots \\ \hat{B}_{n1} & \lambda_n\hat{B}_{n1} & \cdots & \lambda_n^{n-1}\hat{B}_{n1} \end{bmatrix} \tag{2.6.4}$$

because $\hat{A}^m = \Lambda^m = \text{diag}[\lambda_i^m]$, a diagonal matrix for all integers $m \geq 0$. Otherwise stated, each zero kth row element $\hat{B}_{k1}$ of $\hat{B}$ implies an uncontrollable state $\hat{x}_k(t)$, whose time behavior is characterized by the

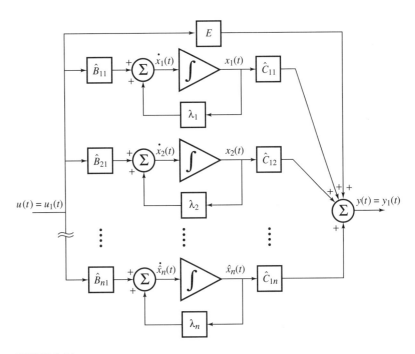

● **FIGURE 2.14**
A State-Space System in Modal Canonical Form

uncontrollable mode $e^{\lambda_k t}$, as well as a completely zero kth row of the controllability matrix $\hat{C}$.[4]

On the other hand, each nonzero kth row element of $\hat{B}$ implies a direct influence of $u(t)$ on $\hat{x}_k(t)$, hence a **controllable** state $\hat{x}_k(t)$ or mode $e^{\lambda_k t}$, and a corresponding nonzero kth row of $\hat{C}$ defined by $\hat{B}_{k1}[1, \lambda_k, \lambda_k^2, \ldots, \lambda_k^{n-1}]$. In the case assumed here of distinct eigenvalues, each such nonzero row of $\hat{B}$ increases the rank of $\hat{C}$ by one. Therefore, the rank of $\hat{C}$ corresponds to the total number of states or modes that are controllable by the input $u(t)$, which is termed the **controllability rank** of the system.

Fortunately, it is not necessary to transform a given state-space system to modal canonical form to determine its controllability rank. In particular, Eq. (2.6.2) implies that $B = Q^{-1}\hat{B}$, $AB = Q^{-1}\hat{A}QQ^{-1}\hat{B} = Q^{-1}\hat{A}\hat{B}$, or that $A^m B = Q^{-1}\hat{A}^m \hat{B}$ in general, so that the controllability matrix of the system defined by Eq. (2.4.8), namely

$$C \stackrel{\text{def}}{=} [B, AB, \ldots, A^{n-1}B] = Q^{-1}\hat{C} \qquad (2.6.5)$$

---

[4]The reader should be careful not to confuse the controllability matrix $\hat{C}$ of Eq. (2.6.4) with the output matrix $\hat{C}$ of Eq. (2.6.2).

with $Q^{-1} = V$ nonsingular. Therefore, *the rank of* C (which is equal to the rank of $\hat{C}$) *equals the controllability rank of the system.* It is important to note that this result holds in the case of nondistinct eigenvalues, as well as the multi-input case where B has m columns, so that

$$B = [B_1, \ B_2, \ \ldots, \ B_m] \tag{2.6.6}$$

and the controllability matrix C, as defined by Eq. (2.6.5), is an $(n \times nm)$ matrix. Therefore, any state-space system defined by Eq. (2.4.8) will be *completely (state or modal) controllable* if its $(n \times nm)$ controllability matrix C has full rank n.

In both the single-input and multi-input cases, it can be shown [39] that **complete controllability** implies that any initial state $\mathbf{x}_0$ at any initial time t_0 can be transferred to any final state $\mathbf{x}_f$ at any final time $t_f > t_0$ through the employment of an appropriate control input defined over the time interval $[t_0, \ t_f]$.

Observability

In view of Figure 2.14, we next note that if $\hat{C}_{1i} = 0$ for any $i = 1, 2, \ldots, n$, then the state $\hat{x}_i(t)$ is **unobservable** at the output $y(t) = y_1(t)$ because the mode $e^{\lambda_i t}$, that defines the time behavior of

$$\hat{x}_i(t) = e^{\lambda_i(t-t_0)}\hat{x}_i(t_0) \tag{2.6.7}$$

will not appear at the output $y(t)$. This lack of observability of the state $\hat{x}_i(t)$ (or the mode $e^{\lambda_i t}$) at $y(t)$ is reflected by a completely zero ith column of the so-called **observability matrix** of the system, namely, the $(n \times n)$ matrix

$$\hat{O} \overset{\text{def}}{=} \begin{bmatrix} \hat{C} \\ \hat{C}\hat{A} \\ \vdots \\ \hat{C}\hat{A}^{n-1} \end{bmatrix} = \begin{bmatrix} \hat{C}_{11} & \hat{C}_{12} & \cdots & \hat{C}_{1n} \\ \lambda_1\hat{C}_{11} & \lambda_2\hat{C}_{12} & \cdots & \lambda_n\hat{C}_{1n} \\ \vdots & \vdots & \ddots & \vdots \\ \lambda_1^{n-1}\hat{C}_{11} & \lambda_2^{n-1}\hat{C}_{12} & \cdots & \lambda_n^{n-1}\hat{C}_{1n} \end{bmatrix} \tag{2.6.8}$$

analogous to a completely zero kth row of $\hat{C}$ in Eq. (2.6.4), which corresponds to an uncontrollable state $\hat{x}_k(t)$.

On the other hand, each nonzero ith column element $\hat{C}_{1i}$ of $\hat{C}$ implies a direct influence of $\hat{x}_i(t)$ on $y(t)$, hence an **observable** state $\hat{x}_i(t)$ or mode $e^{\lambda_i t}$, and a corresponding nonzero ith column of $\hat{O}$ defined by $[1, \lambda_i, \lambda_i^2, \ldots, \lambda_i^{n-1}]^T \hat{C}_{1i}$. In the case assumed here of distinct eigenvalues, each such nonzero element of $\hat{C}$ increases the rank of $\hat{O}$ by one. Therefore, the rank of $\hat{O}$ corresponds to the total number of states or modes that are observable at the output $y(t)$, which is termed the **observability rank** of the system.

As in the case of controllability, it is not necessary to transform a given state-space system to modal canonical form to determine its observability rank. In particular, Eq. (2.6.2) implies that $C = \hat{C}Q$, $CA =$

$\hat{C}QQ^{-1}\hat{A}Q = \hat{C}\hat{A}Q$, or that $CA^m = \hat{C}\hat{A}^m Q$ in general, so that the observability matrix of the system defined by Eq. (2.4.8),

$$\mathcal{O} \overset{\text{def}}{=} \begin{bmatrix} C \\ CA \\ \vdots \\ CA^{n-1} \end{bmatrix} = \hat{\mathcal{O}}Q \tag{2.6.9}$$

with $Q = V^{-1}$ nonsingular. Therefore, *the rank of $\mathcal{O}$* (which is equal to the rank of $\hat{\mathcal{O}}$) *equals the observability rank of the system.* It is important to note that this result holds in the case of nondistinct eigenvalues, as well the multi-output case in which C has p rows, so that

$$C = \begin{bmatrix} C_1 \\ C_2 \\ \vdots \\ C_p \end{bmatrix} \tag{2.6.10}$$

and the observability matrix $\mathcal{O}$, as defined by Eq. (2.6.9), is a $(pn \times n)$ matrix. In view of this, a state-space system defined by Eq. (2.4.8) will be *completely (state or modal) observable* if its $(pn \times n)$ observability matrix $\mathcal{O}$ has full rank n.

In both the single-output and multi-output cases, it can be shown [39] that **complete observability** implies that knowledge of the input and the output over any finite time interval $[t_0, t_f]$ is sufficient to determine the entire initial state $\mathbf{x}(t_0) = \mathbf{x}_0$ and, as a consequence, $\mathbf{x}(t)$, for all t between t_0 and t_f.[5]

. . . .

EXAMPLE 2.6.11 Consider a SISO state-space system defined by the A matrix of Example 2.5.18, with

$$B = \begin{bmatrix} 0 \\ 0 \\ 1 \end{bmatrix}, \qquad C = [0 \quad 1 \quad -1], \quad \text{and} \quad E = 0$$

Using Eq. (2.6.5), the controllability matrix

$$\mathcal{C} = [B, \ AB, \ A^2 B] = \begin{bmatrix} 0 & 0 & 1 \\ 0 & 1 & -4 \\ 1 & -4 & 15 \end{bmatrix}$$

is nonsingular, since $|\mathcal{C}| = -1$. Therefore, the system is completely (state or modal) controllable.

However, using Eq. (2.6.9), the observability matrix

$$\mathcal{O} = \begin{bmatrix} C \\ CA \\ CA^2 \end{bmatrix} = \begin{bmatrix} 0 & 1 & -1 \\ -6 & 1 & 5 \\ 30 & -11 & -19 \end{bmatrix} \quad \text{with} \quad |\mathcal{O}| = -66 + 150 + 30 - 114 = 0$$

---

[5]In light of Eq. (2.5.9), knowledge of $\mathbf{x}(t_0)$ and $\mathbf{u}(t)$ for $t \in [t_0, t_f]$ directly implies knowledge of $\mathbf{x}(t)$ for all $t \in [t_0, t_f]$.

so that $\mathcal{O}$ is a singular matrix of rank $2 < n = 3$. Therefore, two of the system modes are observable and one is unobservable.

To determine the observable and unobservable modes,[6] the system can be transformed to modal canonical form by means of the state transformation matrix

$$Q = V^{-1} = \begin{bmatrix} 1 & 1 & 1 \\ 1 & -2 & -3 \\ 1 & 4 & 9 \end{bmatrix}^{-1} = \begin{bmatrix} 0.5 & 0.417 & 0.083 \\ 1 & -0.667 & -0.333 \\ -0.5 & 0.25 & 0.25 \end{bmatrix}$$

as in Example 2.5.18. In this particular case,

$$\hat{A} = V^{-1}AV = \begin{bmatrix} 1 & 0 & 0 \\ 0 & -2 & 0 \\ 0 & 0 & -3 \end{bmatrix} \quad \text{and} \quad \hat{C} = CQ^{-1} = CV = [0 \ -6 \ -12]$$

Since $\hat{C}_{11}$ is the zero element of $\hat{C}$, the state $\hat{x}_1(t)$, characterized by the mode $e^{\lambda_1 t} = e^t$, is unobservable, while the states $\hat{x}_2(t)$ and $\hat{x}_3(t)$, characterized by the modes $e^{\lambda_2 t} = e^{-2t}$ and $e^{\lambda_3 t} = e^{-3t}$, respectively, are observable. Using Eq. (2.6.1), note that $\hat{x}_1(t)$ is equal to the first row of Q times $\mathbf{x}(t)$. Therefore, a linear combination of the state variables of the original system,

$$\hat{x}_1(t) = [0.5 \quad 0.417 \quad 0.083]\mathbf{x}(t) = 0.5x_1(t) + 0.417x_2(t) + 0.083x_3(t)$$

represents the unobservable "state" of the given system.

· ·

Component Controllability and Observability

In the multi-input and multi-output cases, it is often useful to determine the controllability and observability rank of a system relative to the individual components of its input and output. Such a determination would be important, for example, if one or more of the input components (actuators) or output components (sensors) were to fail.

Suppose the system of Eq. (2.4.8) has $m > 1$ inputs, $u_1(t), u_2(t), \ldots,$ $u_m(t)$, so that the input matrix B has m columns, as in Eq. (2.6.6). If we disregard all inputs except $u_j(t)$, the resulting **controllability matrix associated with input** $u_j(t)$ is defined as the $(n \times n)$ matrix

$$\mathcal{C}_j \stackrel{\text{def}}{=} [B_j, \ AB_j, \ \ldots, \ A^{n-1}B_j] \tag{2.6.12}$$

In light of this observation, it follows that the rank of each such $\mathcal{C}_j$ determines the number of states or modes (but not their identity) that are controllable by $u_j(t)$, the jth component of the input.

In a "dual" manner, suppose the given state-space system has $p > 1$ outputs, $y_1(t), y_2(t), \ldots, y_p(t)$, so that the output matrix C has p rows,

---

[6]Note that for a general state-space system, the rank of $\mathcal{O}$ tells us only the *number* of observable (and unobservable) modes and not their identity, an observation that also holds with respect to the controllability properties of the system, as defined by the rank of $\mathcal{C}$.

as in Eq. (2.6.10). If we disregard all outputs except $y_q(t)$, the resulting **observability matrix associated with output** $y_q(t)$ is defined as the $(n \times n)$ matrix

$$
\mathcal{O}_q \stackrel{\text{def}}{=} \begin{bmatrix} C_q \\ C_q A \\ \vdots \\ C_q A^{n-1} \end{bmatrix}
\tag{2.6.13}
$$

As in the case of controllability, the rank of each such $\mathcal{O}_q$ determines the number of states or modes that are observable by $y_q(t)$, the qth component of the output.

- -

EXAMPLE 2.6.14 To illustrate the preceding discussion, recall the linearized equations of motion of the orbiting satellite of Example 2.4.7 with the $\delta x_i(t)$ replaced by $x_i(t)$ and $d = 1$, so that

$$
\underbrace{\begin{bmatrix} \dot{x}_1(t) \\ \dot{x}_2(t) \\ \dot{x}_3(t) \\ \dot{x}_4(t) \end{bmatrix}}_{\dot{\mathbf{x}}(t)} = \underbrace{\begin{bmatrix} 0 & 1 & 0 & 0 \\ 3\omega^2 & 0 & 0 & 2\omega \\ 0 & 0 & 0 & 1 \\ 0 & -2\omega & 0 & 0 \end{bmatrix}}_{A} \underbrace{\begin{bmatrix} x_1(t) \\ x_2(t) \\ x_3(t) \\ x_4(t) \end{bmatrix}}_{\mathbf{x}(t)} + \underbrace{\begin{bmatrix} 0 & 0 \\ 1 & 0 \\ 0 & 0 \\ 0 & 1 \end{bmatrix}}_{B} \underbrace{\begin{bmatrix} u_1(t) \\ u_2(t) \end{bmatrix}}_{\mathbf{u}(t)}
$$

and

$$
\underbrace{\begin{bmatrix} y_1(t) \\ y_2(t) \end{bmatrix}}_{\mathbf{y}(t)} = \underbrace{\begin{bmatrix} 1 & 0 & 0 & 0 \\ 0 & 0 & 1 & 0 \end{bmatrix}}_{C} \mathbf{x}(t)
$$

The reader can verify (see Problem 2-9) that the $(n \times nm = 4 \times 8)$ controllability matrix $\mathcal{C} = [B, AB, A^2 B, A^3 B]$ has full rank $4 = n$ in this case, so that the entire state is controllable using both inputs. However, since

$$
\mathcal{C}_1 = [B_1, AB_1, A^2 B_1, A^3 B_1] = \begin{bmatrix} 0 & 1 & 0 & -\omega^2 \\ 1 & 0 & -\omega^2 & 0 \\ 0 & 0 & -2\omega & 0 \\ 0 & -2\omega & 0 & 2\omega^3 \end{bmatrix}
$$

is singular, that is, $|\mathcal{C}_1| = 4\omega^4 - 4\omega^4 = 0$, with rank $\mathcal{C}_1 = 3 < 4 = n$, it follows that one of the "states" cannot be controlled by the radial thruster $u_1(t)$ alone, which would be unfortunate if the tangential thruster $u_2(t)$ were to fail.

We next note that

$$
\mathcal{C}_2 = [B_2, AB_2, A^2 B_2, A^3 B_2] = \begin{bmatrix} 0 & 0 & 2\omega & 0 \\ 0 & 2\omega & 0 & -2\omega^3 \\ 0 & 1 & 0 & -4\omega^2 \\ 1 & 0 & -4\omega^2 & 0 \end{bmatrix}
$$

is nonsingular, since $|\mathcal{C}_2| = 4\omega^4 - 16\omega^4 = -12\omega^4 \neq 0$. Thus, complete state control is possible by the tangential thruster $u_2(t)$ alone if the radial thruster $u_1(t)$ were to fail.

As far as observability is concerned, $y_1(t) = C_1\mathbf{x}(t) = x_1(t) = r(t) - d$ represents the radial deviation of $r(t)$ from a nominal value of $d = 1$, while output $y_2(t) = C_2\mathbf{x}(t) = x_3(t) = \alpha(t) - \omega t$ represents the tangential deviation of $\alpha(t)$ from ωt. The reader can verify (see Problem 2-10) that the $(pn \times n = 8 \times 4)$ observability matrix $\mathcal{O}$, as given by Eq. (2.6.9), has full rank $n = 4$ in this case, so that the entire state is observable using both outputs. However, since

$$\mathcal{O}_1 = \begin{bmatrix} C_1 \\ C_1 A \\ C_1 A^2 \\ C_1 A^3 \end{bmatrix} = \begin{bmatrix} 1 & 0 & 0 & 0 \\ 0 & 1 & 0 & 0 \\ 3\omega^2 & 0 & 0 & 2\omega \\ 0 & -\omega^2 & 0 & 0 \end{bmatrix}$$

is clearly singular (because its third column is zero), with rank $\mathcal{O}_1 = 3 < 4 = n$, it follows that one of the "states" cannot be observed by $y_1(t)$ alone.

We note finally that

$$\mathcal{O}_2 = \begin{bmatrix} C_2 \\ C_2 A \\ C_2 A^2 \\ C_2 A^3 \end{bmatrix} = \begin{bmatrix} 0 & 0 & 1 & 0 \\ 0 & 0 & 0 & 1 \\ 0 & -2\omega & 0 & 0 \\ -6\omega^3 & 0 & 0 & -4\omega^2 \end{bmatrix}$$

is nonsingular, since $|\mathcal{O}_2| = -12\omega^4 \neq 0$, so that the entire state can be observed by $y_2(t)$ alone.

The Multi-Input/Multi-Output (MIMO) Case

In the general MIMO case, the explicit modal controllability and observability properties of a system with distinct eigenvalues can be determined by transforming the system to modal canonical form. In particular, a zero in any kth row of column $\hat{B}_j$ of the input matrix $\hat{B}$ implies the uncontrollability of state $\hat{x}_k(t)$ (or the mode $e^{\lambda_k t}$) by $u_j(t)$. Furthermore, a completely zero kth row of $\hat{B}$ implies the complete uncontrollability of state $\hat{x}_k(t)$ (or the mode $e^{\lambda_k t}$) with respect to the entire vector input $\mathbf{u}(t)$. Each such zero row of $\hat{B}$ implies a corresponding zero row of $\hat{C}$, thereby reducing the rank of the $(n \times nm)$ controllability matrices $\hat{C}$ and C by one. The number of controllable modes therefore is given by the rank of $\hat{C}$ or C, the controllability rank of the system.

Dual results hold with respect to the observability properties of a system. A zero in any ith column of row $\hat{C}_q$ of the output matrix $\hat{C}$ implies the unobservability of state $\hat{x}_i(t)$ (or the mode $e^{\lambda_i t}$) by $y_q(t)$. Further, a completely zero ith column of $\hat{C}$ implies the complete unobservability of state $\hat{x}_i(t)$ (or the mode $e^{\lambda_i t}$) with respect to the entire vector output $\mathbf{y}(t)$. Each such zero column of $\hat{C}$ implies a corresponding zero column of $\hat{O}$, thereby reducing the rank of the $(pn \times n)$ observability matrices $\hat{O}$ and $\mathcal{O}$ by one. The number of observable modes therefore is given by the rank of $\hat{O}$ or $\mathcal{O}$, the observability rank of the system.

EXAMPLE 2.6.15 To illustrate the preceding discussion, consider a two-input, two-output, fourth-order state-space system

$$
\begin{bmatrix} \dot{x}_1(t) \\ \dot{x}_2(t) \\ \dot{x}_3(t) \\ \dot{x}_4(t) \end{bmatrix} = \underbrace{\begin{bmatrix} 0 & 0 & 0 & 15 \\ 1 & 0 & 0 & 22 \\ 0 & 1 & 0 & 6 \\ 0 & 0 & 1 & -2 \end{bmatrix}}_{A} \begin{bmatrix} x_1(t) \\ x_2(t) \\ x_3(t) \\ x_4(t) \end{bmatrix} + \underbrace{\begin{bmatrix} 5 & -1 \\ 4 & 0 \\ 1 & 1 \\ 0 & 0 \end{bmatrix}}_{B} \underbrace{\begin{bmatrix} u_1(t) \\ u_2(t) \end{bmatrix}}_{\mathbf{u}(t)}
$$

with

$$
\underbrace{\begin{bmatrix} y_1(t) \\ y_2(t) \end{bmatrix}}_{\mathbf{y}(t)} = \underbrace{\begin{bmatrix} 0 & 0 & -2 & 10 \\ 0 & 0 & -1 & 1 \end{bmatrix}}_{C} \mathbf{x}(t)
$$

The reader can verify, for example, using MATLAB [46], that the ($n \times nm = 4 \times 8$) controllability matrix

$$
\mathcal{C} = [B,\ AB,\ A^2B,\ A^3B] = \begin{bmatrix} 5 & -1 & 0 & 0 & 15 & 15 & 30 & -30 \\ 4 & 0 & 5 & -1 & 22 & 22 & 59 & -29 \\ 1 & 1 & 4 & 0 & 11 & 5 & 34 & 10 \\ 0 & 0 & 1 & 1 & 2 & -2 & 7 & 9 \end{bmatrix}
$$

has full rank $4 = n$ in this case, so that the system is completely state controllable by both inputs. However, the rank of the controllability matrix associated with input $u_1(t)$

$$
\mathcal{C}_1 = [B_1,\ AB_1,\ A^2B_1,\ A^3B_1] = \begin{bmatrix} 5 & 0 & 15 & 30 \\ 4 & 5 & 22 & 59 \\ 1 & 4 & 11 & 34 \\ 0 & 1 & 2 & 7 \end{bmatrix}
$$

is only 2, while the rank of the controllability matrix associated with input $u_2(t)$

$$
\mathcal{C}_2 = [B_2,\ AB_2,\ A^2B_2,\ A^3B_2] = \begin{bmatrix} -1 & 0 & 15 & -30 \\ 0 & -1 & 22 & -29 \\ 1 & 0 & 5 & 10 \\ 0 & 1 & -2 & 9 \end{bmatrix}
$$

is 3 (see Problem 2-13). We therefore conclude that while all four system modes are controllable by both inputs, two of the system modes are uncontrollable by $u_1(t)$, while one mode is uncontrollable by $u_2(t)$.

We observe next that the ($pn \times n = 8 \times 4$) observability matrix

$$
\mathcal{O} = \begin{bmatrix} C \\ CA \\ CA^2 \\ CA^3 \end{bmatrix} = \begin{bmatrix} 0 & 0 & -2 & 10 \\ 0 & 0 & -1 & 1 \\ 0 & -2 & 10 & -32 \\ 0 & -1 & 1 & -8 \\ -2 & 10 & -32 & 80 \\ -1 & 1 & -8 & 0 \\ 10 & -32 & 80 & -162 \\ 1 & -8 & 0 & -41 \end{bmatrix}
$$

has full rank $4 = n$, so that the system is completely state observable by both outputs. However, the rank of the observability matrix associated with output $y_1(t)$

$$\mathcal{O}_1 = \begin{bmatrix} C_1 \\ C_1 A \\ C_1 A^2 \\ C_1 A^3 \end{bmatrix} = \begin{bmatrix} 0 & 0 & -2 & 10 \\ 0 & -2 & 10 & -32 \\ -2 & 10 & -32 & 80 \\ 10 & -32 & 80 & -162 \end{bmatrix}$$

is only 3 (see Problem 2-14), while the rank of the observability matrix associated with output $y_2(t)$

$$\mathcal{O}_2 = \begin{bmatrix} C_2 \\ C_2 A \\ C_2 A^2 \\ C_2 A^3 \end{bmatrix} = \begin{bmatrix} 0 & 0 & -1 & 1 \\ 0 & -1 & 1 & -8 \\ -1 & 1 & -8 & 0 \\ 1 & -8 & 0 & -41 \end{bmatrix}$$

is also 3. We therefore conclude that while all four system modes are observable by both outputs, one mode is unobservable by $y_1(t)$, and one mode is unobservable by $y_2(t)$.

To determine these modes, we can now transform the system to modal canonical form by the complex similarity transformation matrix[7]

$$Q = \begin{bmatrix} -0.111 & -0.331 & -0.994 & -2.983 \\ -0.553 + 0.419j & 0.687 - 1.39j & 0.017 + 3.467j & -3.5 - 6.917j \\ -0.553 - 0.419j & 0.687 + 1.39j & 0.017 - 3.467j & -3.5 + 6.917j \\ -2.077 & 2.077 & -2.077 & 2.077 \end{bmatrix}$$

to obtain the equivalent state-space system

$$\underbrace{\begin{bmatrix} \dot{\hat{x}}_1(t) \\ \dot{\hat{x}}_2(t) \\ \dot{\hat{x}}_3(t) \\ \dot{\hat{x}}_4(t) \end{bmatrix}}_{\dot{\mathbf{\hat{x}}}(t)} = \underbrace{\begin{bmatrix} 3 & 0 & 0 & 0 \\ 0 & -2+j & 0 & 0 \\ 0 & 0 & -2-j & 0 \\ 0 & 0 & 0 & -1 \end{bmatrix}}_{\hat{A} = QAQ^{-1} = \Lambda} \underbrace{\begin{bmatrix} \hat{x}_1(t) \\ \hat{x}_2(t) \\ \hat{x}_3(t) \\ \hat{x}_4(t) \end{bmatrix}}_{\mathbf{\hat{x}}(t)} + \underbrace{\begin{bmatrix} -2.87 & -0.88 \\ 0 & 0.57+3.05 \\ 0 & 0.57-3.05j \\ -4.15 & 0 \end{bmatrix}}_{\hat{B} = QB} \underbrace{\begin{bmatrix} u_1(t) \\ u_2(t) \end{bmatrix}}_{\mathbf{u}(t)}$$

with output

$$\underbrace{\begin{bmatrix} y_1(t) \\ y_2(t) \end{bmatrix}}_{\mathbf{y}(t)} = \underbrace{\begin{bmatrix} 0 & -1.01 + 0.139j & -1.01 - 0.139j & 0.482 \\ 0.348 & -0.106 + 0.094j & -0.106 - 0.094j & 0 \end{bmatrix}}_{\hat{C} = CQ^{-1}} \begin{bmatrix} \hat{x}_1(t) \\ \hat{x}_2(t) \\ \hat{x}_3(t) \\ \hat{x}_4(t) \end{bmatrix}$$

By associating the three zero elements of $\hat{B} = QB$, namely, $\hat{B}_{21} = \hat{B}_{31} = \hat{B}_{42} = 0$, with the corresponding, diagonal elements of $\hat{A}$, it is immediately apparent that the states $\hat{x}_2(t)$ and $\hat{x}_3(t)$, which are characterized by the modes[8] $e^{(-2+j)t}$ and $e^{(-2-j)t}$, are uncontrollable by $u_1(t)$, while the

---

[7]This particular matrix Q was obtained using MATLAB [46].

[8]As we will show in the next chapter, complex modes are always "paired" with their conjugates, so that if any complex mode $e^{(\alpha+j\beta)t}$ is uncontrollable, then its complex conjugate mode $e^{(\alpha-j\beta)t}$ will also be uncontrollable.

state $\hat{x}_4(t)$, or the mode e^{-t}, is uncontrollable by $u_2(t)$. The state $\hat{x}_1(t)$, or the mode e^{3t}, therefore is the only one that is controllable by both inputs. We further observe that the rows of Q will define linear functions of the original state $\mathbf{x}(t)$ that are either controllable or uncontrollable by the defined inputs (see Problem 2-15).

By associating the two zero elements of $\hat{C}$, namely, $\hat{C}_{11} = \hat{C}_{24} = 0$, with the corresponding, diagonal elements of $\hat{A}$, it is immediately apparent that the state $\hat{x}_1(t)$, or the mode e^{3t}, is unobservable by output $y_1(t)$, while the state $\hat{x}_4(t)$, or the mode e^{-t}, is unobservable by $y_2(t)$. The other two states, $\hat{x}_2(t)$ and $\hat{x}_3(t)$, or the modes $e^{(-2+j)t}$ and $e^{(-2-j)t}$, are observable by both outputs.

We have now shown that a system whose state matrix A has n distinct eigenvalues λ_i will have n corresponding modes $e^{\lambda_i t}$. Moreover, each of these modes can be classified as either *controllable* or *uncontrollable* by each component of the input, as well as either *observable* or *unobservable* by each component of the output. If a system is in modal canonical form, then the n components of its state can also be classified in the same manner, which usually is not the case for general state-space representations. Therefore, the concepts of controllability and observability are more appropriately associated with the modes, rather than the states, of a given dynamic system.

2.7 DIFFERENTIAL OPERATOR REPRESENTATIONS

We now show how the controllability and observability properties of a dynamic system can be determined directly from the differential equations that describe its behavior, independent of any state-space representation. The *RLC* network and the spring-mass-damper system of Section 2.2 represent two such systems whose dynamic behavior is specified by differential operator equations. Although it is possible to express the dynamic behavior of these two systems in state-space form, this is not necessary whenever the defining differential equations can be placed in the **differential operator form**:

$$a(D)z(t) = b(D)u(t); \qquad y(t) = c(D)z(t) + e(D)u(t) \qquad (2.7.1)$$

where $a(D)$, $b(D)$, $c(D)$, and $e(D)$ are polynomials[9] in the **differential operator** $D = \frac{d}{dt}$, with $a(D)$ a monic polynomial of degree n, which defines the **order** of this representation, and $z(t)$ is a single-valued function of time that is called the **partial state.** We will often find it convenient to "abbreviate" Eq. (2.7.1) by the polynomial quadruple $\{a(D), b(D), c(D), e(D)\}$, that is, $\{a(D), b(D), c(D), e(D)\} \iff$ Eq. (2.7.1).

---

[9]We will later allow both $b(D)$ and $c(D)$ to be polynomial vectors, thereby enlarging the class of systems considered beyond the SISO case defined by Eq. (2.7.1).

An Equivalent State-Space Representation

We first show that Eq. (2.7.1) has an equivalent state-space representation that can be determined directly by inspection of $a(D)$ and $b(D)$ when the degree of $b(D)$, $\deg[b(D)] < n = \deg[a(D)]$. Suppose we employ the coefficients of

$$a(D) = D^n + a_{n-1}D^{n-1} + \cdots + a_1 D + a_0$$

and

$$b(D) = b_{n-1}D^{n-1} + \cdots + b_1 D + b_0$$

to define the following state-space system:

$$\underbrace{\begin{bmatrix} \dot{x}_1(t) \\ \dot{x}_2(t) \\ \vdots \\ \dot{x}_n(t) \end{bmatrix}}_{\dot{\mathbf{x}}(t)} = \underbrace{\begin{bmatrix} 0 & 0 & 0 & \cdots & -a_0 \\ 1 & 0 & 0 & \cdots & -a_1 \\ 0 & 1 & 0 & \cdots & \\ \vdots & \vdots & \vdots & \ddots & \vdots \\ 0 & 0 & \cdots & 1 & -a_{n-1} \end{bmatrix}}_{A} \underbrace{\begin{bmatrix} x_1(t) \\ x_2(t) \\ \vdots \\ x_n(t) \end{bmatrix}}_{\mathbf{x}(t)} + \underbrace{\begin{bmatrix} b_0 \\ b_1 \\ \vdots \\ b_{n-1} \end{bmatrix}}_{B} u(t)$$

$$(2.7.2)$$

with

$$z(t) \overset{\text{def}}{=} x_n(t) = [0 \quad 0 \quad \cdots \quad 0 \quad 1]\mathbf{x}(t) \overset{\text{def}}{=} C_z\mathbf{x}(t) \qquad (2.7.3)$$

Since A is a right-column companion matrix, it follows (see Appendix A) that the characteristic polynomial of A, is given by

$$|\lambda I - A| = \lambda^n + a_{n-1}\lambda^{n-1} + \cdots + a_1\lambda + a_0 = a(\lambda) \qquad (2.7.4)$$

Therefore, the n zeros of $a(\lambda)$ will correspond to the n eigenvalues λ_i of A, which define the system modes $e^{\lambda_i t}$. As in the previous section, we will assume that these n eigenvalues of A are distinct.

In terms of the differential operator D, Eq. (2.7.2) can be written as

$$\underbrace{\begin{bmatrix} D & 0 & 0 & \cdots & a_0 \\ -1 & D & 0 & \cdots & a_1 \\ 0 & -1 & D & \cdots & a_2 \\ \vdots & \vdots & & \ddots & \vdots \\ 0 & 0 & \cdots & -1 & D + a_{n-1} \end{bmatrix}}_{(DI - A)} \underbrace{\begin{bmatrix} x_1(t) \\ x_2(t) \\ \vdots \\ x_n(t) \end{bmatrix}}_{\mathbf{x}(t)} = \underbrace{\begin{bmatrix} b_0 \\ b_1 \\ \vdots \\ b_{n-1} \end{bmatrix}}_{B} u(t) \quad (2.7.5)$$

If we now premultiply Eq. (2.7.5) by the row vector $[1 \quad D \quad D^2 \quad \cdots \quad D^{n-1}]$, noting that $x_n(t) = z(t)$, we obtain the relationship:

$$[0 \quad 0 \quad \cdots \quad 0 \quad a(D)] \begin{bmatrix} x_1(t) \\ x_2(t) \\ \vdots \\ x_n(t) \end{bmatrix} = a(D)z(t) = b(D)u(t) \qquad (2.7.6)$$

thereby establishing the equivalence of the state-space system defined by Eq. (2.7.2) and the partial state/input relation $a(D)z(t) = b(D)u(t)$ of Eq. (2.7.1).

Since $x_n(t) = z(t)$ in light of Eq. (2.7.3), Eq. (2.7.2) implies that

$$Dz(t) = \dot{x}_n(t) = x_{n-1}(t) - a_{n-1}x_n(t) + b_{n-1}u(t)$$

$$D^2z(t) = \dot{x}_{n-1}(t) - a_{n-1}\dot{x}_n(t) + b_{n-1}\dot{u}(t)$$

$$= x_{n-2}(t) - a_{n-2}x_n(t) + b_{n-2}u(t)$$

$$-a_{n-1}[x_{n-1}(t) - a_{n-1}x_n(t) + b_{n-1}u(t)] + b_{n-1}\dot{u}(t)$$

and so on, which enables us to express the output relationship of Eq. (2.7.1)

$$y(t) = c(D)z(t) + e(D)u(t) = c(D)x_n(t) + e(D)u(t)$$

as a function of $\mathbf{x}(t)$ and $u(t)$ and its derivatives. As a consequence,

$$y(t) = C\mathbf{x}(t) + E(D)u(t) \tag{2.7.7}$$

for some constant $(1 \times n)$ vector C and a corresponding polynomial $E(D)$.

We have therefore established a **complete equivalence relationship** between the differential operator representation of Eq. (2.7.1) and the state-space representation defined by Eqs. (2.7.2) and (2.7.7), with E "expanded" to $E(D)$ (if necessary) to include derivatives of the input. We denote this equivalence relationship as

$$\underbrace{\{A,\ B,\ C,\ E(D)\}}_{\text{of Eqs.(2.7.2) and (2.7.7)}} \overset{\text{equiv}}{\Longleftrightarrow} \underbrace{\{a(D),\ b(D),\ c(D),\ e(D)\}}_{\text{of Eq.(2.7.1)}} \tag{2.7.8}$$

Observable Canonical Forms

If $c(D) = 1$ in Eq. (2.7.1), so that

$$a(D)z(t) = b(D)u(t); \qquad y(t) = z(t) + e(D)u(t) \tag{2.7.9}$$

the equivalent state-space system defined by Eqs. (2.7.2) and (2.7.7) is characterized by an output matrix $C = C_z = [0 \quad 0 \quad \cdots \quad 0 \quad 1]$ and $E(D) = e(D)$ in Eq. (2.7.7), that is,

$$y(t) = \underbrace{[0 \quad 0 \quad \cdots \quad 0 \quad 1]}_{C = C_z}\mathbf{x}(t) + \underbrace{E(D)}_{e(D)}u(t) \tag{2.7.10}$$

Therefore, Eqs. (2.7.2) and (2.7.10) represent a state-space system equivalent to the differential operator system defined by Eq. (2.7.9). We denote this equivalence relationship as

$$\underbrace{\{A,\ B,\ C,\ E(D)\}}_{\text{of Eqs. (2.7.2) and (2.7.10)}} \overset{\text{equiv}}{\Longleftrightarrow} \underbrace{\{a(D),\ b(D),\ c(D) = 1,\ e(D)\}}_{\text{of Eq. (2.7.9)}} \tag{2.7.11}$$

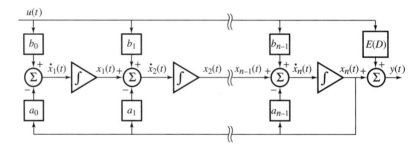

● **FIGURE 2.15**
A State-Space System in Observable Canonical Form

Moreover, *both of these representations are completely observable.* In particular, the differential operator representation is observable because $a(D)$ and $c(D) = 1$ are coprime,[10] and the state-space representation is observable because its observability matrix

$$\mathcal{O} = \begin{bmatrix} C \\ CA \\ CA^2 \\ \vdots \\ CA^{n-1} \end{bmatrix} = \begin{bmatrix} 0 & \cdots & 0 & 1 \\ 0 & \cdots & 1 & * \\ \vdots & & * & \vdots \\ 0 & 1 & \cdots & * \\ 1 & * & \cdots & * \end{bmatrix} \qquad (2.7.12)$$

is nonsingular (the $*$'s denoting irrelevant scalars).

Note further that the $\{A, C\}$ pair of Eqs. (2.7.2) and (2.7.10) is in a special "canonical" form. In particular, A is a right-column companion matrix and C is identically zero except for a 1 in the right column. In light of these observations, we say that both of the representations defined by Eq. (2.7.11) are in **observable canonical form**. Figure 2.15 depicts a block diagram of a state-space system in observable canonical form, as defined by Eqs. (2.7.2) and (2.7.10).

Differential Operator Controllability

Because of the right-column companion form structure of A in Eq. (2.7.2), it follows that in the case assumed here of distinct eigenvalues,[11] the vector $[1 \quad \lambda_i \quad \lambda_i^2 \quad \cdots \quad \lambda_i^{n-1}]$ is a row eigenvector of A in the sense that

$$[1 \quad \lambda_i \quad \lambda_i^2 \quad \cdots \quad \lambda_i^{n-1}]A = \lambda_i[1 \quad \lambda_i \quad \lambda_i^2 \quad \cdots \quad \lambda_i^{n-1}] \qquad (2.7.13)$$

---

[10]We will formally establish this condition for differential operator observability later in this section.

[11]However, the results presented do hold in the case of nondistinct eigenvalues as well.

for each $i = 1, 2, \ldots, n$. Therefore, the transpose of a Vandermonde matrix V of n column eigenvectors of A

$$
V^T = \begin{bmatrix} 1 & 1 & \cdots & 1 \\ \lambda_1 & \lambda_2 & \cdots & \lambda_n \\ \lambda_1^2 & \lambda_2^2 & \cdots & \lambda_n^2 \\ \vdots & \vdots & & \vdots \\ \lambda_1^{n-1} & \lambda_2^{n-1} & \cdots & \lambda_n^{n-1} \end{bmatrix}^T = \begin{bmatrix} 1 & \lambda_1 & \lambda_1^2 & \cdots & \lambda_1^{n-1} \\ 1 & \lambda_2 & \lambda_2^2 & \cdots & \lambda_2^{n-1} \\ \vdots & \vdots & \vdots & & \vdots \\ 1 & \lambda_n & \lambda_n^2 & \cdots & \lambda_n^{n-1} \end{bmatrix}
$$

$$(2.7.14)$$

diagonalizes A. Stated otherwise, a transformation of state defined by $\hat{\mathbf{x}}(t) = V^T \mathbf{x}(t)$ reduces the state-space system defined by Eq. (2.7.2) to the following modal canonical form:

$$
\begin{bmatrix} \dot{\hat{x}}_1(t) \\ \dot{\hat{x}}_2(t) \\ \vdots \\ \dot{\hat{x}}_n(t) \end{bmatrix} = \underbrace{\begin{bmatrix} \lambda_1 & 0 & 0 & \cdots \\ 0 & \lambda_2 & 0 & \cdots \\ \vdots & & \ddots & \\ 0 & 0 & \cdots & \lambda_n \end{bmatrix}}_{V^T A V^{-T} = \hat{A} = \mathrm{diag}[\lambda_i]} \begin{bmatrix} \hat{x}_1(t) \\ \hat{x}_2(t) \\ \vdots \\ \hat{x}_n(t) \end{bmatrix} + \underbrace{\begin{bmatrix} b(\lambda_1) \\ b(\lambda_2) \\ \vdots \\ b(\lambda_n) \end{bmatrix}}_{V^T B} u(t);
$$

$$(2.7.15)$$

with the elements of $V^T B$ given by $b(\lambda_i)$ because

$$[1 \quad \lambda_i \quad \lambda_i^2 \quad \cdots \quad \lambda_i^{n-1}]B = b(\lambda_i) \quad \text{for} \quad i = 1, 2, \ldots, n \quad (2.7.16)$$

Figure 2.14 and the results presented in the previous section show that each $\hat{x}_i(t)$ will be controllable if and only if $b(\lambda_i) \neq 0$. Therefore, the state-space system defined by Eq. (2.7.2) will be completely (state or modal) controllable if and only if the polynomials $a(\lambda)$ and $b(\lambda)$, or the differential operator pair $a(D)$ and $b(D)$, are coprime.

When this is not the case, every zero λ_k of $a(\lambda)$, which also is a zero of $b(\lambda)$, implies an uncontrollable state $\hat{x}_k(t) = [1 \quad \lambda_k \quad \lambda_k^2 \quad \cdots \quad \lambda_k^{n-1}]\mathbf{x}(t)$, characterized by an uncontrollable mode $e^{\lambda_k t}$. Moreover, each such λ_k reduces the controllability rank of the system by one. The controllability properties of a dynamic system in differential operator form therefore can be completely specified by the zeros of the polynomials $a(D)$ and $b(D)$ of Eq. (2.7.1), independent of any state-space representation.

Controllable Canonical Forms

When $b(D) = 1$ and $\deg[c(D)] < n = \deg[a(D)]$, the differential operator system defined by Eq. (2.7.1)

$$\underbrace{(D^n + a_{n-1}D^{n-1} + \cdots + a_1 D + a_0)}_{a(D)} z(t) = u(t);$$

$$y(t) = \underbrace{(c_{n-1}D^{n-1} + \cdots + c_1 D + c_0)}_{c(D)} z(t) + e(D)u(t) \qquad (2.7.17)$$

has an alternative, equivalent state-space representation that can be determined directly by inspection of $a(D)$ and $c(D)$.

Suppose we employ the coefficients of $a(D)$ and $c(D)$ to define the following state-space system:

$$\underbrace{\begin{bmatrix} \dot{x}_1(t) \\ \dot{x}_2(t) \\ \vdots \\ \dot{x}_n(t) \end{bmatrix}}_{\dot{\mathbf{x}}(t)} = \underbrace{\begin{bmatrix} 0 & 1 & 0 & \cdots & 0 \\ 0 & 0 & 1 & \cdots & 0 \\ \vdots & \vdots & & \ddots & \vdots \\ -a_0 & -a_1 & & \cdots & -a_{n-1} \end{bmatrix}}_{A} \underbrace{\begin{bmatrix} x_1(t) \\ x_2(t) \\ \vdots \\ x_n(t) \end{bmatrix}}_{\mathbf{x}(t)} + \underbrace{\begin{bmatrix} 0 \\ \vdots \\ 0 \\ 1 \end{bmatrix}}_{B} u(t);$$

$$y(t) = \underbrace{[c_0 \quad c_1 \quad \cdots \quad c_{n-1}]}_{C} \begin{bmatrix} x_1(t) \\ x_2(t) \\ \vdots \\ x_n(t) \end{bmatrix} + \underbrace{E(D)}_{e(D)} u(t) \qquad (2.7.18)$$

Since A is a bottom-row companion matrix, it follows (see Appendix A) that the characteristic polynomial of A is given by

$$|\lambda I - A| = \lambda^n + a_{n-1}\lambda^{n-1} + \cdots + a_1\lambda + a_0 = a(\lambda) \qquad (2.7.19)$$

as in Eq. (2.7.4). Therefore, the n zeros of $a(\lambda)$ correspond to the n (assumed distinct) eigenvalues λ_i of A that define the system modes $e^{\lambda_i t}$.

If $z(t) \stackrel{\text{def}}{=} x_1(t)$ in Eq. (2.7.18), it follows that $Dz(t) = \dot{x}_1(t) = x_2(t)$, $D^2 z(t) = \dot{x}_2(t) = x_3(t), \ldots, D^{n-1}z(t) = \dot{x}_{n-1}(t) = x_n(t)$, or that

$$\begin{bmatrix} 1 \\ D \\ \vdots \\ D^{n-1} \end{bmatrix} z(t) = \begin{bmatrix} x_1(t) \\ x_2(t) \\ \vdots \\ x_n(t) \end{bmatrix} = \mathbf{x}(t) \qquad (2.7.20)$$

The substitution of Eq. (2.7.20) for $\mathbf{x}(t)$ in Eq. (2.7.18) therefore implies that

$$\begin{bmatrix} D & -1 & 0 & \cdots & 0 \\ 0 & D & -1 & \cdots & 0 \\ \vdots & \vdots & & \ddots & \vdots \\ a_0 & a_1 & & \cdots & D+a_{n-1} \end{bmatrix} \begin{bmatrix} 1 \\ D \\ \vdots \\ D^{n-1} \end{bmatrix} z(t) = \begin{bmatrix} 0 \\ \vdots \\ 0 \\ 1 \end{bmatrix} u(t)$$

or that

$$a(D)z(t) = u(t); \quad y(t) = C\mathbf{x}(t) + E(D)u(t) = c(D)z(t) + e(D)u(t) \qquad (2.7.21)$$

thus establishing the equivalence of the two representations. We will denote this equivalence relationship as

$$\underbrace{\{A, B, C, E(D)\}}_{\text{of Eq. (2.7.18)}} \stackrel{\text{equiv}}{\Longleftrightarrow} \underbrace{\{a(D), b(D) = 1, c(D), e(D)\}}_{\text{of Eq. (2.7.17)}} \qquad (2.7.22)$$

Note that both of the representations defined by Eq. (2.7.22) are completely controllable. The differential operator representation is controllable because $a(D)$ and $b(D) = 1$ are coprime, and the state-space representation is controllable because its controllability matrix

$$\mathcal{C} = [B,\ AB,\ \cdots,\ A^{n-1}B] = \begin{bmatrix} 0 & \cdots & 0 & 1 \\ 0 & \cdots & 1 & * \\ \vdots & & * & \vdots \\ 0 & 1 & \cdots & * \\ 1 & * & \cdots & * \end{bmatrix} \qquad (2.7.23)$$

is nonsingular (the $*$'s denoting irrelevant scalars).

Further, the $\{A,\ B\}$ pair of Eq. (2.7.18) is in a special "canonical" form: A is a bottom-row companion matrix and B is identically zero except for the 1 in its bottom row. Hence, we say that both of the representations defined by Eq. (2.7.22) are in **controllable canonical form.** Figure 2.16 depicts a block diagram of a state-space system in controllable canonical form, as defined by Eq. (2.7.18).

Differential Operator Observability

Because of the bottom-row companion form structure of A in Eq. (2.7.18), it follows that for each $i = 1, 2, \ldots, n$, $\begin{bmatrix} 1 \\ \lambda_i \\ \vdots \\ \lambda_i^{n-1} \end{bmatrix}$ is a column eigenvector of A in the sense that

$$A \begin{bmatrix} 1 \\ \lambda_i \\ \vdots \\ \lambda_i^{n-1} \end{bmatrix} = \begin{bmatrix} 1 \\ \lambda_i \\ \vdots \\ \lambda_i^{n-1} \end{bmatrix} \lambda_i \qquad (2.7.24)$$

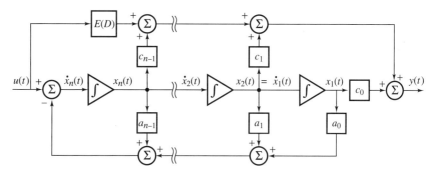

● **FIGURE 2.16**
A State-Space System in Controllable Canonical Form

Therefore, if V is a Vandermonde matrix of n column eigenvectors of A, as in Eq. (2.7.14), it follows that its inverse V^{-1} diagonalizes A. Stated otherwise, a transformation of state defined by $\hat{\mathbf{x}}(t) = V^{-1}\mathbf{x}(t)$ reduces the state-space system defined by Eq. (2.7.18) to the following modal canonical form:

$$
\begin{bmatrix} \dot{\hat{x}}_1(t) \\ \dot{\hat{x}}_2(t) \\ \vdots \\ \dot{\hat{x}}_n(t) \end{bmatrix} = \underbrace{\begin{bmatrix} \lambda_1 & 0 & 0 & \cdots \\ 0 & \lambda_2 & 0 & \cdots \\ \vdots & & \ddots & \cdots \\ 0 & 0 & \cdots & \lambda_n \end{bmatrix}}_{V^{-1}AV = \hat{A} = \mathrm{diag}[\lambda_i]} \begin{bmatrix} \hat{x}_1(t) \\ \hat{x}_2(t) \\ \vdots \\ \hat{x}_n(t) \end{bmatrix} + \underbrace{\begin{bmatrix} \hat{b}_0 \\ \hat{b}_1 \\ \vdots \\ \hat{b}_{n-1} \end{bmatrix}}_{V^{-1}B} u(t);
$$

$$
y(t) = \underbrace{[c(\lambda_1),\ c(\lambda_2),\ \cdots,\ c(\lambda_n)]}_{CV} \begin{bmatrix} \hat{x}_1(t) \\ \hat{x}_2(t) \\ \vdots \\ \hat{x}_n(t) \end{bmatrix} + E(D)u(t) \qquad (2.7.25)
$$

with the elements of CV given by $c(\lambda_i)$ because

$$
C \begin{bmatrix} 1 \\ \lambda_i \\ \lambda_i^2 \\ \vdots \\ \lambda_i^{n-1} \end{bmatrix} = c(\lambda_i) \quad \text{for} \quad i = 1, 2, \ldots, n \qquad (2.7.26)
$$

Figure 2.14 and the results presented in the previous section show that each $\hat{x}_i(t)$ will be observable if and only if $c(\lambda_i) \neq 0$. Therefore, the state-space system defined by Eqs. (2.7.2) and (2.7.7) will be completely (state or modal) observable if and only if the polynomials $a(\lambda)$ and $c(\lambda)$, or the differential operator pair $a(D)$ and $c(D)$, are coprime.

When this is not the case, every zero λ_k of $a(\lambda)$, which is also a zero of $c(\lambda)$, implies an unobservable state $\hat{x}_k(t)$ characterized by an uncontrollable mode $e^{\lambda_k t}$. Moreover, each such λ_k reduces the observability rank of the system by one. The observability properties of a dynamic system in differential operator form can therefore be completely specified by the zeros of the polynomials $a(D)$ and $c(D)$ of Eq. (2.7.1), independent of any state-space representation.

The Multi-Input/Multi-Output Case

Although we initially assumed that Eq. (2.7.1) defines a SISO system, it can be modified to include certain MIMO systems as well. In particular, a vector input

$$\mathbf{u}(t) = \begin{bmatrix} u_1(t) \\ u_2(t) \\ \vdots \\ u_m(t) \end{bmatrix} \qquad (2.7.27)$$

can be accommodated by allowing the polynomial $b(D)$ in Eq. (2.7.1) to be a row vector of polynomials

$$b(D) = [b_1(D),\ b_2(D),\ \ldots,\ b_m(D)] \qquad (2.7.28)$$

Each polynomial element of $b(D)$ defines a corresponding, real $(n \times 1)$ column of the input matrix B of an equivalent state-space system, analogous to that defined by Eq. (2.7.2).

In a dual manner, a vector output

$$\mathbf{y}(t) = \begin{bmatrix} y_1(t) \\ y_2(t) \\ \vdots \\ y_p(t) \end{bmatrix} \qquad (2.7.29)$$

can be accommodated by allowing the polynomial $c(D)$ in Eq. (2.7.1) to be a column vector of polynomials

$$c(D) = \begin{bmatrix} c_1(D) \\ c_2(D) \\ \vdots \\ c_p(D) \end{bmatrix} \qquad (2.7.30)$$

Of course, $e(D)$ would also be a vector or matrix of polynomials in these cases. Each polynomial element of $c(D)$ then defines a corresponding real $(1 \times n)$ row of the output matrix C of an equivalent state-space system, analogous to that defined by Eq. (2.7.7). A block diagram of such a MIMO system is depicted in Figure 2.17.

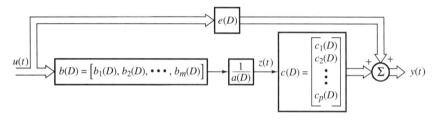

● **FIGURE 2.17**
A MIMO Differential Operator System

. . . .

EXAMPLE 2.7.31 Consider a dynamic system defined by the two-input/two-output differential equation:

$$\frac{d^4 z(t)}{dt^4} + 2\frac{d^3 z(t)}{dt^3} - 6\frac{d^2 z(t)}{dt^2} - 22\frac{dz(t)}{dt} - 15z(t) =$$

$$\frac{d^2 u_1(t)}{dt^2} + 4\frac{du_1(t)}{dt} + 5u_1(t) + \frac{d^2 u_2(t)}{dt^2} - u_2(t)$$

with

$$y_1(t) = -2\frac{dz(t)}{dt} + 6z(t) \text{ and } y_2(t) = -\frac{dz(t)}{dt} - z(t)$$

This system can readily be placed in a MIMO differential operator form analogous to that defined by Eq. (2.7.1)

$$\underbrace{(D^4 + 2D^3 - 6D^2 - 22D - 15)}_{a(D)} z(t) = \underbrace{[D^2 + 4D + 5, \ D^2 - 1]}_{b(D) = [b_1(D), \ b_2(D)]} \underbrace{\begin{bmatrix} u_1(t) \\ u_2(t) \end{bmatrix}}_{\mathbf{u}(t)};$$

$$\mathbf{y}(t) = \begin{bmatrix} y_1(t) \\ y_2(t) \end{bmatrix} = \begin{bmatrix} c_1(D) \\ c_2(D) \end{bmatrix} z(t) = \underbrace{\begin{bmatrix} -2D+6 \\ -D-1 \end{bmatrix}}_{c(D)} z(t)$$

Since $a(D)$ can be factored as

$$a(D) = (D+1)(D-3)(D^2 + 4D + 5) = (D+1)(D-3)(D+2-j)(D+2+j)$$

the system modes are defined by the $n = 4$ zeros of $a(D)$: $-1, +3$, and $-2 \pm j$.

We next note that $b_1(D) = D^2 + 4D + 5$, which is a factor of $a(D)$ as well. Therefore, the modes $e^{(-2+j)t}$ and $e^{(-2-j)t}$, which imply the real-valued modes $e^{-2t} \sin t$ and $e^{-2t} \cos t$, are uncontrollable by $u_1(t)$. Moreover, since $b_2(D) = (D+1)(D-1)$, the mode e^{-t} is uncontrollable by $u_2(t)$. Therefore, the remaining mode e^{3t} is the only one that is controllable by both inputs. Since all of the modes are controllable by at least one of the inputs, the system is completely controllable by the vector input $\mathbf{u}(t)$. This latter observation also holds because $a(D)$ and the polynomial vector $b(D) = [b_1(D), \ b_2(D)]$ are coprime, that is, none of the zeros of $a(D)$ are also zeros of *both* $b_1(D)$ and $b_2(D)$.

We further note that $c_1(D) = -2(D-3)$ while $c_2(D) = -(D+1)$. Therefore, the mode e^{3t} is unobservable by $y_1(t)$, while e^{-t} is unobservable by $y_2(t)$. Since all of the modes are observable by at least one of the outputs, the system is completely observable by the vector output $\mathbf{y}(t)$. This latter observation also holds because $a(D)$ and the polynomial vector $c(D)$ are coprime.

These observations can now be verified if we transform the given differential operator system to the equivalent state-space system in Eq. (2.7.8). In this case, the direct employment of the coefficients of $a(D)$ and $b(D)$, as in Eq. (2.7.2), implies that

$$\underbrace{\begin{bmatrix} \dot{x}_1(t) \\ \dot{x}_2(t) \\ \dot{x}_3(t) \\ \dot{x}_4(t) \end{bmatrix}}_{} = \underbrace{\begin{bmatrix} 0 & 0 & 0 & 15 \\ 1 & 0 & 0 & 22 \\ 0 & 1 & 0 & 6 \\ 0 & 0 & 1 & -2 \end{bmatrix}}_{A} \begin{bmatrix} x_1(t) \\ x_2(t) \\ x_3(t) \\ x_4(t) \end{bmatrix} + \underbrace{\begin{bmatrix} 5 & -1 \\ 4 & 0 \\ 1 & 1 \\ 0 & 0 \end{bmatrix}}_{B} \begin{bmatrix} u_1(t) \\ u_2(t) \end{bmatrix}$$

which defines the state A and input B matrices of the equivalent state-space system, with $z(t) = x_4(t)$.

Since $y_1(t) = c_1(D)z(t) = (-2D + 6)z(t) = -2\dot{z}(t) + 6z(t) = -2\dot{x}_4(t) + 6x_4(t)$ and $\dot{x}_4(t) = x_3(t) - 2x_4(t)$, it follows that $y_1(t) = -2x_3(t) + 10x_4(t)$. In an analogous manner, $y_2(t) = (-D - 1)z(t) = -\dot{z}(t) - z(t) = -\dot{x}_4(t) - x_4(t) = -x_3(t) + x_4(t)$, so that

$$\mathbf{y}(t) = \begin{bmatrix} y_1(t) \\ y_2(t) \end{bmatrix} = \underbrace{\begin{bmatrix} 0 & 0 & -2 & 10 \\ 0 & 0 & -1 & 1 \end{bmatrix}}_{C} \begin{bmatrix} x_1(t) \\ x_2(t) \\ x_3(t) \\ x_4(t) \end{bmatrix} + \underbrace{\begin{bmatrix} 0 & 0 \\ 0 & 0 \end{bmatrix}}_{E} \begin{bmatrix} u_1(t) \\ u_2(t) \end{bmatrix}$$

Note that this resulting, equivalent state-space system is identical to the one defined in Example 2.6.15. As expected, the controllability and observability observations made in that earlier example are identical to those made here.

··

2.8   SUMMARY

We have now shown how the dynamic behavior of many different physical systems can be mathematically modeled by either linear differential (operator) equations or equivalent state-space representations, which are useful for a number of reasons: They are relatively simple to implement using basic analog elements, and the state transition matrix e^{At} implies complete response solutions in the linear, time-invariant case.

In very common situations, when the state matrix A has n distinct eigenvalues, λ_1, λ_2, ..., λ_n, the elements of e^{At} were shown to consist of linear combinations of the natural modes of the system, namely, $e^{\lambda_i t}$, for $i = 1, 2, ..., n$. Moreover, each of these distinct modes can be characterized as either controllable or uncontrollable (observable or unobservable) by each component of the input $\mathbf{u}(t)$ (output $\mathbf{y}(t)$). Matrix rank conditions were presented for directly determining the controllability (observability) rank of a given state-space system from knowledge of the defining vectors and matrices.

The dual notions of controllability and observability were shown to extend naturally to systems defined in the differential operator form:

$$a(D)z(t) = b(D)u(t) \qquad y(t) = c(D)z(t) + e(D)u(t) \qquad (2.7.1)$$

with the common roots of $a(D)$ and $b(D)$ ($c(D)$) directly defining any and all of the uncontrollable (unobservable) modes.

In the special case in which $b(D) = 1$ ($c(D) = 1$), Eq. (2.7.1) was shown to be equivalent to a special state-space system in controllable (observable) canonical form. The controllability (observability) properties of SISO systems were shown to directly apply to the individual components of the input (output) in both the state-space and differential operator MIMO cases.

PROBLEMS [12]

2-1. Determine a state-space representation for the series RLC circuit that was defined in Section 2.2, whose dynamic behavior is given by Eq. (2.2.1). (*Hint*: You may wish to define one of your state variables as the charge across the capacitor.)

2-2. Determine a state-space representation for the Section 2.2 system consisting of two coupled masses, whose dynamic behavior is defined by Eqs. (2.2.4) and (2.2.5).

2-3. Determine a state-space representation for the liquid level system of Section 2.2, whose dynamic behavior is defined by Eqs. (2.2.6) and (2.2.7).

2-4. What terms in the differential equations Eqs. (2.3.10) and (2.3.11), which define the dynamic behavior of the orbiting satellite, account for its nonlinear behavior?

2-5. Consider the inverted pendulum on a moving cart, whose dynamic behavior is defined by Eqs. (2.3.6) and (2.3.7).

(a) Verify that the pendulum in a vertically balanced or **stabilized position**, as defined when $\theta(t) = 0°$, $\dot{x}(t) = 0$ and $F(t) = 0$, represents a nominal solution.

(b) Determine a linear state-space description of the system around this nominal solution assuming that $\cos\theta \approx 1$ and $\sin\theta \approx \theta$.

2-6.* Determine the eigenvalues of the state matrix associated with the DC servomotor of Example 2.4.10 if $R_a = 0.1$, $L_a = 0.0095$, $K_b = 0.505$, $K_m = 10$, $J = 105.3$, and $D = 49.5$ in compatible units. If we now assume that only D can vary, for what positive values of D, if any, will there be repeated eigenvalues?

2-7. Consider a state-space system defined by the state matrix

$$A = \begin{bmatrix} 0 & 1 \\ -4 & -4 \end{bmatrix}$$

(a) Show that A has (two) repeated eigenvalues at -2 in the complex plane.

(b) Verify that in this case (setting $t_0 = 0$ for convenience)

$$e^{At} = \begin{bmatrix} e^{-2t} + 2te^{-2t} & te^{-2t} \\ -4te^{-2t} & e^{-2t} - 2te^{-2t} \end{bmatrix}$$

---

[12] An asterisk (*) placed next to a problem number indicates that a computer-aided design package, such as MATLAB or MATRIX$_x$, would be particularly useful in minimizing the numerical computations required to solve the problem.

2-8. Note that the terms that comprise the successive rows 2 and 3 of the state transition matrix e^{At} of Example 2.5.18 are time derivatives of the corresponding terms that comprise the preceding rows 1 and 2, respectively. State why this is true, in general, whenever A is an $(n \times n)$ matrix in bottom-row companion form.

2-9. Verify that the $\{A, B\}$ pair associated with the linearized equations of motion of the orbiting satellite of Example 2.6.14

$$A = \begin{bmatrix} 0 & 1 & 0 & 0 \\ 3\omega^2 & 0 & 0 & 2\omega \\ 0 & 0 & 0 & 1 \\ 0 & -2\omega & 0 & 0 \end{bmatrix}; \qquad B = \begin{bmatrix} 0 & 0 \\ 1 & 0 \\ 0 & 0 \\ 0 & 1 \end{bmatrix}$$

is controllable since $\mathcal{C} = [B, \ AB, \ A^2B, \ A^3B]$ has full rank $4 = n$.

2-10. Verify that the linearized state-space representation of the orbiting satellite also is observable if

$$C = \begin{bmatrix} 1 & 0 & 0 & 0 \\ 0 & 0 & 1 & 0 \end{bmatrix}$$

as in Example 2.6.14.

2-11.* Consider the state-space system: $\dot{\mathbf{x}}(t) = A\mathbf{x}(t) + bu(t); \ y(t) = c\mathbf{x}(t) + eu(t)$, with

$$A = \begin{bmatrix} 0 & 0 & 6 \\ 1 & 0 & -1 \\ 0 & 1 & -4 \end{bmatrix} \qquad b = \begin{bmatrix} 6 \\ -5 \\ -1 \end{bmatrix} \qquad c = [0 \ 3 \ -3] \quad \text{and} \quad e = [1]$$

    (a) Determine the eigenvalues of the state matrix A.

    (b) Determine any and all uncontrollable modes.

    (c) Determine any and all unobservable modes.

(*Hint*: You may wish to use the results of Example 2.5.18.)

2-12. The **dual** of a state-space representation $\{A, B, C, E\}$, is defined by transposing all four matrices and then interchanging B^T and C^T.

$$\{A, \ B, \ C, \ D\} \overset{\text{dual}}{\Longleftrightarrow} \{A^T, \ C^T, \ B^T, \ E^T\}$$

Using this definition, prove that a given state-space system is controllable if and only if its dual is observable.

2-13.* Verify in Example 2.6.15 that the rank of the controllability matrix associated with input $u_1(t)$

$$\mathcal{C}_1 = [B_1, \ AB_1, \ A^2B_1, \ A^3B_1] = \begin{bmatrix} 5 & 0 & 15 & 30 \\ 4 & 5 & 22 & 59 \\ 1 & 4 & 11 & 34 \\ 0 & 1 & 2 & 7 \end{bmatrix}$$

is only 2, while the rank of the controllability matrix associated with input $u_2(t)$

$$C_2 = [B_2, \ AB_2, \ A^2 B_2, \ A^3 B_2] = \begin{bmatrix} -1 & 0 & 15 & -30 \\ 0 & -1 & 22 & -29 \\ 1 & 0 & 5 & 10 \\ 0 & 1 & -2 & 9 \end{bmatrix}$$

is 3.

2-14.* Verify that in Example 2.6.15 the rank of the observability matrix associated with output $y_1(t)$

$$\mathcal{O}_1 = \begin{bmatrix} C_1 \\ C_1 A \\ C_1 A^2 \\ C_1 A^3 \end{bmatrix} = \begin{bmatrix} 0 & 0 & -2 & 10 \\ 0 & -2 & 10 & -32 \\ -2 & 10 & -32 & 80 \\ 10 & -32 & 80 & -162 \end{bmatrix}$$

is 3.

2-15. Consider the MIMO state-space system of Example 2.6.15.

   (a) Determine the explicit linear functions of the original state $\mathbf{x}(t)$ that are completely controllable and completely uncontrollable by each of the defined inputs $u_1(t)$ and $u_2(t)$.

   (b) Determine the explicit linear functions of the state $\mathbf{x}(t)$ that are completely observable and completely unobservable by each of the defined outputs $y_1(t)$ and $y_2(t)$.

2-16. Determine the four eigenvalues of the state matrix A associated with the linearized equations of motion of the orbiting satellite, as given in Example 2.6.14.

   (a) Which of these eigenvalues defines a corresponding mode that is uncontrollable whenever the tangential thruster fails?

   (b) Which of these eigenvalues defines a corresponding mode that is unobservable whenever the $y_2(t)$ sensor fails?

2-17. If the two, single-input, completely controllable state-space systems $\{A, B, C, E\}$ and $\{\hat{A}, \hat{B}, \hat{C}, \hat{E}\} = \{QAQ^{-1}, QB, CQ^{-1}, E\}$ are equivalent, show how their controllability matrices $\mathcal{C}$ and $\hat{\mathcal{C}}$ can be used to determine the state transformation matrix Q.

2-18. Assume that the "state" of the electrical network depicted in Figure 2.18 is defined by the current through the inductor and the charge across the capacitor, that is, $x_1(t) = i(t)$ and $x_2(t) = q(t)$, with input $u(t) = v_i(t)$ and output $y(t) = v_o(t)$, as shown.

   (a) Determine the corresponding state-space representation.

   (b) Under what conditions (on L, R_1, R_2, and C), if any, is the network uncontrollable?

   (c) Under what conditions, if any, is the network unobservable?

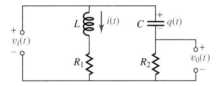

• **FIGURE 2.18**
The Electrical Network of Problem 2-18

2-19. A square $(n \times n)$ matrix A is said to be **cyclic** if a corresponding n-dimensional column vector b can be found that makes the $\{A, b\}$ pair controllable, thus implying the nonsingularity of the $(n \times n)$ controllability matrix

$$C = [b, \ Ab, \ A^2b, \ \ldots, \ A^{n-1}b]$$

   (a) Display an A matrix that is not cyclic.

   (b) Prove that if A has n distinct eigenvalues, then it is cyclic.

2-20.* It can be shown [62] that if A is cyclic (see the previous problem) and q_1 is defined as the last row of C^{-1}, then the $(n \times n)$ state transformation matrix

$$Q \stackrel{\text{def}}{=} \begin{bmatrix} q_1 \\ q_1 A \\ \vdots \\ q_1 A^{n-1} \end{bmatrix}$$

will transform the $\{A, b\}$ pair to an equivalent $\{\hat{A}, \hat{b}\}$ pair in the controllable canonical form defined by Eq. (2.7.18).

$$\hat{A} = QAQ^{-1} = \begin{bmatrix} 0 & 1 & 0 & \cdots & 0 \\ 0 & 0 & 1 & \cdots & 0 \\ \vdots & \vdots & & \ddots & \vdots \\ -a_0 & -a_1 & & \cdots & -a_{n-1} \end{bmatrix} \quad \text{and } \hat{b} = Qb = \begin{bmatrix} 0 \\ \vdots \\ 0 \\ 1 \end{bmatrix}$$

Verify this observation by transforming the $\{A, B_2\}$ pair of Example 2.6.14 to controllable companion form when $\omega = 1$.

2-21. Explain why the polynomials $a(s) = s^n + a_{n-1}s^{n-1} + \cdots + a_1s + a_0$ and $b(s) = b_{n-1}s^{n-1} + \cdots + b_1s + b_0$ are coprime if and only if the state-space pair

$$A = \begin{bmatrix} 0 & 0 & 0 & \cdots & -a_0 \\ 1 & 0 & 0 & \cdots & -a_1 \\ 0 & 1 & 0 & \cdots & \\ \vdots & \vdots & \ddots & & \vdots \\ 0 & 0 & \cdots & 1 & -a_{n-1} \end{bmatrix} \quad \text{and} \quad b = \begin{bmatrix} b_0 \\ b_1 \\ \vdots \\ b_{n-1} \end{bmatrix}$$

is controllable.

2-22. * Consider a two-input, two-output MIMO system defined by the differential operator representation:

$$\underbrace{(D^5 + 6D^4 + 11D^3 + 16D^2 + 70D + 100)}_{a(D) = (D+2)(D+3\pm j)(D-1\pm j2)} z(t);$$

$$= \underbrace{[D^2 + 6D + 10,\ 2D - 4]}_{b(D) = [b_1(D),\ b_2(D)]} \overbrace{\begin{bmatrix} u_1(t) \\ u_2(t) \end{bmatrix}}^{\mathbf{u}(t)}$$

$$\mathbf{y}(t) = \begin{bmatrix} y_1(t) \\ y_2(t) \end{bmatrix} = \begin{bmatrix} c_1(D) \\ c_2(D) \end{bmatrix} z(t) = \underbrace{\begin{bmatrix} D^2 + D - 2 \\ 2D^3 + 2D + 20 \end{bmatrix}}_{c(D)} z(t)$$

(a) Determine the system modes that are uncontrollable by $u_1(t)$ alone and by $u_2(t)$ alone.

(b) Determine the system modes that are unobservable by $y_1(t)$ alone and by $y_2(t)$ alone.

(c) Determine an equivalent state-space representation and verify that the ranks of the appropriate controllability and observability matrices are consistent with your findings in parts (a) and (b) of this problem.

2-23. * Consider the depth control of the submarine depicted in Figure 2.19, where $\theta(t)$ defines its directional velocity $v(t)$ (angle of attack) relative to its heading $\alpha(t)$. Assume that at a constant velocity of 20 knots, the dynamic behavior of the submarine can be defined by the nominal, linearized, state-space representation:

$$\begin{bmatrix} \dot{x}_1(t) \\ \dot{x}_2(t) \\ \dot{x}_3(t) \end{bmatrix} = \begin{bmatrix} 0 & 1 & 0 \\ -0.008 & -0.15 & 0.12 \\ 0 & 0.06 & -0.4 \end{bmatrix} \begin{bmatrix} x_1(t) \\ x_2(t) \\ x_3(t) \end{bmatrix} + \begin{bmatrix} 0 \\ -0.1 \\ 0.8 \end{bmatrix} u(t)$$

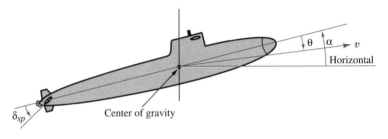

● **FIGURE 2.19**
The Depth Control of the Submarine of Problem 2-23

where $x_1(t) = \alpha(t)$, $x_2(t) = \dot{x}_1(t) = \dot{\alpha}(t)$, $x_3(t) = \theta(t)$, and $u(t) = \delta_{sp}(t)$, the angular deflection of the stern plane control surface.

(a) Determine the modes of the system.

(b) Determine any uncontrollable modes.

(c) Express the complete dynamic behavior of the system by a single third-order differential equation with partial state $z(t)$.

(d) Express $z(t)$ as a function of $\alpha(t)$ and $\theta(t)$.

3

THE TRANSFER
FUNCTION/MATRIX

3.1   THE DEFINITION

The so-called "transfer function" $G(s)$ represents one of the most useful descriptions we will employ in the analysis and design of dynamic systems. By definition, the *transfer function* in the SISO case or the *transfer matrix* (a matrix of transfer functions) in the MIMO case is defined as the "ratio" of the Laplace transform (see Appendix B) of the system output $\mathbf{y}(t)$ to the Laplace transform of the system input $\mathbf{u}(t)$, when all of the initial conditions are zero.

Therefore, in the general MIMO case, the $(p \times m)$ **transfer matrix** $G(s)$ of a dynamic system is defined by the relationship:

$$\underbrace{\mathcal{L}[\mathbf{y}(t)]}_{\mathbf{y}(s)} = G(s) \underbrace{\mathcal{L}[\mathbf{u}(t)]}_{\mathbf{u}(s)} \qquad (3.1.1)$$

so that in the SISO case, the **transfer function**

$$G(s) = \frac{y(s)}{u(s)} \qquad (3.1.2)$$

It may be noted that a transfer matrix is often defined in the linear, time-invariant case in which it is expressed as a matrix of **rational transfer functions,** that is, ratios of two polynomials in the Laplace variable s. Moreover, the transfer matrix represents an external or an input/output description of a dynamic system. As such, it is independent of any particular internal, state-space, or differential operator representation of the system.

The Impulse Response

We now illustrate two alternative ways that Eq. (3.1.1) can be employed to obtain an explicit expression for the transfer matrix $G(s)$ of any state-space system defined by the matrix quadruple $\{A, B, C, E\}$. Consider the output response of a state-space system when a **unit impulse function** $\delta(t)$ is applied to the jth component of the input at $t_0 = 0$, that is, when $u_j(t) = \delta(t)$, with all other input components and initial conditions equal to zero, as depicted in Figure 3.1.

Since $\mathcal{L}[\delta(t)] = 1$, the column vector $\mathcal{L}[\mathbf{u}(t)]$ is identically zero except for a solitary 1 in the jth row. Using Eq. (3.1.1), therefore, the Laplace transform of the resulting impulse response output $\mathbf{y}(t)$ defines the jth column of the transfer matrix of the system:

$$\mathcal{L}[\mathbf{y}(t)]_{u_j(t)=\delta(t)} = \begin{bmatrix} G_{1j}(s) \\ G_{2j}(s) \\ \vdots \\ G_{pj}(s) \end{bmatrix} \tag{3.1.3}$$

In the SISO case, the transfer function is defined as the Laplace transform of the (scalar) unit impulse response, or the so-called **weighting function** of the system, which is the Laplace transform of the system output $y(t)$ when input $u(t) = \delta(t)$.

By setting $\mathbf{x}_0 = 0$ in Eq. (2.5.10), it follows that the complete $(p \times m)$ transfer matrix $G(s)$ of the system is given by the relationship:

$$G(s) = \mathcal{L}\left[C \int_0^t e^{A(t-\tau)} B \operatorname{diag}[\delta(\tau)]\, d\tau + E \operatorname{diag}[\delta(t)] \right]$$

$$= \mathcal{L}\left[Ce^{At} B + E \right] = C\mathcal{L}[e^{At}]B + E \tag{3.1.4}$$

where $Ce^{At} B + E$ is the so-called **impulse response matrix** of the system, and $\operatorname{diag}[\delta(t)]$ is a diagonal $(m \times m)$ matrix of unit impulse functions. Therefore, $G(s)$ is equal to the Laplace transform of the impulse response matrix of a MIMO system.

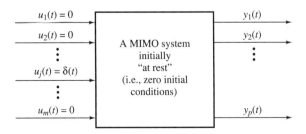

• **FIGURE 3.1**
A Unit Impulse Response

We note next that

$$e^{At} = I + At + \frac{A^2 t^2}{2!} + \frac{A^3 t^3}{3!} + \cdots \qquad (3.1.5)$$

in light of Eq. (2.5.3), so that

$$\mathcal{L}[e^{At}] = \frac{I}{s} + \frac{A}{s^2} + \frac{A^2}{s^3} + \cdots = (sI - A)^{-1} \qquad (3.1.6)$$

since

$$(sI - A)\left[\frac{I}{s} + \frac{A}{s^2} + \frac{A^2}{s^3} + \cdots\right] = I - \frac{A}{s} + \frac{A}{s} - \frac{A^2}{s^3} + \frac{A^2}{s^3} + \cdots = I$$

Equation (3.1.6) thus implies that the state transition matrix e^{At} can be determined by means of the *inverse Laplace transform* of $(sI - A)^{-1}$:

$$e^{At} = \mathcal{L}^{-1}\left\{(sI - A)^{-1}\right\} \qquad (3.1.7)$$

If we now employ Eq. (3.1.6) in Eq. (3.1.4), it follows that the $(p \times m)$ transfer matrix

$$G(s) = C(sI - A)^{-1}B + E \qquad (3.1.8)$$

which establishes the relationship between the external input/output transfer function and a state-space description of the system "at rest."

State-Space Representations

We now present an alternative derivation of Eq. (3.1.8). Recall that a state-space representation $\{A, B, C, E\}$ is defined by the equations:

$$\dot{\mathbf{x}}(t) = A\mathbf{x}(t) + B\mathbf{u}(t) \qquad \mathbf{y}(t) = C\mathbf{x}(t) + E\mathbf{u}(t) \qquad (3.1.9)$$

with $(n \times m)$ input matrix B and $(p \times n)$ output matrix C. If we denote $\mathcal{L}[\mathbf{x}(t)]$ as $\mathbf{x}(s)$, noting that $\mathcal{L}[\dot{\mathbf{x}}(t)] = s\mathbf{x}(s)$ when the initial state $\mathbf{x}_0 = 0$, it follows that

$$\mathcal{L}[\dot{\mathbf{x}}(t) = A\mathbf{x}(t) + B\mathbf{u}(t)] = s\mathbf{x}(s) = A\mathbf{x}(s) + B\mathbf{u}(s)$$

or that

$$(sI - A)\mathbf{x}(s) = B\mathbf{u}(s) \qquad (3.1.10)$$

Therefore,

$$\mathbf{x}(s) = (sI - A)^{-1}B\mathbf{u}(s) \qquad (3.1.11)$$

and since

$$\mathcal{L}[\mathbf{y}(t) = C\mathbf{x}(t) + E\mathbf{u}(t)] = \mathbf{y}(s) = C\mathbf{x}(s) + E\mathbf{u}(s) \qquad (3.1.12)$$

the substitution of Eq. (3.1.11) for $\mathbf{x}(s)$ in Eq. (3.1.12) implies that

$$\underbrace{\begin{bmatrix} y_1(s) \\ y_2(s) \\ \vdots \\ y_p(s) \end{bmatrix}}_{\mathbf{y}(s)} = \underbrace{[C(sI-A)^{-1}B+E]}_{G(s)} \underbrace{\begin{bmatrix} u_1(s) \\ u_2(s) \\ \vdots \\ u_m(s) \end{bmatrix}}_{\mathbf{u}(s)} \qquad (3.1.13)$$

which verifies Eq. (3.1.8).

EXAMPLE 3.1.14 To illustrate the determination of $G(s)$ using Eq. (3.1.8), consider the two-input, two-output, fourth-order state-space system defined in Example 2.6.15, namely,

$$\begin{bmatrix} \dot{x}_1(t) \\ \dot{x}_2(t) \\ \dot{x}_3(t) \\ \dot{x}_4(t) \end{bmatrix} = \underbrace{\begin{bmatrix} 0 & 0 & 0 & 15 \\ 1 & 0 & 0 & 22 \\ 0 & 1 & 0 & 6 \\ 0 & 0 & 1 & -2 \end{bmatrix}}_{A} \begin{bmatrix} x_1(t) \\ x_2(t) \\ x_3(t) \\ x_4(t) \end{bmatrix} + \underbrace{\begin{bmatrix} 5 & -1 \\ 4 & 0 \\ 1 & 1 \\ 0 & 0 \end{bmatrix}}_{B} \begin{bmatrix} u_1(t) \\ u_2(t) \end{bmatrix};$$

$$\begin{bmatrix} y_1(t) \\ y_2(t) \end{bmatrix} = \underbrace{\begin{bmatrix} 0 & 0 & -2 & 10 \\ 0 & 0 & -1 & 1 \end{bmatrix}}_{C} \begin{bmatrix} x_1(t) \\ x_2(t) \\ x_3(t) \\ x_4(t) \end{bmatrix}$$

Since

$$(sI-A)^{-1} = \begin{bmatrix} s & 0 & 0 & -15 \\ -1 & s & 0 & -22 \\ 0 & -1 & s & -6 \\ 0 & 0 & -1 & s+2 \end{bmatrix}^{-1} = \frac{(sI-A)^+}{|sI-A|} =$$

$$\frac{\begin{bmatrix} s^3+2s^2-6s-22 & 15 & 15s & 15s^2 \\ s^2+2s-6 & s^3+2s^2-6s & 22s+15 & 22s^2+15s \\ s+2 & s^2+2s & s^3+2s^2 & 6s^2+22s+15 \\ 1 & s & s^2 & s^3 \end{bmatrix}}{s^4+2s^3-6s^2-22s-15}$$

it follows that the premultiplication of the above expression by

$$C = \begin{bmatrix} 0 & 0 & -2 & 10 \\ 0 & 0 & -1 & 1 \end{bmatrix}$$

and the subsequent postmultiplication of the resulting expression by

$$B = \begin{bmatrix} 5 & -1 \\ 4 & 0 \\ 1 & 1 \\ 0 & 0 \end{bmatrix}$$

yields the (2×2) transfer matrix

$$G(s) = \frac{\begin{bmatrix} -2s^3 - 2s^2 + 14s + 30 & -2s^3 + 6s^2 + 2s - 6 \\ -s^3 - 5s^2 - 9s - 5 & -s^3 - s^2 + s + 1 \end{bmatrix}}{s^4 + 2s^3 - 6s^2 - 22s - 15}$$

$$= \begin{bmatrix} \dfrac{-2s^3 - 2s^2 + 14s + 30}{s^4 + 2s^3 - 6s^2 - 22s - 15} & \dfrac{-2s^3 + 6s^2 + 2s - 6}{s^4 + 2s^3 - 6s^2 - 22s - 15} \\[2ex] \dfrac{-s^3 - 5s^2 - 9s - 5}{s^4 + 2s^3 - 6s^2 - 22s - 15} & \dfrac{-s^3 - s^2 + s + 1}{s^4 + 2s^3 - 6s^2 - 22s - 15} \end{bmatrix}$$

Note that each of the $(pm = 4)$ elements of the transfer matrix $G(s)$ is a rational transfer function in the Laplace operator s.

....

Proper, Strictly Proper, and Improper Transfer Functions

Any rational transfer function

$$G(s) = \frac{m(s)}{a(s)}$$

whose numerator polynomial $m(s)$ is of degree (deg) no greater than that of its denominator polynomial $a(s)$, so that $\deg[m(s)] \leq \deg[a(s)]$, is termed **proper**, and any such $G(s)$ is characterized by the condition that $\lim_{s \to \infty} G(s) = k < \infty$. In such cases, $\deg[a(s)] - \deg[m(s)]$ is called the **relative degree of the transfer function**, or the **relative order of the system** [11] defined by the transfer function. If the degree of the numerator of $G(s)$ is strictly less than the degree of the denominator, that is, if $\deg[m(s)] < \deg[a(s)]$, the transfer function is termed **strictly proper**, and any such $G(s)$ is characterized by the condition that $\lim_{s \to \infty} G(s) = 0$. If the degree of the numerator of $G(s)$ is greater than the degree of the denominator, $G(s)$ is called **improper**, and $\lim_{s \to \infty} G(s) \to \infty$.

In light of these definitions, recall that

$$(sI - A)^{-1} = \frac{(sI - A)^+}{|sI - A|} \tag{3.1.15}$$

with

$$|sI - A| = a(s) = s^n + a_{n-1}s^{n-1} + \cdots + a_1 s + a_0 \tag{3.1.16}$$

Since each element of $(sI - A)^+ = [(sI - A)^*]^T$ is of lower degree than n; that is, since

$$\deg[(sI - A)^+] < \deg[a(s)] = n \tag{3.1.17}$$

it follows that each rational, transfer function element of $C(sI - A)^{-1}B$ is strictly proper, so that

$$\lim_{s \to \infty} \left[C(sI - A)^{-1}B = \frac{C(sI - A)^+ B}{|sI - A|} \right] = 0 \tag{3.1.18}$$

Therefore, in light of Eqs. (3.1.8) and (3.1.18),

$$\lim_{s \to \infty} G(s) = \lim_{s \to \infty} \left[C(sI - A)^{-1} B + E \right] = E \qquad (3.1.19)$$

We conclude that each transfer function element, $G_{ij}(s) = C_i(sI - A)^{-1} B_j + E_{ij}$, of a state-space transfer matrix $G(s)$ is strictly proper if and only if $E_{ij} = 0$. Moreover, the transfer function elements of a state-space system, as defined by Eq. (3.1.8), will not be improper unless E is "expanded" to $E(D)$, to accommodate derivatives of the input, as in Eq. (2.7.7). In such cases, each nonconstant polynomial element $E_{ij}(D)$ of $E(D)$ implies a corresponding, improper transfer function element $G_{ij}(s)$ of $G(s)$.

Equivalent State-Space Systems

Let us now recall the definition of *state-space equivalence* given by Eq. (2.5.23), namely,

$$\{A, \ B, \ C, \ E\} \overset{\text{equiv}}{\Longleftrightarrow} \{\hat{A}, \ \hat{B}, \ \hat{C}, \ \hat{E}\} = \{QAQ^{-1}, \ QB, \ CQ^{-1}, \ E\}$$

If $\hat{G}(s) = \hat{C}(sI - \hat{A})^{-1} \hat{B} + \hat{E}$ represents the transfer matrix of a state-space system that is equivalent to the system defined by Eq. (3.1.9), it follows that

$$\hat{G}(s) = \hat{C}(sI - \hat{A})^{-1} \hat{B} + \hat{E} = CQ^{-1}(sQQ^{-1} - QAQ^{-1})^{-1}QB + E =$$

$$CQ^{-1}Q(sI - A)^{-1}Q^{-1}QB + E = C(sI - A)^{-1}B + E = G(s) \qquad (3.1.20)$$

as given by Eq. (3.1.8), which establishes the fact that *the transfer matrices of equivalent state-space systems*[1] *are identical.* As a consequence, $G(s)$ is independent of the particular internal state-space description of the system. The fact that a unit impulse output response is independent of the internal system representation is also clear observing Figure 3.1.

Differential Operator Representations

Consider any "sufficiently differentiable" function $f(t)$ of the time t. If

$$\mathcal{L}[f(t)] \overset{\text{def}}{=} f(s)$$

then (see Appendix B) $\mathcal{L}[Df(t)] = sf(s), \mathcal{L}[D^2 f(t)] = s^2 f(s), \ldots$, or, in general,

$$\mathcal{L}[D^k f(t)] = s^k f(s) \qquad (3.1.21)$$

---

[1] Actually, the transfer matrices of any two equivalent systems are identical, regardless of how they are defined.

for all integers $k \geq 0$, assuming zero initial conditions on $f(t)$ and all of its defined derivatives.

In light of Eq. (3.1.21), it follows that the transfer function $G(s)$ of a system defined by the differential operator representation Eq. (2.7.1), namely,

$$a(D)z(t) = b(D)u(t); \qquad y(t) = c(D)z(t) + e(D)u(t)$$

which is initially at rest, can be determined by a simple "substitution" of the Laplace operator s for both the differential operator D and the time t in Eq. (2.7.1). In other words, the Laplace transformation of the differential operator representation Eq. (2.7.1) directly implies that

$$a(s)z(s) = b(s)u(s); \qquad y(s) = c(s)z(s) + e(s)u(s) \qquad (3.1.22)$$

We can now solve Eq. (3.1.22) for $y(s)$ as a function of $u(s)$ to obtain an expression for $G(s)$. Since

$$z(s) = \frac{b(s)}{a(s)} u(s) \qquad (3.1.23)$$

it follows that

$$y(s) = c(s)z(s) + e(s)u(s) = \left[\frac{c(s)b(s)}{a(s)} + e(s) \right] u(s) \qquad (3.1.24)$$

so that the transfer function

$$G(s) = \frac{y(s)}{u(s)} = \frac{c(s)b(s)}{a(s)} + e(s) = \frac{c(s)b(s) + e(s)a(s)}{a(s)} \qquad (3.1.25)$$

It may be noted that Eq. (3.1.25) holds in the multi-input and multi-output cases as well, if $b(D)$ or $c(D)$ become polynomial vectors in order to accommodate a multiple-input $\mathbf{u}(t)$ or a multiple-output $\mathbf{y}(t)$. In such cases, the dimensions of $e(D)$ changes appropriately. If $a(D)$ remains a single polynomial, Eqs. (3.1.23) and (3.1.24) still hold, so that

$$\mathbf{y}(s) = \underbrace{\left[\frac{c(s)b(s)}{a(s)} + e(s) \right]}_{G(s)} \mathbf{u}(s) \qquad (3.1.26)$$

thus defining the **transfer matrix** $G(s)$ of the MIMO differential operator system.

EXAMPLE 3.1.27 Consider the two-input/two-output differential operator system of Example 2.7.31,

$$\underbrace{(D^4 + 2D^3 - 6D^2 - 22D - 15)}_{a(D)} z(t) = \underbrace{[D^2 + 4D + 5, \ D^2 - 1]}_{b(D) = [b_1(D), \ b_2(D)]} \underbrace{\begin{bmatrix} u_1(t) \\ u_2(t) \end{bmatrix}}_{\mathbf{u}(t)};$$

$$\mathbf{y}(t) = \begin{bmatrix} y_1(t) \\ y_2(t) \end{bmatrix} = \begin{bmatrix} c_1(D) \\ c_2(D) \end{bmatrix} z(t) = \underbrace{\begin{bmatrix} -2D + 6 \\ -D - 1 \end{bmatrix}}_{c(D)} z(t)$$

By substituting the Laplace operator s for D in the expressions for $a(D)$, $b(D)$, and $c(D)$, we immediately obtain $a(s)$, $b(s)$, and $c(s)$, respectively. The substitution of these three expressions in Eq. (3.1.25), noting that $e(s) = 0$, subsequently implies that

$$
\begin{aligned}
G(s) = \frac{c(s)b(s)}{a(s)} &= \frac{\begin{bmatrix} -2s + 6 \\ -s - 1 \end{bmatrix} [s^2 + 4s + 5, \ s^2 - 1]}{s^4 + 2s^3 - 6s^2 - 22s - 15} \\[2mm]
&= \frac{\begin{bmatrix} -2s^3 - 2s^2 + 14s + 30 & -2s^3 + 6s^2 + 2s - 6 \\ -s^3 - 5s^2 - 9s - 5 & -s^3 - s^2 + s + 1 \end{bmatrix}}{s^4 + 2s^3 - 6s^2 - 22s - 15}
\end{aligned}
$$

Note that this expression for $G(s)$ is the same expression we obtained earlier in Example 3.1.14. This must be the case, however, because the differential operator system of this example is equivalent to the state-space system of Example 3.1.14, as we noted in Example 2.7.31. Therefore, this example also serves to illustrate that the transfer matrix of a system is independent of the particular representation employed to define its dynamic behavior.

3.2 CONTROLLABLE AND OBSERVABLE REALIZATIONS

As we have now shown, the transfer function/matrix represents another way, in addition to state-space and differential operator representations, of describing the behavior of dynamic systems. We introduced the transfer function/matrix by illustrating how it can be obtained from state-space representations using Eq. (3.1.8) and from differential operator representations using Eq. (3.1.25).

It is often of interest to develop techniques that essentially "reverse" the relations given by Eqs. (3.1.8) and (3.1.25) to determine either state-space or differential operator representations that describe a system that is initially defined by some $G(s)$. This goal is the primary objective of this section.

Consider any SISO system characterized by a rational transfer function

$$G(s) = \frac{m(s)}{a(s)} = \frac{m_r s^r + m_{r-1} s^{r-1} + \cdots + m_1 s + m_0}{a_n s^n + a_{n-1} s^{n-1} + \ldots + a_1 s + a_0} \quad (3.2.1)$$

with both $m_r \neq 0$ and $a_n \neq 0$. Note that we can "replace" $a(s)$ by a monic polynomial of degree n, the defined *order* of $G(s)$, by dividing both $m(s)$ and $a(s)$ by a_n. For convenience, let us assume that this has already been done or, equivalently, that $a_n = 1$ in Eq. (3.2.1), so that

$$a(s) = s^n + a_{n-1} s^{n-1} + \cdots + a_1 s + a_0 \quad (3.2.2)$$

In light of Eq. (3.1.8), any state-space system $\{A, B, C, E(D)\}^2$ that satisfies the relationship

$$C(sI - A)^{-1} B + E(s) = G(s) = \frac{m(s)}{a(s)} = \frac{m_r s^r + \cdots + m_1(s) + m_0}{s^n + \ldots + a_1 s + a_0} \quad (3.2.3)$$

with $|sI - A| = a(s)$ will be called a **state-space realization of** $G(s)$.

Similarly, in light of Eq. (3.1.25), any differential operator system $\{a(D), b(D), c(D), e(D)\}$ that satisfies the relationship

$$\frac{c(s)b(s)}{a(s)} + e(s) = G(s) = \frac{m(s)}{a(s)} = \frac{m_r s^r + \cdots + m_1 s + m_0}{s^n + \cdots + a_1 s + a_0} \quad (3.2.4)$$

will be called a **differential operator realization of** *G(s)*. These two definitions can be extended to the MIMO case, as we will later do in Section 3.4.

The Euclidean Algorithm

If $G(s)$ is not strictly proper, we can recall the **Euclidean algorithm** [38], which states that if we divide a polynomial $m(s) = m_r s^r + m_{r-1} s^{r-1} + \cdots + m_1 s + m_0$ by another polynomial $a(s) = s^n + a_{n-1} s^{n-1} + \cdots + a_1 s + a_0$, we will obtain a unique **quotient** polynomial $g(s)$ and a unique **remainder** polynomial $l(s)$ such that

$$m(s) = a(s)g(s) + l(s) \quad \text{with} \quad \deg[l(s)] < \deg[a(s)] = n \quad (3.2.5)$$

Therefore, in light of Eq. (3.2.5), any rational $G(s)$ can be expressed as a unique sum

$$G(s) = \frac{m(s)}{a(s)} = \frac{l(s)}{a(s)} + g(s) \quad (3.2.6)$$

---

[2] By allowing E to be an $E(D)$, we have again "expanded" the class of state-space systems to include those whose output may involve derivatives of the input.

of its **strictly proper part** $\dfrac{l(s)}{a(s)}$ and its quotient $g(s)$. Note that if $G(s)$ is strictly proper to begin with,[3] then $l(s) = m(s)$ and $g(s) = 0$. Moreover, if $G(s)$ is proper, then $g(s) = g$, a constant that is given by $\lim_{s \to \infty} G(s)$.

....

EXAMPLE 3.2.7 To illustrate the preceding discussion, consider a system defined by the improper transfer function

$$G(s) = \frac{m(s)}{a(s)} = \frac{2s^4 + 9s^3 + 6s^2 - 9s - 2}{s^3 + 4s^2 + s - 6}$$

If we divide $m(s)$ by $a(s)$,

$$(s^3 + 4s^2 + s - 6) \overline{\smash{\big)}\, 2s^4 + 9s^3 + 6s^2 - 9s - 2}$$

we readily determine (see Problem 3-6) that $g(s) = 2s + 1$ and $l(s) = 2s + 4$, so that

$$G(s) = \frac{2s^4 + 9s^3 + 6s^2 - 9s - 2}{s^3 + 4s^2 + s - 6} = \underbrace{\frac{2s + 4}{s^3 + 4s^2 + s - 6}}_{\text{strictly proper part}} + \underbrace{2s + 1}_{\text{quotient}}$$

....

Canonical Realizations

Using Eq. (3.1.25), it is now a relatively simple matter to obtain two "dual" differential operator realizations of a rational $G(s)$ defined by Eq. (3.2.6), namely, the **controllable canonical differential operator realization** $\{a(D), b(D), c(D), e(D)\} = \{a(D), 1, l(D), g(D)\}$, or

$$a(D)z(t) = \underbrace{b(D)}_{1} u(t); \qquad y(t) = \underbrace{c(D)}_{l(D)} z(t) + \underbrace{e(D)}_{g(D)} u(t) \qquad (3.2.8)$$

and the **observable canonical differential operator realization** $\{a(D), b(D), c(D), e(D)\} = \{a(D), l(D), 1, g(D)\}$, or

$$a(D)z(t) = \underbrace{b(D)}_{l(D)} u(t); \qquad y(t) = \underbrace{c(D)}_{1} z(t) + \underbrace{e(D)}_{g(D)} u(t) \qquad (3.2.9)$$

Note that these differential operator realizations imply corresponding, equivalent, canonical state-space realizations, in light of equivalence relationships Eqs. (2.7.22) and (2.7.11), respectively.

---

[3] Although this will be the case for the majority of physical systems we will encounter, dynamic compensators are often characterized by proper, and even improper transfer functions, as we will later show.

EXAMPLE 3.2.10 To illustrate the two controllable canonical realizations, let us consider the system defined in Example 3.2.7 by the transfer function

$$G(s) = \frac{m(s)}{a(s)} = \frac{2s^4 + 9s^3 + 6s^2 - 9s - 2}{s^3 + 4s^2 + s - 6} = \frac{2s + 4}{s^3 + 4s^2 + s - 6} + 2s + 1$$

with $b(s) = 1$, $c(s) = l(s) = 2s + 4$, and $e(s) = g(s) = 2s + 1$. In light of Eq. (3.2.8), it follows that the differential operator system,

$$\underbrace{(D^3 + 4D^2 + D - 6)}_{a(D)} z(t) = u(t); \qquad y(t) = \underbrace{(2D + 4)}_{c(D)} z(t) + \underbrace{(2D + 1)}_{e(D)} u(t) \qquad (3.2.11)$$

with $b(D) = 1$, represents the unique controllable canonical differential operator realization of $G(s)$. Furthermore, in light of equivalence relationship Eq. (2.7.22), the state-space system

$$\begin{bmatrix} \dot{x}_1(t) \\ \dot{x}_2(t) \\ \dot{x}_3(t) \end{bmatrix} = \underbrace{\begin{bmatrix} 0 & 1 & 0 \\ 0 & 0 & 1 \\ 6 & -1 & -4 \end{bmatrix}}_{A} \begin{bmatrix} x_1(t) \\ x_2(t) \\ x_3(t) \end{bmatrix} + \underbrace{\begin{bmatrix} 0 \\ 0 \\ 1 \end{bmatrix}}_{B} u(t); \qquad y(t) = \underbrace{[4 \ \ 2 \ \ 0]}_{C} \begin{bmatrix} x_1(t) \\ x_2(t) \\ x_3(t) \end{bmatrix} + \underbrace{[2D + 1]}_{E(D)} u(t)$$

$$(3.2.12)$$

represents the corresponding equivalent controllable canonical state-space realization of $G(s)$.

It is of interest to note that although the two controllable canonical form realizations are completely controllable, in general, they are not necessarily observable as well. More specifically, the completely controllable differential operator realization defined by Eq. (3.2.8) is observable if and only if $a(D)$ and $c(D)$ are coprime, and this may not be the case. In particular, since $D+2$ is a factor of both $a(D) = D^3 + 4D^2 + D - 6$ and $c(D) = 2D + 4$ in this example, the equivalent controllable realizations of $G(s)$ defined by Eqs. (3.2.11) and (3.2.12) are characterized by the unobservable mode e^{-2t}.

EXAMPLE 3.2.13 To illustrate the two observable canonical realizations, let us again consider the system defined in Example 3.2.7 by the transfer function

$$G(s) = \frac{m(s)}{a(s)} = \frac{2s^4 + 9s^3 + 6s^2 - 9s - 2}{s^3 + 4s^2 + s - 6} = \frac{2s + 4}{s^3 + 4s^2 + s - 6} + 2s + 1$$

with $b(s) = l(s) = 2s + 4$, $c(s) = 1$, and $e(s) = g(s) = 2s + 1$. In light of Eq. (3.2.9), it follows that the differential operator system

$$\underbrace{(D^3 + 4D^2 + D - 6)}_{a(D)} z(t) = \underbrace{(2D + 4)}_{b(D)} u(t); \qquad y(t) = z(t) + \underbrace{(2D + 1)}_{e(D)} u(t) \qquad (3.2.14)$$

with $c(D) = 1$, represents the unique observable canonical differential operator realization of $G(s)$.

Furthermore, in light of equivalence relationship Eq. (2.7.11), the state-space system

$$
\begin{bmatrix} \dot{x}_1(t) \\ \dot{x}_2(t) \\ \dot{x}_3(t) \end{bmatrix} = \underbrace{\begin{bmatrix} 0 & 0 & 6 \\ 1 & 0 & -1 \\ 0 & 1 & -4 \end{bmatrix}}_{A} \begin{bmatrix} x_1(t) \\ x_2(t) \\ x_3(t) \end{bmatrix} + \underbrace{\begin{bmatrix} 4 \\ 2 \\ 0 \end{bmatrix}}_{B} u(t); \qquad y(t) = \underbrace{[0 \ 0 \ 1]}_{C} \begin{bmatrix} x_1(t) \\ x_2(t) \\ x_3(t) \end{bmatrix} + \underbrace{[2D+1]}_{E(D)} u(t)
$$

$$(3.2.15)$$

represents the corresponding equivalent observable canonical state-space realization of $G(s)$.

In general, the two observable canonical form realizations are not necessarily controllable. More specifically, the completely observable differential operator realization defined by Eq. (3.2.9) is controllable if and only if $a(D)$ and $b(D)$ are coprime, and this may not be the case. In particular, since $D+2$ is a factor of both $a(D) = D^3 + 4D^2 + D - 6$ and $b(D) = 2D + 4$ in this example, the equivalent observable realizations of $G(s)$ defined by Eqs. (3.2.14) and (3.2.15) are characterized by the uncontrollable mode e^{-2t}.

....

Dual Systems and Realizations

The controllable canonical realizations and the observable canonical realizations, which we have derived for SISO systems, represent "dual realizations" of the transfer function $G(s)$. To formalize this notion in the MIMO case, we will say that any two state-space systems, $\{A_c, B_c, C_c, E_c(D)\}$ and $\{A_o, B_o, C_o, E_o(D)\}$, are **dual** to one another (see Problem 2-12 as well) if

$$\{A_o^T, C_o^T, B_o^T, E_o^T(D)\} = \{A_c, B_c, C_c, E_c(D)\} \Longleftrightarrow$$

$$\{A_c^T, C_c^T, B_c^T, E_c^T(D)\} = \{A_o, B_o, C_o, E_o(D)\} \qquad (3.2.16)$$

In an analogous manner, we will say that any two differential operator systems, $\{a_c(D), b_c(D), c_c(D), e_c(D)\}$ and $\{a_o(D), b_o(D), c_o(D), e_o(D)\}$, are dual to one another if[4]

$$\{a_o(D), c_o^T(D), b_o^T(D), e_o^T(D)\} = \{a_c(D), b_c(D), c_c(D), e_c(D)\} \Longleftrightarrow$$

$$\{a_c(D), c_c^T(D), b_c^T(D), e_c^T(D)\} = \{a_o(D), b_o(D), c_o(D), e_o(D)\}$$
$$(3.2.17)$$

In light of these definitions, it can be verified that the controllable canonical realizations of $G(s)$ defined by equivalence relationship Eq. (2.7.22), and the observable canonical realizations of the same $G(s)$ defined by equivalence relationship Eq. (2.7.11) are dual to one another, as are the systems defined by Eqs. (3.2.11) and (3.2.14), as well as the systems defined by Eqs. (3.2.12) and (3.2.15).

---

[4]Since $a(D)$ is assumed to be a single polynomial in all of the differential operator systems considered in the text, $a^T(D)$ is always equal to $a(D)$.

Since

$$\mathcal{C}_c = [B_c,\ A_c B_c,\ \cdots,\ A_c^{n-1} B_c] = \begin{bmatrix} C_o \\ C_o A_o \\ \vdots \\ C_o A_o^{n-1} \end{bmatrix}^T = \mathcal{O}_o^T \quad \text{and} \quad \mathcal{C}_o = \mathcal{O}_c^T$$

(3.2.18)

it follows by *duality* that the controllability "properties" of a dynamic system correspond to the observability "properties" of its dual system and vice versa. In particular, if $e^{\lambda t}$ is a controllable or uncontrollable mode of one system, then $e^{\lambda t}$ will be an observable or unobservable mode of its dual system and vice versa.

We further note that the transfer matrix of a dynamic system is equal to the transpose of the transfer matrix of its dual system; that is, the reader can directly verify (see Problem 3-13) that in light of Eqs. (3.1.8) and (3.1.26),

$$G_o^T(s) = G_c(s) \iff G_c^T(s) = G_o(s)$$

(3.2.19)

One immediate and useful consequence of this fact has already been observed: Whenever $G(s) = G^T(s)$, which is always the case for SISO systems, the determination of any one realization of $G(s)$, such as Eqs. (3.2.11) or (3.2.12) in Example 3.2.10, immediately implies a **dual realization** of the same $G(s)$, such as Eqs. (3.2.14) or (3.2.15) in Example 3.2.13. Therefore, the same algorithm can be used to obtain either pair of $G(s)$ realizations, because the other pair of realizations can be obtained using the duality relations Eqs. (3.2.16) and (3.2.17).

3.3   MINIMAL REALIZATIONS

We have now shown how dual controllable and observable realizations of a rational transfer function

$$G(s) = \frac{m(s)}{a(s)} = \frac{m_r s^r + m_{r-1} s^{r-1} + \cdots + m_1 s + m_0}{a_n s^n + a_{n-1} s^{n-1} + \cdots + a_1 s + a_0}$$

(3.3.1)

can be directly determined, once $G(s)$ is expressed as the sum of its strictly proper part and quotient, as in Eq. (3.2.6); that is, when

$$G(s) = \frac{m(s)}{a(s)} = \frac{l(s)}{a(s)} + g(s)$$

(3.3.2)

It may now be asked, "Which of these dual realizations is the 'appropriate one' to use, and when will dual realizations of the same $G(s)$ be equivalent to one another?" In order to resolve such questions, as well as provide additional insight into the relationships among state-space, differential operator, and transfer matrix representations, we now introduce the concepts of the "poles" and "zeros" of a rational transfer function.

Poles and Zeros

The **poles** of a system defined by a rational transfer function, such as the $G(s)$ of Eq. (3.3.1), are defined as the n roots p_i for $i = 1, 2, \ldots, n$, of the denominator polynomial $a(s)$.[5] In light of Eqs. (3.2.3) and (3.2.4), the poles p_i of a system correspond to the eigenvalues λ_i of the state matrix A of a state-space realization of its transfer function $G(s)$, as well as the roots of the $a(D)$ polynomial of a differential operator realization of $G(s)$, any of which define the system modes $e^{p_i t} = e^{\lambda_i t}$.

The **zeros** of $G(s)$ are defined as the r roots z_j for $j = 1, 2, \ldots, r$ of the numerator polynomial $m(s)$. The zeros of $G(s)$ are so-named because they normally "zero" $G(s)$; that is, $G(s = z_j) = 0$ provided $a(s = z_j) \neq 0$. Generally speaking, both the poles and zeros of $G(s)$ can be complex quantities, which will always appear as complex conjugate pairs, because the coefficients of both $m(s)$ and $a(s)$ are real.

Partial Fraction Modal Representations

Note that $a(s)$ can always be written as the product

$$a(s) = (s - p_1)(s - p_2) \cdots (s - p_n) = \prod_{i=1}^{n} (s - p_i) \tag{3.3.3}$$

Moreover, if the poles are distinct,[6] the strictly proper part of $G(s)$ can be expressed by a **partial fraction expansion**, or a sum of simple, first-order transfer functions, namely,

$$\frac{l(s)}{a(s)} = \frac{r_1}{s - p_1} + \frac{r_2}{s - p_2} + \cdots + \frac{r_n}{s - p_n} \tag{3.3.4}$$

The numerical values of each of the n numerator **residues** r_i can be determined by multiplying Eq. (3.3.4) by $(s - p_i)$ and evaluating the resulting expression at $s = p_i$. For example, to find r_2, we could first multiply Eq. (3.3.4) by $(s - p_2)$ to obtain

$$\frac{(s - p_2)l(s)}{a(s)} = \frac{(s - p_2)r_1}{s - p_1} + \frac{(s - p_2)r_2}{s - p_2} + \cdots + \frac{(s - p_2)r_n}{s - p_n}$$

which, when evaluated at $s = p_2$, implies that

$$\left. \frac{l(s)}{(s - p_1)(s - p_3) \cdots (s - p_n)} \right|_{s = p_2} = r_2$$

---

[5]These are the values of s at which the function $G(s)$ is not analytic, since $G(p_i) \to \infty$. The p_i are so-named because they are analogous to the supporting poles of a tent which cause its height to peak at their locations.

[6]We again note that this will be the case in the vast majority of physical systems. However, when certain of the poles are repeated, a "modified" partial fraction expansion can be employed (see Problem 3-15).

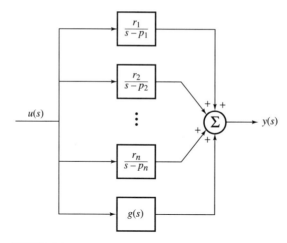

● **FIGURE 3.2**
A Modal Transfer Function Representation of $G(s)$

In general, therefore,

$$r_i = \left. \frac{l(s)}{\prod_{\substack{j=1 \\ j \neq i}}^{n} (s - p_j)} \right|_{s=p_i} \qquad \text{for } i = 1, 2, \ldots, n \qquad (3.3.5)$$

Since $y(s) = G(s)u(s)$, Eqs. (3.3.2) and (3.3.4) imply that

$$y(s) = \left[\frac{r_1}{s - p_1} + \frac{r_2}{s - p_2} + \cdots + \frac{r_n}{s - p_n} + g(s) \right] u(s) \qquad (3.3.6)$$

which represents a **modal transfer function representation** of the system, as depicted in Figure 3.2. Note that the output $y(s)$ is equal to the sum of each of the partial fraction terms multiplied by the input $u(s)$, in addition to the "feedforward" term defined by $g(s)u(s)$.

. . . .

EXAMPLE 3.3.7 Consider a system defined by the transfer function of Example 3.2.7,

$$G(s) = \frac{2s^4 + 9s^3 + 6s^2 - 9s - 2}{s^3 + 4s^2 + s - 6} = \underbrace{\frac{2s + 4}{s^3 + 4s^2 + s - 6}}_{\text{strictly proper part}} + \underbrace{2s + 1}_{\text{quotient}}$$

Since $a(s) = (s - 1)(s + 2)(s + 3)$, the poles of the system lie at $s = +1 \stackrel{\text{def}}{=} p_1$, $s = -2 \stackrel{\text{def}}{=} p_2$, and $s = -3 \stackrel{\text{def}}{=} p_3$ in the complex plane. Equation (3.3.5), therefore, implies that

$$r_1 = \left. \frac{2s + 4}{(s + 2)(s + 3)} \right|_{s=1} = \frac{6}{12} = 0.5 \qquad r_2 = \left. \frac{2s + 4}{(s - 1)(s + 3)} \right|_{s=-2} = \frac{0}{-3} = 0$$

$$r_3 = \left. \frac{2s + 4}{(s - 1)(s + 2)} \right|_{s=-3} = -\frac{2}{4} = -0.5$$

so that

$$\frac{l(s)}{a(s)} = \frac{2s + 4}{(s-1)(s+2)(s+3)} = \frac{0.5}{s-1} + \frac{0}{s+2} + \frac{-0.5}{s+3}$$

as in Eq. (3.3.4).

It is of interest to note that $r_2 = 0$ in this example because $l(s) = 2(s+2)$, so that there is a system zero at the same $s = -2$ location as the pole p_2. Therefore, the transfer function is characterized by a **common pole-zero pair**, as represented by the $(s+2)$ factor in *both* the numerator and denominator of $G(s)$. The presence of such a common factor implies the potential to "cancel" this term, thereby obtaining a transfer function of lower order (2) than $n = 3$, namely,

$$G(s) = \frac{2\cancel{(s+2)}}{(s-1)\cancel{(s+2)}(s+3)} + 2s + 1 = \frac{2s^3 + 5s^2 - 4s - 1}{s^2 + 2s - 3}$$

Controllability and Observability Consequences

We will now establish some useful facts regarding the controllability and observability properties of systems characterized by common pole-zero pairs. Note that each partial fraction term

$$G_i(s) = \frac{r_i}{s - p_i} \tag{3.3.8}$$

in a modal representation of $G(s)$ can be *realized* by the first-order state-space system

$$\dot{x}_i(t) = p_i x_i(t) + b_i u(t) \qquad y_i(t) = c_i x_i(t) \tag{3.3.9}$$

provided $b_i c_i = r_i$, because

$$G_i(s) = \frac{y_i(s)}{u(s)} = c_i(s - p_i)^{-1} b_i = \frac{c_i b_i = r_i}{s - p_i} \tag{3.3.10}$$

Therefore, a modal transfer function representation of $G(s)$ can be realized by the nth order modal state-space system with A in a diagonal Jordan form,

$$\begin{bmatrix} \dot{x}_1(t) \\ \dot{x}_2(t) \\ \vdots \\ \dot{x}_n(t) \end{bmatrix} = \underbrace{\begin{bmatrix} p_1 & 0 & 0 & \cdots \\ 0 & p_2 & 0 & \cdots \\ \vdots & & \ddots & \\ 0 & 0 & \cdots & p_n \end{bmatrix}}_{A} \begin{bmatrix} x_1(t) \\ x_2(t) \\ \vdots \\ x_n(t) \end{bmatrix} + \underbrace{\begin{bmatrix} b_1 \\ b_2 \\ \vdots \\ b_n \end{bmatrix}}_{B} u(t);$$

$$y(t) = \sum_{i=1}^{n} y_i(t) + g(D)u(t) = \underbrace{[c_1, c_2, \ldots, c_n]}_{C} \begin{bmatrix} x_1(t) \\ x_2(t) \\ \vdots \\ x_n(t) \end{bmatrix} + \underbrace{E(D)}_{g(D)} u(t) \tag{3.3.11}$$

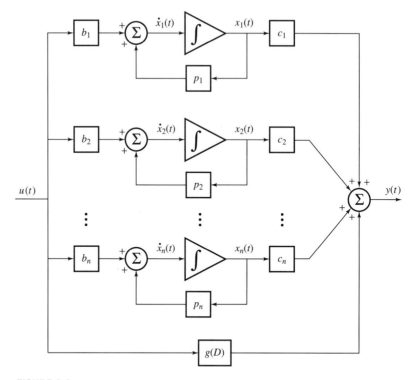

• **FIGURE 3.3**
A Modal State-Space Realization of $G(s)$

provided $b_i c_i = r_i$, for $i = 1, 2, \ldots, n$. This modal realization of $G(s)$, which is analogous to the modal canonical form depicted in Figure 2.14 of Section 2.6, is depicted here in Figure 3.3.

If $G(s)$ is characterized by a common pole-zero pair at $s = p_i$, it follows that $l(s = p_i) = 0$ in Eq. (3.3.5), so that $r_i = c_i b_i = 0$ in Eq. (3.3.10). As a consequence, the state $x_i(t)$, hence the mode $e^{p_i t} = e^{\lambda_i t}$ is uncontrollable if $b_i = 0$, unobservable if $c_i = 0$, or both uncontrollable and unobservable if both $b_i = 0$ and $c_i = 0$.

Now consider any differential operator realization $\{a(D), b(D), c(D), e(D)\}$ of a transfer function $G(s)$ that satisfies Eq. (3.2.4), that is,

$$\frac{c(s)b(s)}{a(s)} + e(s) = G(s) = \frac{m(s)}{a(s)} \qquad (3.3.12)$$

so that

$$c(s)b(s) = m(s) - e(s)a(s) \iff c(D)b(D) = m(D) - e(D)a(D) \qquad (3.3.13)$$

If p_i is a zero of both $a(s)$ and $m(s)$, it will also be a zero of $a(D)$ and either $c(D), b(D)$, or both $c(D)$ and $b(D)$. Therefore, the mode $e^{p_i t} = e^{\lambda_i t}$ is uncontrollable if $b(D = p_i) = 0$, unobservable if $c(D = p_i) = 0$, or both uncontrollable and unobservable if both $b(D = p_i) = 0$ and $c(D = p_i) = 0$.

When a transfer function $G(s)$ is characterized by a common pole-zero pair, then knowledge of $G(s)$ alone does *not* enable us to determine whether the system is uncontrollable, unobservable, or both. For example, the presence of a common pole-zero factor $(s - p_i)$ in $G(s)$ implies only that the input does not affect the output by way of the path from $u(t)$ to $y_i(t) = c_i x_i(t)$ in Figure 3.3. This could be due to an "open" input or output connection. Therefore, the three possibilities are that $b_i = 0$, $c_i = 0$, or both $b_i = 0$ and $c_i = 0$, which imply that the mode $e^{p_i t}$ is uncontrollable, unobservable, or both uncontrollable and unobservable, respectively.

Prime Transfer Functions and Minimal Realizations

In light of the preceding discussion, it is often useful to deal only with **prime** or **minimal transfer functions**, which are defined as those transfer functions characterized by coprime numerator and denominator polynomials, because they represent the only "part" of a system that can be both observed and controlled.

Using these observations, a **minimal state-space** or **differential operator realization** of a rational transfer function $G(s)$ is defined as one that realizes only the "prime part" of $G(s)$, after all common pole-zero factors have been cancelled. Therefore, a minimal realization of $G(s)$ is both completely controllable and completely observable.

For example, if A is the state matrix of any minimal state-space realization of the $G(s)$ of Example 3.3.7, then $|sI - A| = s^2 + 2s - 3 = (s - 1)(s + 3)$. We note further that any two minimal realizations of a rational transfer function are equivalent, although this fact will not be formally established here. In light of this observation, however, two differential operator systems are defined to be *equivalent* to one another if they are equivalent to equivalent state-space systems.

A completely controllable, completely observable system can be characterized by any one of an infinite number of equivalent, minimal representations, all of which realize $G(s)$. However, the controllable canonical state-space and differential operator representations of Eq. (2.7.22) and the *dual* observable canonical state-space and differential operator representations of Eq. (2.7.11) are particularly useful in control system analysis and design because of their special structure. Table 3.1 displays all five of these equivalent system representations in the minimal case, and Example 3.3.14 illustrates these five representations for a particular system. Note that none of these representations require explicit knowledge of the system modes, nor do they rely on any assumptions of distinct modes. These observations will prove useful in certain of our subsequent discussions when we develop control system design algorithms that do not depend on explicit knowledge of the zeros and poles (modes) of the system.

TABLE 3.1 Equivalent Minimal System Representations

The Minimal Transfer Function

$$G(s) = \frac{l(s)}{a(s)} + g(s) \quad \text{with} \quad a(s) = \underbrace{a_0 + a_1 s + \cdots a_{n-1}s^{n-1} + s^n}_{a(D = s)} \tag{3.2.6}$$

and $\quad l(s) = \underbrace{b_0 + b_1 s + \cdots + b_{n-1}s^{n-1}}_{b(D = s)} \quad \text{or} \quad \underbrace{c_0 + c_1 s + \cdots + c_{n-1}s^{n-1}}_{c(D = s)} \quad$ prime.

The Two Minimal Controllable Canonical Representations

$$a(D)z(t) = u(t); \quad y(t) = c(D)z(t) + [e(D) = g(D)]u(t) \tag{2.7.17}$$

and

$$\begin{bmatrix} \dot{x}_1(t) \\ \dot{x}_2(t) \\ \vdots \\ \dot{x}_n(t) \end{bmatrix} = \underbrace{\begin{bmatrix} 0 & 1 & 0 & \cdots & 0 \\ 0 & 0 & 1 & \cdots & 0 \\ \vdots & \vdots & & \ddots & \vdots \\ -a_0 & -a_1 & & \cdots & -a_{n-1} \end{bmatrix}}_{A} \begin{bmatrix} x_1(t) \\ x_2(t) \\ \vdots \\ x_n(t) \end{bmatrix} + \underbrace{\begin{bmatrix} 0 \\ \vdots \\ 0 \\ 1 \end{bmatrix}}_{B} u(t);$$

$$y(t) = \underbrace{[c_0 \ c_1 \ \cdots \ c_{n-1}]}_{C} \begin{bmatrix} x_1(t) \\ x_2(t) \\ \vdots \\ x_n(t) \end{bmatrix} + \underbrace{g(D)}_{E(D)} u(t) \tag{2.7.18}$$

The Two Minimal Observable Canonical Representations

$$a(D)z(t) = b(D)u(t); \quad y(t) = z(t) + [e(D) = g(D)]u(t) \tag{2.7.9}$$

and

$$\begin{bmatrix} \dot{x}_1(t) \\ \dot{x}_2(t) \\ \vdots \\ \dot{x}_n(t) \end{bmatrix} = \underbrace{\begin{bmatrix} 0 & 0 & 0 & \cdots & -a_0 \\ 1 & 0 & 0 & \cdots & -a_1 \\ 0 & 1 & 0 & \cdots & \\ \vdots & \vdots & \ddots & & \vdots \\ 0 & 0 & \cdots & 1 & -a_{n-1} \end{bmatrix}}_{A} \begin{bmatrix} x_1(t) \\ x_2(t) \\ \vdots \\ x_n(t) \end{bmatrix} + \underbrace{\begin{bmatrix} b_0 \\ b_1 \\ \vdots \\ b_{n-1} \end{bmatrix}}_{B} u(t); \tag{2.7.2}$$

$$y(t) = \underbrace{[0 \ 0 \ \cdots \ 0 \ 1]}_{C} \begin{bmatrix} x_1(t) \\ x_2(t) \\ \vdots \\ x_n(t) \end{bmatrix} + \underbrace{g(D)}_{E(D)} u(t) \tag{2.7.10}$$

EXAMPLE 3.3.14 To illustrate the five equivalent ways that the dynamic behavior of a minimal system can be defined, as outlined in Table 3.1, let us consider the particular prime transfer function:

$$G(s) = \frac{l(s)}{a(s)} + g(s) = \frac{5s^2 - 7}{s^3 + 6s^2 - 2s + 8} - 4s + 3$$

In light of Table 3.1, the minimal controllable canonical differential operator representation is given by Eq. (2.7.17), namely,

$$[D^3 + 6D^2 - 2D + 8]z(t) = u(t); \qquad y(t) = [5D^2 - 7]z(t) + [-4D + 3]u(t)$$

and the minimal controllable canonical state-space representation is given by Eq. (2.7.18), namely

$$\begin{bmatrix} \dot{x}_1(t) \\ \dot{x}_2(t) \\ \dot{x}_3(t) \end{bmatrix} = \begin{bmatrix} 0 & 1 & 0 \\ 0 & 0 & 1 \\ -8 & 2 & -6 \end{bmatrix} \begin{bmatrix} x_1(t) \\ x_2(t) \\ x_3(t) \end{bmatrix} + \begin{bmatrix} 0 \\ 0 \\ 1 \end{bmatrix} u(t);$$

$$y(t) = [-7 \ 0 \ 5] \begin{bmatrix} x_1(t) \\ x_2(t) \\ x_3(t) \end{bmatrix} + [-4D + 3]u(t)$$

Moreover, the dual minimal observable canonical differential operator representation is given by Eq. (2.7.9), namely,

$$[D^3 + 6D^2 - 2D + 8]z(t) = [5D^2 - 7]u(t); \qquad y(t) = z(t) + [-4D + 3]u(t)$$

and the dual minimal observable canonical state-space representation is given by Eqs. (2.7.2) and (2.7.10), namely,

$$\begin{bmatrix} \dot{x}_1(t) \\ \dot{x}_2(t) \\ \dot{x}_3(t) \end{bmatrix} = \begin{bmatrix} 0 & 0 & -8 \\ 1 & 0 & 2 \\ 0 & 1 & -6 \end{bmatrix} \begin{bmatrix} x_1(t) \\ x_2(t) \\ x_3(t) \end{bmatrix} + \begin{bmatrix} -7 \\ 0 \\ 5 \end{bmatrix} u(t);$$

$$y(t) = [0 \ 0 \ 1] \begin{bmatrix} x_1(t) \\ x_2(t) \\ x_3(t) \end{bmatrix} + [-4D + 3]u(t)$$

Note that each of these five representations directly implies the other four.

3.4   TRANSFER MATRIX REALIZATIONS

The notion of a state-space or a differential operator realization of a rational transfer function will now be extended to include the matrix case as well. If $G(s)$ is a $(p \times m)$ matrix of transfer functions, that is, a transfer matrix, then any state-space system $\{A, B, C, E(D)\}$ that satisfies Eq. (3.1.8), so that

$$C(sI - A)^{-1}B + E(s) = G(s) \tag{3.4.1}$$

is called a **state-space realization of G(s)**. Moreover, any differential operator system $\{a(D), b(D), c(D), e(D)\}$ that satisfies Eq. (3.1.26), so that

$$\frac{c(s)b(s)}{a(s)} + e(s) = G(s) \tag{3.4.2}$$

is called a **differential operator realization of G(s)**.

If a given rational $G(s)$ has any improper elements, then the Euclidean algorithm can be applied to each of these individually, as in Eq. (3.2.6), to determine the $(p \times m)$ matrix quotient $E(s) = e(s)$ and, as a consequence, the **strictly proper part of G(s)**,

$$\hat{G}(s) \stackrel{\text{def}}{=} G(s) - E(s) = C(sI - A)^{-1}B = G(s) - e(s) = \frac{c(s)b(s)}{a(s)}$$
$$\tag{3.4.3}$$

Since $c(s)$ and $b(s)$ are restricted to be polynomial vectors,[7] the determination of an appropriate differential operator realization or an equivalent state-space realization of $G(s)$ is restricted to those systems for which $\hat{G}(s)$ has rank one, in the sense that $a(s)\hat{G}(s)$ can be expressed as the product of two polynomial vectors, namely a $(p \times 1)$ column vector $c(s)$ and a $(1 \times m)$ row vector $b(s)$.

The actual determination of a pair of polynomial vectors $c(s)$ and $b(s)$ so that

$$c(s)b(s) = a(s)\hat{G}(s) = a(s)[G(s) - e(s)] \tag{3.4.4}$$

then implies a differential operator realization $\{a(D), b(D), c(D), e(D)\}$ of $G(s)$ since Eq. (3.4.4) implies Eq. (3.4.2). A corresponding, equivalent, state-space realization of $G(s)$ can then be determined by the equivalence relationship Eq. (2.7.8), as in Example 2.7.31.

It is a relatively simple matter to determine when $\hat{G}(s)$ has rank one and, in such cases, to subsequently determine a $c(s)$ and a $b(s)$ that satisfy Eq. (3.4.4). In particular, if we express both $c(s)$ and $b(s)$ in terms of their individual polynomial components, so that

$$c(s) = \begin{bmatrix} c_1(s) \\ c_2(s) \\ \vdots \\ c_p(s) \end{bmatrix} \quad \text{and} \quad b(s) = [b_1(s) \ b_2(s) \ \cdots b_m(s)]$$

---

[7]This restriction can be removed by allowing $a(D)$, hence $a(s)$, to be a polynomial matrix [62]. However, such an assumption would extend our coverage beyond the intended scope of this text.

it follows that

$$c(s)b(s) = \begin{bmatrix} c_1(s)b_1(s) & c_1(s)b_2(s) & \cdots & c_1(s)b_m(s) \\ c_2(s)b_1(s) & c_2(s)b_2(s) & \cdots & c_2(s)b_m(s) \\ \vdots & \vdots & & \vdots \\ c_p(s)b_1(s) & c_p(s)b_2(s) & \cdots & c_p(s)b_m(s) \end{bmatrix} \qquad (3.4.5)$$

In light of Eqs. (3.4.4) and (3.4.5), note that each $c_i(s)$ divides the ith row of $a(s)\hat{G}(s)$, while each $b_j(s)$ divides the jth column of $a(s)\hat{G}(s)$. Appropriate elements of $c(s)$ and $b(s)$ can therefore be found by determining the greatest common divisors of the p rows and m columns of $a(s)\hat{G}(s)$, once its elements are expressed in factored form.

Note further that any transfer matrix pole p_i that zeros $c_j(s)$, so that $c_j(p_i) = 0$, will also zero the entire jth row of the numerator matrix $c(s)b(s) = a(s)\hat{G}(s)$, hence $a(s)G(s)$ as well. Any such p_i implies a corresponding system mode $e^{p_i t}$ that is unobservable by the output component $y_j(t)$. Conversely, any transfer matrix pole p_i that zeros $b_k(s)$, so that $b_k(p_i) = 0$, also zeros the entire kth column of the numerator matrix $c(s)b(s) = a(s)\hat{G}(s)$, hence $a(s)G(s)$ as well. Any such p_i implies a corresponding system mode $e^{p_i t}$ that is uncontrollable by the input component $u_k(t)$. Therefore, the controllability and observability "properties" of a rank one MIMO system can be determined directly from knowledge of its transfer matrix $G(s)$.

. . . .

EXAMPLE 3.4.6 Consider the strictly proper, rank one rational transfer matrix

$$G(s) = \frac{\begin{bmatrix} -2s^3 - 2s^2 + 14s + 30 & -2s^3 + 6s^2 + 2s - 6 \\ -s^3 - 5s^2 - 9s - 5 & -s^3 - s^2 + s + 1 \end{bmatrix}}{s^4 + 2s^3 - 6s^2 - 22s - 15 = (s+1)(s-3)(s^2+4s+5)}$$

which is equivalent to a strictly proper $\hat{G}(s)$ in this case because

$$\lim_{s \to \infty} G(s) = g(s) = e(s) = \begin{bmatrix} 0 & 0 \\ 0 & 0 \end{bmatrix}$$

We first factor each element of $a(s)\hat{G}(s) = a(s)G(s)$, thereby obtaining the expression:

$$c(s)b(s) = a(s)G(s) = \begin{bmatrix} -2(s-3)(s^2+4s+5) & -2(s-3)(s+1)(s-1) \\ -(s+1)(s^2+4s+5) & -(s+1)^2(s-1) \end{bmatrix}$$

Clearly, $(s-3)$ and $(s+1)$ represent greatest common divisors of rows 1 and 2 of $a(s)G(s)$. Therefore, the corresponding modes, e^{3t} and e^{-t}, are unobservable by the output components $y_1(t)$ and $y_2(t)$, respectively. Since (s^2+4s+5) and $(s+1)(s-1) = (s^2-1)$ represent greatest common divisors of columns 1 and 2 of $a(s)G(s)$, it follows that the modes $e^{-2t}\sin t$ and e^{-t} are uncontrollable by the input components $u_1(t)$ and $u_2(t)$, respectively.

Since the product

$$\underbrace{\begin{bmatrix} -2(s-3) \\ -(s+1) \end{bmatrix}}_{c(s)} \underbrace{[s^2 + 4s + 5, \quad s^2 - 1]}_{b(s)} = a(s)G(s)$$

in this case, it follows that

$$\underbrace{(D^4 + 2D^3 - 6D^2 - 22D - 15)}_{a(s\,=\,D)} z(t) = \underbrace{[D^2 + 4D + 5, \quad D^2 - 1]}_{b(s\,=\,D)} \mathbf{u}(t);$$

$$\mathbf{y}(t) = \begin{bmatrix} y_1(t) \\ y_2(t) \end{bmatrix} = \underbrace{\begin{bmatrix} -2D+6 \\ -D-1 \end{bmatrix}}_{c(s\,=\,D)} z(t)$$

represents a differential operator realization of $G(s)$.

It may be noted that that we have essentially "reversed" the algorithm used to determine the transfer matrix of Example 3.1.27, which was based on knowledge of the differential operator representation of Example 2.7.31. Therefore, the equivalent $\{A, B, C\}$ system of that example

$$\begin{bmatrix} \dot{x}_1(t) \\ \dot{x}_2(t) \\ \dot{x}_3(t) \\ \dot{x}_4(t) \end{bmatrix} = \underbrace{\begin{bmatrix} 0 & 0 & 0 & 15 \\ 1 & 0 & 0 & 22 \\ 0 & 1 & 0 & 6 \\ 0 & 0 & 1 & -2 \end{bmatrix}}_{A} \begin{bmatrix} x_1(t) \\ x_2(t) \\ x_3(t) \\ x_4(t) \end{bmatrix} + \underbrace{\begin{bmatrix} 5 & -1 \\ 4 & 0 \\ 1 & 1 \\ 0 & 0 \end{bmatrix}}_{B} \begin{bmatrix} u_1(t) \\ u_2(t) \end{bmatrix};$$

$$\mathbf{y}(t) = \begin{bmatrix} y_1(t) \\ y_2(t) \end{bmatrix} = \underbrace{\begin{bmatrix} 0 & 0 & -2 & 10 \\ 0 & 0 & -1 & 1 \end{bmatrix}}_{C} \begin{bmatrix} x_1(t) \\ x_2(t) \\ x_3(t) \\ x_4(t) \end{bmatrix}$$

represents an equivalent state-space realization of $G(s)$ in this case.

Minimal Transfer Matrix Realizations

The notion of a minimal realization of a rational transfer function, which was introduced in the previous section, readily extends to the matrix case if we restrict our discussion to those situations where $\hat{G}(s) = G(s) - e(s)$ has rank one, so that it can be expressed as

$$\frac{c(s)b(s)}{a(s)} = \hat{G}(s) \tag{3.4.7}$$

A corresponding differential operator realization $\{a(D), b(D), c(D)\}$ of $\hat{G}(s)$ is controllable (or observable) if and only if $a(D)$ and $b(D)$ (or $c(D)$) are coprime. As a consequence, the realization is *minimal* (both controllable and observable) if and only if $a(D)$ and *all* of the (pm)

elements of the product $c(D)b(D)$ are coprime. Therefore, in the rank one case, any state-space or differential operator realization of $G(s) = \hat{G}(s) + e(s)$, with

$$|sI - A| = a(D = s) = a(s) \tag{3.4.8}$$

is minimal if and only if there are no pole-zero factors common to both $a(s)$ and *all* of the elements of $a(s)b(s) + e(s)a(s) = a(s)G(s)$, which represents the numerator matrix of

$$G(s) = \frac{c(s)b(s) + e(s)a(s)}{a(s)}$$

This is clearly the case in Example 3.4.6, so that the equivalent, fourth-order differential operator and state-space realizations derived in that example are both controllable and observable, hence minimal, which verifies the conclusions that were made earlier in Example 2.7.31.

. . . .

EXAMPLE 3.4.9 Consider a system defined by the $(p \times m = 2 \times 3)$ rank one rational transfer matrix

$$G(s) = \frac{\begin{bmatrix} 3s + 6 & s^2 + 9s + 14 & 0 \\ s^3 + 2s^2 - 2s - 4 & -4s - 4 & -2s^2 - 6s - 4 \end{bmatrix}}{s^2 + 3s + 2}$$

Since

$$\frac{s^2 + 9s + 14}{s^2 + 3s + 2} = \frac{6s + 12}{s^2 + 3s + 2} + 1, \qquad \frac{-2s^2 - 6s - 4}{s^2 + 3s + 2} = -2$$

and

$$\frac{s^3 + 2s^2 - 2s - 4}{s^2 + 3s + 2} = \frac{-s - 2}{s^2 + 3s + 2} + s - 1$$

using the Euclidean algorithm, it follows that the matrix quotient

$$E(s) = e(s) = \begin{bmatrix} 0 & 1 & 0 \\ s - 1 & 0 & -2 \end{bmatrix}$$

and the strictly proper part

$$\hat{G}(s) = \frac{\begin{bmatrix} 3s + 6 & 6s + 12 & 0 \\ -s - 2 & -4s - 4 & 0 \end{bmatrix}}{s^2 + 3s + 2} = \frac{\begin{bmatrix} 3(s + 2) & 6(s + 2) & 0 \\ -(s + 2) & -2(s + 2) & 0 \end{bmatrix}}{(s + 1)(s + 2)}$$

The presence of a factor $(s + 2)$ common to $a(s)$ and all of the $(pm = 6)$ polynomial elements of the numerator matrix $a(s)\hat{G}(s)$ and, therefore, all of the polynomial elements of $a(s)G(s)$ as well, implies that the mode e^{-2t} is either uncontrollable, unobservable, or both. Additional information about the system would be required in order to determine which of these three possibilities holds in this particular case.

Note that such a determination need not be made to define a minimal realization of $G(s)$, since the "prime part" of $\hat{G}(s)$

$$\hat{G}(s) = \frac{\begin{bmatrix} 3 & 6 & 0 \\ -1 & -2 & 0 \end{bmatrix}}{s + 1}$$

can be realized directly by any number of first ($n = 1$) order state-space or differential operator systems. For example, both the minimal state-space representation

$$\{A,\ B,\ C\} = \left\{ -1,\ [1 \quad 2],\ \begin{bmatrix} 3 \\ -1 \end{bmatrix} \right\}$$

and the equivalent, minimal differential operator representation

$$\{a(D),\ b(D),\ c(D)\} = \left\{ D + 1,\ [1 \quad 2],\ \begin{bmatrix} 3 \\ -1 \end{bmatrix} \right\}$$

realize the prime $\hat{G}(s)$, irrespective of the controllability and observability properties of the system defined by the original $G(s)$.

3.5   SUMMARY

We have now illustrated three alternative ways of representing the dynamic behavior of a linear time-invariant, dynamic system: the internal state-space representation, the differential operator representation, and the transfer matrix/function, which defines only the external input/output behavior of a system.

A transfer matrix is easily obtained from a state-space representation by Eq. (3.1.8),

$$G(s) = C(sI - A)^{-1}B + E$$

Two different derivations of this relationship were presented. A transfer matrix also can be obtained directly from a differential operator system by Eq. (3.1.26),

$$G(s) = \frac{c(s)b(s)}{a(s)} + e(s)$$

Proper, strictly proper, and improper transfer functions were defined. Dual, differential operator realizations of a strictly proper transfer function were shown to be directly obtainable from $G(s)$, and the Euclidean algorithm can be used when this is not the case to "separate" the quotient $g(s) = e(s)$ from $G(s)$. The dual, differential operator representations

were then shown to directly imply corresponding, dual, state-space realizations of $G(s)$.

A modal representation of $G(s)$ was then employed to illustrate that
when a transfer function has common pole-zero factors, a corresponding
realization will be uncontrollable, unobservable, or both. When there are
no such common factors, all realizations will be minimal, that is, both
controllable and observable, as well as equivalent. Table 3.1 displayed
five equivalent, canonical system representations in the minimal case, and
Example 3.3.14 served to illustrate these five representations for a particular system. Many of the SISO results were extended to the rank one
MIMO case in Section 3.4.

PROBLEMS

3-1. Determine the (3×1) **transfer vector** (a vector of transfer functions)
of the state-space system:

$$\underbrace{\begin{bmatrix} 0 & 1 & 0 \\ 0 & 0 & 1 \\ -5 & -2 & -2 \end{bmatrix}}_{A}, \quad \underbrace{\begin{bmatrix} 0 \\ 0 \\ 1 \end{bmatrix}}_{B}, \quad \underbrace{\begin{bmatrix} 1 & 0 & 0 \\ 0 & 1 & -2 \\ 3 & 0 & 2 \end{bmatrix}}_{C}$$

3-2. Determine the (1×3) transfer vector of the state-space system (which
is dual to that defined in Problem 3-1):

$$\underbrace{\begin{bmatrix} 0 & 0 & -5 \\ 1 & 0 & -2 \\ 0 & 1 & -2 \end{bmatrix}}_{A}, \quad \underbrace{\begin{bmatrix} 1 & 0 & 3 \\ 0 & 1 & 0 \\ 0 & -2 & 2 \end{bmatrix}}_{B}, \quad \underbrace{[0 \ 0 \ 1]}_{C}$$

3-3. Consider a system defined by the prime transfer function

$$G(s) = \frac{s^3 + s^2 + 4s - 2}{s^2 + 1}$$

(a) Determine both the controllable and the observable canonical differential operator realizations of this $G(s)$.

(b) Determine both the controllable and the observable canonical state-
space realizations of this $G(s)$.

(c) Determine the unique equivalence transformation matrix Q that
transforms the observable canonical state-space realization to the
controllable canonical state-space realization.

3-4. Consider a system defined by the following differential equations:

$$\ddot{q}(t) - q(t) = \dot{r}(t) + 2r(t) \qquad w(t) = \dot{q}(t) + 3q(t) - r(t)$$

Determine an equivalent state-space representation with output $y(t) =$

$w(t)$ and input $u(t) = r(t)$. Verify your findings by showing the transfer function equivalence of the two representations.

3-5. Using Eq. (3.2.6),

$$G(s) = \frac{l(s)}{a(s)} + g(s) = \frac{l(s) + g(s)a(s)}{a(s)} = \frac{m(s)}{a(s)}$$

prove that $l(s)$ and $a(s)$ are coprime if and only if $m(s) = l(s) + g(s)a(s)$ and $a(s)$ are coprime.

3-6. Show that the division of $m(s) = 2s^4 + 9s^3 + 6s^2 - 9s - 2$ by $a(s) = s^3 + 4s^2 + s - 6$ yields the unique quotient $q(s) = 2s + 1$ and the unique remainder $l(s) = 2s + 4$, as stated in Example 3.2.7.

3-7. Consider a system defined by the transfer function

$$G(s) = \frac{s + 2}{s^2 + 3s + 2} = \frac{s + 2}{(s + 1)(s + 2)}$$

Determine a second order state-space realization of $G(s)$, with

which is

$$A = \begin{bmatrix} 0 & -2 \\ 1 & -3 \end{bmatrix}$$

    (a) controllable, but unobservable;

    (b) observable, but uncontrollable;

    (c) both uncontrollable and unobservable.

Determine an equivalent differential operator representation in all three cases.

3-8. Consider a dynamic system defined by the state-space representation: $\dot{\mathbf{x}}(t) = A\mathbf{x}(t) + B\mathbf{u}(t);\ \mathbf{y}(t) = C\mathbf{x}(t)$, when

$$A = \begin{bmatrix} 0 & 0 \\ 1 & -1 \end{bmatrix} \qquad B = \begin{bmatrix} 1 & 0 \\ 1 & 1 \end{bmatrix} \qquad C = \begin{bmatrix} 1 & 0 \\ -1 & 1 \end{bmatrix}$$

    (a) Transform the system to modal form and identify the modes that are uncontrollable and unobservable by the individual components of the input and output.

    (b) Now use Eq. (2.7.8) to determine an equivalent differential operator system $\{a(D),\ b(D),\ c(D),\ e(D)\}$. Using Eq. (3.4.5), verify that the controllability and observability properties of the differential operator system are identical to those of the state-space system.

    (c) Verify the equivalence of the transfer matrix $G(s)$, as determined by *both* Eqs. (3.1.8) and (3.1.26), that is, show that in this particular case,

$$G(s) = C(sI - A)^{-1}B = \frac{c(s)b(s)}{a(s)} + e(s) = \begin{bmatrix} \frac{1}{s} & 0 \\ 0 & \frac{1}{s+1} \end{bmatrix}$$

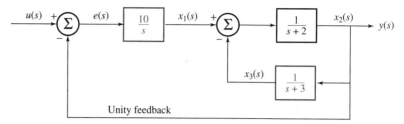

• **FIGURE 3.4**
The Transfer Function Block Diagram for Problem 3-9

3-9. Consider the third-order unity feedback system depicted in Figure 3.4.

    (a) Determine a state-space representation of this system assuming that the outputs of the three transfer function "blocks" are $x_1(s)$, $x_2(s) = y(s)$, and $x_3(s)$.

    (b) Determine the closed loop transfer function $G(s)$ between $y(s)$ and $u(s)$.

    (c) Determine the transfer function $G_e(s)$ between the error $e(s) = u(s) - y(s)$ and $u(s)$.

3-10. Consider a fourth-order system consisting of a series connection of

$$H(s) = \frac{s+1}{s(s+2)} \quad \text{followed by} \quad G(s) = \frac{s+2}{s(s+1)}$$

as depicted in Figure 3.5(a).

    (a) Show that such a connection implies a single uncontrollable mode e^{-t} and a single unobservable mode e^{-2t}.

    (b) Show that a reversed series connection of $G(s)$ followed by $H(s)$, as depicted in Figure 3.5(b), implies a single uncontrollable mode e^{-2t} and a single unobservable mode e^{-t}.

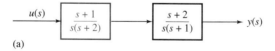

(a)

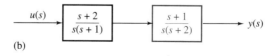

(b)

• **FIGURE 3.5**
The Series Systems of Problem 3-10

(c) How may this observation be generalized to series connections of systems described by arbitrary transfer functions?

3-11. Determine the unit impulse response of the system defined in Example 3.3.7.

3-12. Consider the second-order system depicted by the transfer function block diagram of Figure 3.6. Determine an equivalent state-space system and show that the mode e^{-at} will be unobservable by the output $y(t)$ regardless of whether the "feedback switch" is open or closed. It might be noted that this problem serves to motivate the so-called *internal model principle*, which will be presented in Section 7.3.

3-13. In light of Eqs. (3.2.16) and (3.1.8), verify that Eq. (3.2.19) holds for dual state-space systems. In light of Eqs. (3.2.17) and (3.1.26), verify that Eq. (3.2.19) holds for dual differential operator systems.

3-14. Use Eq. (3.1.7) to determine the state transition matrix e^{At} when

$$A = \begin{bmatrix} 0 & 1 & 0 \\ 0 & 0 & 1 \\ 6 & -1 & -4 \end{bmatrix}$$

and verify that it is identical to that given in Example 2.5.18.

3-15. Consider the strictly proper transfer function

$$G(s) = \frac{y(s)}{u(s)} = \frac{s+4}{s^2 + 4s + 4} = \frac{s+4}{(s+2)^2}$$

which has a repeated pole at $s = -2$. In such a case, a partial fraction expansion of $G(s)$ will include a second-order term, since

$$G(s) = \frac{s+4}{(s+2)^2} = \frac{r_1}{s+2} + \frac{r_2}{(s+2)^2} = \frac{r_1(s+2) + r_2}{(s+2)^2}$$

so that $r_1 = 1$ and $r_2 = 2$ (because $2r_1 + r_2 = 4$).

(a) Determine the unit impulse output response of the system.

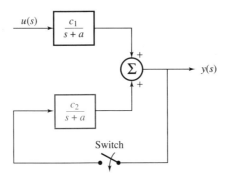

● **FIGURE 3.6**
The Transfer Function System of Problem 3-12

(b) Determine e^{At} using Eq. (3.1.7) if

$$A = \begin{bmatrix} 0 & 1 \\ -4 & -4 \end{bmatrix}$$

and verify that it is identical to that given in Problem 2-7.

3-16. Determine the transfer matrix $G(s) = C(sI - A)^{-1}B$ of the orbiting satellite, as defined by the linearized state-space equations of Example 2.6.14.

(a) Which pole implies a corresponding mode that is uncontrollable by the input component $u_1(t)$?

(b) Which pole implies a corresponding mode that is unobservable by the output component $y_1(t)$?

3-17. Consider a system whose dynamic behavior is defined by the differential equations:

$$\frac{d^3z(t)}{dt^3} - \frac{d^2z(t)}{dt^2} - 2\frac{dz(t)}{dt} = \frac{du(t)}{dt} - u(t) \quad \text{and} \quad y(t) = -2\frac{dz(t)}{dt} + 4z(t)$$

(a) Determine an equivalent state-space system.

(b) Determine the transfer function between $y(s)$ and $u(s)$.

(c) Determine all uncontrollable and unobservable modes.

(d) Determine an equivalent state-space system in modal canonical form.

3-18. Show that the differential operator system

$$\underbrace{(D^2 + 6.5D + 22.1)}_{a(D)} z(t) = \underbrace{[-36D - 160.8, \ -279.4D - 896.73]}_{b(D) = [b_1(D), \ b_2(D)]} \begin{bmatrix} u_1(t) \\ u_2(t) \end{bmatrix};$$

$$y(t) = z(t) + \underbrace{[8, \ 37.3]}_{e(D) = e} \begin{bmatrix} u_1(t) \\ u_2(t) \end{bmatrix}$$

and the equivalent state-space system:

$$\begin{bmatrix} \dot{x}_1(t) \\ \dot{x}_2(t) \end{bmatrix} = \underbrace{\begin{bmatrix} 0 & -22.1 \\ 1 & -6.5 \end{bmatrix}}_{A} \begin{bmatrix} x_1(t) \\ x_2(t) \end{bmatrix} + \underbrace{\begin{bmatrix} -160.8 & -279.4 \\ -36 & -896.73 \end{bmatrix}}_{B} \begin{bmatrix} u_1(t) \\ u_2(t) \end{bmatrix};$$

$$y(t) = \underbrace{x_2(t)}_{z(t)} + [8, \ 37.3] \begin{bmatrix} u_1(t) \\ u_2(t) \end{bmatrix}$$

both realize the (1×2) transfer vector depicted in Figure 3.7.

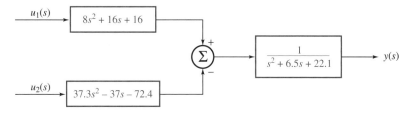

● **FIGURE 3.7**
The (1×2) Transfer Vector of Problem 3-18

3-19.* Verify that the nominal transfer function of the DC servomotor of Example 2.4.10 is given by

$$G(s) = \frac{y(s)}{u(s)} = \frac{\theta(s)}{v_a(s)} = \frac{10}{s^3 + 11s^2 + 10s}$$

if $R_a = 0.1$, $L_a = 0.0095$, $K_b = 0.505$, $K_m = 10$, $J = 105.3$, and $D = 49.5$, as in Problem 2-6.

3-20.* Consider a dynamic system defined by the state-space representation: $\dot{\mathbf{x}}(t) = A\mathbf{x}(t) + B\mathbf{u}(t)$; $\mathbf{y}(t) = C\mathbf{x}(t)$, when

$$A = \begin{bmatrix} 0 & -4 & 3 & -4 \\ 2 & -9 & -2 & -8 \\ 0 & 0 & 1 & 0 \\ -2 & 7 & 5 & 6 \end{bmatrix}, \quad B = \begin{bmatrix} 1 & -1 \\ -1 & -15 \\ 1 & 3 \\ 2 & 19 \end{bmatrix}$$

$$C = \begin{bmatrix} -1 & -1 & 7 & -2 \\ 2 & 1 & -5 & 2 \\ 0 & -3 & 11 & -4 \end{bmatrix}$$

(a) Transform the system to a diagonal, modal form and identify the modes that are uncontrollable and unobservable by the individual components of the input and output.

(b) Determine the transfer matrix $G(s)$ using both the given and the equivalent state-space representations.

(c) Verify the controllability/observability properties determined in Part (a) by "inspection" of $G(s)$.

3-21.* The pitch angle $\rho(t)$ of the helicopter depicted in Figure 3.8 can be controlled by adjusting the angle $\alpha(t)$ of the main rotor blades. At a nominal hovering position, the dynamic behavior of this system can be defined by the differential equations:

$$\frac{d^2\rho(t)}{dt^2} = -0.65\frac{d\rho(t)}{dt} - 0.02\frac{dx(t)}{dt} + 5.4\alpha(t)$$

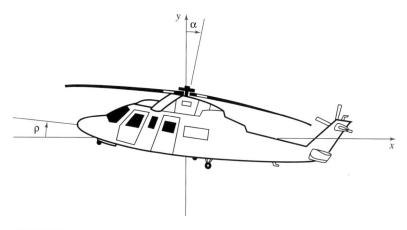

• **FIGURE 3.8**
Pitch Control of the Hovering Helicopter of Problem 3-21

$$\frac{d^2x(t)}{dt^2} = -1.57\frac{d\rho(t)}{dt} - 0.03\frac{dx(t)}{dt} + 9.8[\rho(t) + \alpha(t)]$$

with $x(t)$ representing the horizontal position.

(a) Determine a state-space representation for this system.

(b) Determine the system modes.

(c) Calculate the transfer function relationship between the output $\rho(s)$ and the input $\alpha(s)$.

3-22. In 1953, S. J. Mason introduced a general procedure for analyzing block diagrams.[8] His algorithm has found widespread use in control system analysis and design, primarily in determining the transfer function relationship between defined input/output pairs, such as output $y(s)$ and input $r(s)$ in the block diagram of Figure 3.9. A simplified version of Mason's algorithm follows:

• Determine the number m of closed *loops*, and the **loop gain** L_j of each. (*Note*: This step is independent of the defined input/output pair.)

• Determine the number k of direct *paths* between the defined input and the defined output, and the **path gain** P_i of each. (*Note*: In both of these first two steps, the same point in the block diagram cannot be passed through more than once.)

[8]S. J. Mason, "Feedback Theory—Some Properties of Signal Flow Graphs," *Proc. IRE*, Vol. 44 (7), Sept., 1953.

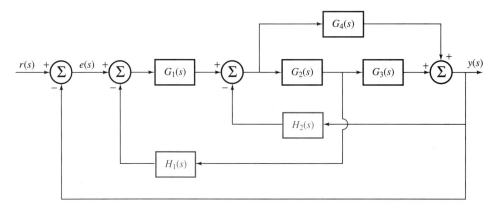

● **FIGURE 3.9**
A Block Diagram for Problem 3-22

● Assuming that each loop touches every other loop, as well as all of the forward paths, a simplified version of **Mason's formula** states that

$$\frac{\text{output}}{\text{input}} = \frac{\sum_{i=1}^{k} P_i}{1 - \sum_{j=1}^{m} L_j \overset{\text{def}}{=} \Delta} \qquad (3.4.10)$$

(a) Verify that there are two direct paths from $r(s)$ to $y(s)$ in Figure 3.9, and determine both of the path gains. (*Hint*: $G_1(s)G_4(s)$ represents one of these path gains.)

(b) Verify that there are five closed loops, and determine all five of the loop gains. (*Hint*: $-G_1(s)G_4(s)$ and $-G_2(s)G_3(s)H_2(s)$ represent two of these loop gains.)

(c) Verify that each loop touches every other loop, as well as both of the forward paths. As a consequence, use Eq. (3.4.10) to determine the closed-loop transfer function between output $y(s)$ and input $r(s)$, that is, determine

$$G_{yr}(s) = \frac{y(s)}{r(s)}$$

If each loop in a block diagram touches every other loop, but *not* all of the forward paths, Mason's formula can be modified as follows:

$$\frac{\text{output}}{\text{input}} = \frac{\sum_{i=1}^{k} P_i \Delta_i}{1 - \sum_{j=1}^{m} L_j = \Delta} \qquad (3.4.11)$$

where each $\Delta_i = \Delta +$ those L_j that touch P_i.

(d) Verify that there is only one direct path from input $r(s)$ to output $e(s)$ in Figure 3.9, and determine its path gain.

(e) Use Eq. (3.4.11) to determine the closed-loop transfer function between $e(s)$ and $r(s)$, that is, determine

$$G_{er}(s) = \frac{e(s)}{r(s)}$$

4

DYNAMIC
RESPONSE

4.1   THE IMPULSE RESPONSE

In the previous two chapters, we focused our attention on mathematical descriptions of dynamic systems. We showed how the dynamic behavior of a linear system can be *explicitly* described by either the internal state-space and differential operator representations or the external transfer function. We also established a notion of "equivalence" among these three types of representations.

In this chapter, we will present an alternative way of *implicitly* describing the behavior of a dynamic system through the physical observation of its dynamic response, which is defined as the time-varying behavior of its state or output to a known set of nonzero initial conditions or a known input. As we will show, various plots of the dynamic response can tell us a great deal about an otherwise unknown system.

The reader may recall that we have already derived some analytical expressions for the dynamic response of a known state-space system in terms of its state transition matrix e^{At}. The complete state response was given by Eq. (2.5.9),

$$\mathbf{x}(t) = \underbrace{e^{A(t-t_0)}\mathbf{x}_0}_{\text{zero-input response}} + \underbrace{\int_{t_0}^{t} e^{A(t-\tau)}B\mathbf{u}(\tau)d\tau}_{\text{zero-state response}}$$

with

$$\mathbf{y}(t) = \underbrace{Ce^{A(t-t_0)}\mathbf{x}_0}_{\text{zero-input response}} + \underbrace{C\int_{t_0}^{t} e^{A(t-\tau)}B\mathbf{u}(\tau)d\tau + E\mathbf{u}(t)}_{\text{zero-state response}}$$

defining the complete output response. The zero-input response of a system is often termed a **natural response**, because it is dependent only on the internal or natural modes of the system and not on the external input $\mathbf{u}(t)$.

In this section, we will obtain an analytical expression for the zero-state **impulse response** of a linear time-invariant SISO system, which is defined as the output $y(t)$ that results from the application of the unit impulse input $u(t) = \delta(t)$ to a system initially at rest. In particular, if we now recall that the *transfer function* $G(s)$ of a SISO system represents the Laplace transform of its impulse response, it follows that the inverse Laplace transform of $G(s)$ yields an analytical expression for its impulse response.

More specifically, consider a system defined by the transfer function

$$G(s) = \frac{m(s)}{a(s)} = \frac{l(s)}{a(s)} + g(s)$$

as defined by Eq. (3.3.2), so that

$$y(s) = \underbrace{\left[\frac{r_1}{s - p_1} + \frac{r_2}{s - p_2} + \cdots + \frac{r_n}{s - p_n} + g(s) \right]}_{G(s)} u(s)$$

in light of Eq. (3.3.6). If we take the inverse Laplace transform of $y(s)$ when $u(s) = \mathcal{L}[\delta(t)] = 1$, it follows that the impulse response

$$y(t) = \mathcal{L}^{-1}[y(s)] = \mathcal{L}^{-1}[G(s)\,\underbrace{u(s)}_{1}] = r_1 e^{p_1 t} + r_2 e^{p_2 t} + \cdots r_n e^{p_n t} + g(D)\delta(t)$$

$$\text{(4.1.1)}$$

Therefore, the impulse response of a linear time-invariant dynamic system is comprised of a "weighted sum" of its modes, as well as $\delta(t)$ and certain of its derivatives.

If any poles of $G(s)$ are complex valued, that is, if $p_i = \alpha + j\beta$ for some value of i between 1 and n, then such poles appear as complex conjugate pairs, so that $p_k = \alpha - j\beta$ for some $k \neq i$, because the coefficients of $a(s)$ are real. In such cases, the corresponding residues, r_i and r_k, also appear as a complex conjugate pair. As a consequence, the impulse response terms in Eq. (4.1.1) that correspond to such a complex conjugate pair, namely,

$$r_i e^{p_i t} + r_k e^{p_k t} = (\gamma + j\rho)e^{(\alpha+j\beta)t} + (\gamma - j\rho)e^{(\alpha-j\beta)t}$$

$$= 2e^{\alpha t}[\gamma \cos \beta t - \rho \sin \beta t] = 2\sqrt{\gamma^2 + \rho^2}\ e^{\alpha t} \cos\left(\beta t + \tan^{-1}\rho/\gamma\right)$$

$$\text{(4.1.2)}$$

as the reader can verify (see Problem 4-1). It may be noted that Eq. (4.1.2) mathematically establishes the physically obvious fact that the impulse response of a dynamic system can always be expressed as a real-valued function of time.

EXAMPLE 4.1.3 Consider a system defined by the differential operator representation:

$$\underbrace{(D^3 + 5D^2 + 11D + 15)}_{a(D)}\, y(t) = \underbrace{(7D^2 + 18D + 15)}_{b(D)}\, u(t)$$

as in Eq. (2.7.1), with $c(D) = 1$ and $e(D) = 0$. To obtain an analytical expression for the unit impulse response of this system by Eq. (4.1.1), we first compute its transfer function using Eq. (3.1.25)

$$G(s) = \frac{c(s)b(s)}{a(s)} = \frac{b(s)}{a(s)} = \frac{7s^2 + 18s + 15}{s^3 + 5s^2 + 11s + 15} = \frac{m(s)}{a(s)} = \frac{l(s)}{a(s)}$$

We next determine that the poles of the system, as given by the roots of $a(s)$, lie at -3, $-1 + j2$, and $-1 - j2$ in the complex s-plane, so that a partial fraction expansion of $G(s)$ is given by

$$G(s) = \frac{3}{s+3} + \frac{2+j}{s+1-j2} + \frac{2-j}{s+1+j2}$$

In light of Eqs. (4.1.1) and (4.1.2), it then follows that the impulse response of the system is given by

$$y(t) = 3e^{-3t} + (2+j)e^{(-1+j2)t} + (2-j)e^{(-1-j2)t} = 3e^{-3t} + 2\sqrt{5}\, e^{-t}\cos(2t + 0.464)$$

as depicted in Figure 4.1.

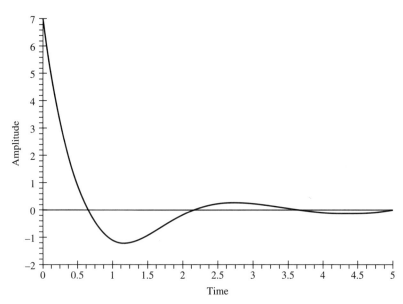

• **FIGURE 4.1**

The Unit Impulse Response of $G(s) = \dfrac{7s^2 + 18s + 15}{s^3 + 5s^2 + 11s + 15}$

Stable and Unstable Responses

The impulse response of a dynamic system is comprised of a weighted sum of its modes. Moreover, the modes of a system are directly dependent on the location of its poles in the complex s (or $\sigma + j\omega$)-plane. Therefore, the locations of the poles of a dynamic system directly imply the time-varying behavior of its impulse response.

A pole at a negative real location, $s = p_i = \alpha$, for some real $\alpha < 0$, contributes an exponentially decaying modal term $r_i e^{\alpha t}$ to the impulse response. Moreover, the closer α is to zero, the slower the time-varying decay of $r_i e^{\alpha t}$ to zero. A similar situation occurs with respect to complex-valued poles. A complex conjugate pair of poles at $s = \alpha \pm j\beta$, with $\alpha < 0$, contributes an exponentially decaying sinusoidal term to the impulse response, and the closer α is to zero, the slower the decay of this term to zero.

In general, any pole with a negative real (Re) part or, equivalently, any pole in the **stable half-plane** $Re(s = \sigma + j\omega) = \sigma < 0$, and its corresponding exponentially decaying mode, is called a **stable pole** and a **stable mode,** respectively. A system with all of its poles in the half-plane $Re(s) < 0$, so that its impulse response is characterized solely by exponentially decaying terms, is called a **stable system,**[1] and the polynomial $a(s)$ whose roots define the poles is called a **stable polynomial**. Note that the system defined in Example 4.1.3 is stable in light of this definition.

Any poles in the **unstable half-plane** $Re(s) > 0$, either at $s = p_i = \alpha$ or at $s = \alpha \pm j\beta$, with $\alpha > 0$, contribute exponentially growing terms to the impulse response that are characterized by an $e^{\alpha t}$ multiplier. Any such pole and its corresponding mode is called an **unstable pole** and an **unstable mode**, respectively. Any system with one or more unstable poles or modes is called an **unstable system.** The closer the poles are to the imaginary axis, the slower the exponential growth of the response toward infinity.

A pair of poles on the imaginary axis (at $s = \pm j\omega_i$) contribute a pure sinusoidal term (of frequency ω_i) to the impulse response of the system, whose amplitude neither grows nor decays with increasing time, reflecting what is sometimes termed **marginal stability.** Figure 4.2 displays some "typical" pole locations along with their corresponding impulsive output responses for comparison purposes.

Initial and Final Values

We now employ two Laplace transform theorems (see Appendix B) that extend our ability to analyze linear time-invariant dynamic systems. For any given function $g(t)$ of time, the **initial value theorem** states that if $g(s) = \mathcal{L}[g(t)]$ can be expressed as a strictly proper transfer function,

---

[1]System stability will be discussed further in Chapter 5.

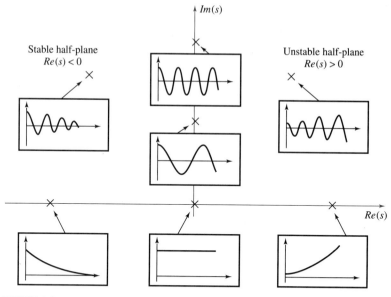

- **FIGURE 4.2**
  Pole Locations and the Corresponding Impulsive Behavior

then the initial value of $g(t)$, namely

$$g_0 \overset{\text{def}}{=} g(t_0 = 0) = \lim_{s \to \infty} s g(s) \tag{4.1.4}$$

In an analogous manner, the **final value theorem** states that if $sg(s) = s \times \mathcal{L}[g(t)]$ can be expressed as a stable transfer function, then the final value of $g(t)$, namely

$$g_\infty \overset{\text{def}}{=} \lim_{t \to \infty} g(t) = \lim_{s \to 0} s g(s) \tag{4.1.5}$$

EXAMPLE 4.1.6 We now illustrate these two theorems using Example 4.1.3, where $y(s)$, the Laplace transform of the impulse response $y(t)$, is defined by the strictly proper transfer function $G(s)$, that is, in light of Eq. (4.1.1), $u(s) = 1$ so that

$$y(s) = G(s) = \frac{7s^2 + 18s + 15}{s^3 + 5s^2 + 11s + 15}$$

Therefore, Eq. (4.1.4) implies that the initial value of the impulse response is given by

$$y_0 = \lim_{s \to \infty} \frac{s(7s^2 + 18s + 15)}{s^3 + 5s^2 + 11s + 15} = \frac{7s^3}{s^3} = 7$$

Since $sG(s)$ is a stable transfer function, the final value of the impulse response is given by

$$y_\infty = \lim_{t \to \infty} y(t) = \lim_{s \to 0} sG(s) = \lim_{s \to 0} \frac{s(7s^2 + 18s + 15)}{s^3 + 5s^2 + 11s + 15} = \frac{0}{15} = 0$$

using Eq. (4.1.5). Note that both of these relationships are verified by the Figure 4.1 impulse response plot of $y(t)$.

4.2   THREE COMMON REFERENCE INPUT RESPONSES

We have now shown that we can obtain an analytical expression for the impulse response of a linear time-invariant dynamic system using the inverse Laplace transform of its transfer function. However, a physical impulse response is practically impossible to achieve, since the required input signal $\delta(t)$ is defined by a limiting process in which its amplitude grows without bound, a physical impossibility. Furthermore, in many practical applications, a common control system "design goal"[2] is to have the physical output $y(t)$ of a system "track" a **command** or **reference input signal** $r(t)$ with as small an **error**

$$e(t) = r(t) - y(t) \tag{4.2.1}$$

as possible. It is difficult to imagine a physical situation in which the output $y(t)$ would be required to track a unit impulse function. Therefore, although the impulse response is a useful mathematical concept, it does not represent a very practical response goal.

In many practical cases, the output $y(t)$ of a SISO dynamic system corresponds to some physical value, such as the heading of an aircraft, the fluid level in a container, or the angular velocity of a radar antenna. Moreover, the reference input $r(t)$ often represents a desired change in $y(t)$, such as a new, time-invariant position or a different velocity or acceleration.

When the desired change in the positional value of $y(t)$ is a constant K, the reference input can be represented by a **step function**, namely,

$$Kr_s(t) = \begin{cases} 0 & \text{for} \quad t < 0 \\ K & \text{for} \quad t > 0 \end{cases} \tag{4.2.2}$$

An example of this function is when a radar antenna is automatically moved from some fixed initial position to another fixed position. The unit step function $r_s(t)$, which is defined when $K = 1$, is often called a *unit position input*.

---

[2]Various control system design goals will be discussed in Part II of this text.

When the desired change in the positional value of $y(t)$ has a constant velocity, the reference input can be represented by a **ramp function**, namely,

$$Kr_r(t) = \begin{cases} 0 & \text{for} \quad t < 0 \\ Kt & \text{for} \quad t \geq 0 \end{cases} \qquad (4.2.3)$$

An example of this situation is using a radar antenna to automatically track an aircraft flying at a constant velocity. The *unit ramp function* $r_r(t)$ is often called a *unit velocity input*.

Finally, in the constant acceleration case, the reference input can be represented by a **parabolic function**, namely,

$$Kr_p(t) = \begin{cases} 0 & \text{for} \quad t < 0 \\ Kt^2 & \text{for} \quad t \geq 0 \end{cases} \qquad (4.2.4)$$

This would be the case when a radar antenna automatically tracks the constant acceleration lift-off of a missile. The *unit parabolic function* $r_p(t)$ is often called a *unit acceleration input*. These three common reference inputs are depicted in Figure 4.3.

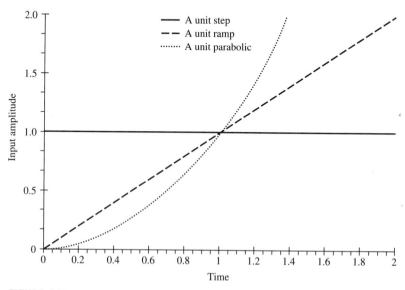

• **FIGURE 4.3**
Three Common Reference Input Signals

Laplace Transform Determination

Although exact output tracking is a common design goal, the three common reference inputs can also be employed as "test signals" that are directly applied as the input $u(t)$ to a system at rest to analyze the resulting output response. This particular "analysis application" is emphasized in this section, mainly with respect to the response of systems to step function inputs.

If the transfer function $G(s)$ of a SISO system is known, then its output response to any one of the common reference inputs can be obtained by the inverse Laplace transform. In particular, if

$$y(s) = G(s)u(s) \qquad (4.2.5)$$

with $G(s)$ a known rational transfer function, then

$$y(t) = \mathcal{L}^{-1}[y(s)] = \mathcal{L}^{-1}[G(s)u(s)] \qquad (4.2.6)$$

Since the Laplace transform of $t^{\bar{n}}$ for $t \geq 0$ is $\dfrac{\bar{n}!}{s^{\bar{n}+1}}$, for all integers $\bar{n} \geq 0$ (see Appendix B), it follows that when the input $u(t)$ is the step function of Eq. (4.2.2), then

$$u(s) = \mathcal{L}[Kr_s(t)] = Kr_s(s) = K\frac{1}{s} \qquad (4.2.7)$$

Furthermore, when $u(t)$ is the ramp function of Eq. (4.2.3), then

$$u(s) = \mathcal{L}[Kr_r(t)] = Kr_r(s) = K\frac{1}{s^2} \qquad (4.2.8)$$

and when $u(t)$ is the parabolic function of Eq. (4.2.4), then

$$u(s) = \mathcal{L}[Kr_p(t)] = Kr_p(s) = K\frac{2}{s^3} \qquad (4.2.9)$$

In any one of these cases, the substitution of the appropriate expression for $u(s)$ in Eq. (4.2.6), and a subsequent inverse Laplace transform evaluation of the resulting rational transfer function $G(s)u(s)$, directly implies the corresponding step, ramp, or parabolic response.

Natural and Forced Responses

Whenever a dynamic system that is initially at rest at $t = t_0$ (which we equate to 0 for convenience) is subjected to an external input, or "forcing" term, its resulting output response can be separated into two mutually exclusive parts: the natural response $y_n(t)$ and the "forced response" $y_f(t)$,

$$y(t) = y_n(t) + y_f(t) \quad \text{for all} \quad t \geq 0 \qquad (4.2.10)$$

The **natural response** $y_n(t)$ is defined as that part of the complete response that consists of the natural modes of the system, while the **forced response** $y_f(t)$ consists of the additional modal terms that are defined by the input $u(t)$.

Consider an output $y(t)$ defined by Eq. (4.2.6). If both $G(s)$ and $u(s)$ can be expressed as rational functions, so that

$$G(s) = \frac{m(s)}{a(s)} \qquad (4.2.11)$$

and

$$u(s) = \frac{m_u(s)}{p_u(s)} \qquad (4.2.12)$$

with $G(s)u(s)$ strictly proper, and if $a(s)$ and $p_u(s)$ have no common roots, then it is possible to separate the Laplace transforms of the natural and forced responses as follows:

$$y(s) = G(s)u(s) = \frac{m(s)m_u(s)}{a(s)p_u(s)} = \underbrace{\frac{\hat{m}(s)}{a(s)}}_{y_n(s)} + \underbrace{\frac{\hat{m}_u(s)}{p_u(s)}}_{y_f(s)} \qquad (4.2.13)$$

with both $y_n(s)$ and $y_f(s)$ strictly proper.

Since

$$\frac{\hat{m}(s)}{a(s)} + \frac{\hat{m}_u(s)}{p_u(s)} = \frac{\hat{m}(s)p_u(s) + \hat{m}_u(s)a(s)}{a(s)p_u(s)} = \frac{m(s)m_u(s)}{a(s)p_u(s)} \qquad (4.2.14)$$

$\hat{m}(s)$ and $\hat{m}_u(s)$ in Eq. (4.2.13) can be determined uniquely using the relationship:

$$\hat{m}(s)p_u(s) + \hat{m}_u(s)a(s) = m(s)m_u(s) \qquad (4.2.15)$$

by equating polynomial coefficients of equal degree (see Problem 4-6). Once such a determination has been made, Eqs. (4.2.6) and (4.2.13) imply that

$$y(t) = \mathcal{L}^{-1}[G(s)u(s)] = \underbrace{\mathcal{L}^{-1}\left[\frac{\hat{m}(s)}{a(s)}\right]}_{y_n(t)} + \underbrace{\mathcal{L}^{-1}\left[\frac{\hat{m}_u(s)}{p_u(s)}\right]}_{y_f(t)} \qquad (4.2.16)$$

as in Eq. (4.2.10).

An alternative way of determining $y_n(s)$ and $y_f(s)$, hence $y_n(t)$ and $y_f(t)$, is by a partial fraction expansion of the $G(s)u(s)$ product. The resulting expression separates all of the modal terms that comprise the output $y(s) = G(s)u(s)$. Those corresponding to the roots of $a(s)$ define $y_n(s)$, hence the natural modes that comprise $y_n(t)$, while those corresponding to the roots of $p_u(s)$ define $y_f(s)$, hence the **forced modes** that comprise $y_f(t.)$

In those cases when $G(s)$ is the transfer function of a stable system, so that $\lim_{t \to \infty} y_n(t) = 0$, the explicit determination of $y_n(s)$, hence $y_n(t)$, may not be relevant. This is particularly true in the case of the "frequency response" of a dynamic system, as we will show in the next section.

. . . .

EXAMPLE 4.2.17 To illustrate the preceding, let us determine the **unit ramp response** of the system defined earlier in Example 4.1.3, whose transfer function

$$G(s) = \frac{7s^2 + 18s + 15}{s^3 + 5s^2 + 11s + 15} = \frac{m(s)}{a(s)}$$

is characterized by three poles at $s = -3$ and $-1 \pm j2$. In light of Eq. (4.2.8),

$$u(s) = r_r(s) = \frac{1}{s^2} = \frac{m_u(s)}{p_u(s)}$$

so that Eq. (4.2.6) implies a unit ramp response defined by

$$y(t) = \mathcal{L}^{-1}\left[G(s)u(s) = \frac{7s^2 + 18s + 15}{s^5 + 5s^4 + 11s^3 + 15s^2}\right]$$

A partial fraction expansion of this $G(s)u(s)$ product now produces the $y_n(s)$, $y_f(s)$ separation, that is,

$$G(s)u(s) = \underbrace{\frac{0.333}{s+3} + \frac{-0.4+j.2}{s+1-j2} + \frac{-0.4-j.2}{s+1+j2}}_{y_n(s)} + \underbrace{\frac{0.467}{s} + \frac{1}{s^2}}_{y_f(s)}$$

which implies a natural response

$$y_n(t) = \mathcal{L}^{-1}[y_n(s)] = 0.333e^{-3t} - 0.894e^{-t}\cos(2t - 0.464)$$

and a forced response

$$y_f(t) = \mathcal{L}^{-1}[y_f(s)] = 0.467 + t$$

for all $t \geq 0$. This unit ramp output response is depicted in Figure 4.4, together with $r_r(t)$. Note that the output response "parallels" (but does not equal) the unit ramp reference input $r_r(t)$ in the limit as $t \to \infty$.

. . . .

The Steady-State Error

The notion of "steady-state" can define the behavior of any time-varying function as $t \to \infty$. However, it usually is restricted to those cases in which a function exhibits some constant or repetitive behavior with increasing time. Therefore, using Eq. (4.2.1), and motivated by the previous example, the **steady-state error** of a dynamic system is defined, in general, by the relationship:

$$e_{ss}(t) \overset{\text{def}}{=} \lim_{t \to \infty} e(t) = \lim_{t \to \infty} [r(t) - y(t)] = \lim_{t \to \infty} r(t) - \lim_{t \to \infty} y(t) \quad (4.2.18)$$

A steady-state error usually is associated with a stable system driven by a nondiminishing reference input. In such cases, the natural response $y_n(t)$ is of a transient nature only, since $\lim_{t\to\infty} y_n(t) = 0$, so that

$$\lim_{t\to\infty} [y(t) = y_n(t) + y_f(t)] = \lim_{t\to\infty} y_f(t) \qquad (4.2.19)$$

As a consequence,

$$e_{ss}(t) = \lim_{t\to\infty} r(t) - \lim_{t\to\infty} y_f(t) \qquad (4.2.20)$$

This latter relationship can be illustrated using Example 4.2.17 where $\lim_{t\to\infty} r(t) = r_r(t) = t$, and $\lim_{t\to\infty} y(t) = y_f(t) = 0.467+t$. Therefore, in light of Eq. (4.2.20),

$$e_{ss}(t) = t - (0.467 + t) = -0.467$$

a result that is verified by Figure 4.4, where the output $y(t)$ exceeds the ramp input $u(t) = r_r(t)$ by a steady-state value of 0.467 as $t \to \infty$.

In this case, the final value theorem could also have been employed to establish the fact that $e_{ss}(t) = e_\infty = -0.467$. In particular, the Laplace transform of Eq. (4.2.1) implies that

$$e(s) = r(s) - y(s) \qquad (4.2.21)$$

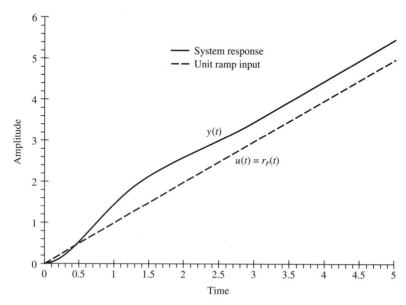

• FIGURE 4.4

The Unit Ramp Response of $G(s) = \dfrac{7s^2 + 18s + 15}{s^3 + 5s^2 + 11s + 15}$

and since $r(s) = u(s)$ and $y(s) = G(s)u(s)$, it follows that

$$e(s) = [1 - G(s)]u(s) \qquad (4.2.22)$$

Therefore, for the particular $G(s)$ of Example 4.2.17,

$$se(s) = s[1 - G(s)]u(s)$$

$$= s \underbrace{\left[\frac{s^3 + 5s^2 + 11s + 15 - (7s^2 + 18s + 15)}{s^3 + 5s^2 + 11s + 15} \right]}_{[1 - G(s)]} \underbrace{\frac{1}{s^2}}_{u(s)}$$

$$= \frac{s^2 - 2s - 7}{s^3 + 5s^2 + 11s + 15}$$

a stable transfer function. As a consequence, Eq. (4.1.5) implies that

$$e_{ss}(t) = e_\infty = \lim_{s \to 0} se(s) = \frac{-7}{15} = -0.467$$

Step Input Test Signals

When a step input is applied to a physical system, the output usually does not change instantaneously, because system inertia must be overcome by a "drive motor" with limited power. For example, the actual position of a large radar antenna will not change instantaneously in response to a step input command. As a consequence, a "typical" output response that would be obtained in the case of a *unit* ($K = 1$) *step input* may resemble the $y(t)$ depicted in Figure 4.5.

Note that there are certain **step response parameters** associated with the step response of Figure 4.5 that can be directly identified by inspection of the response. In particular,

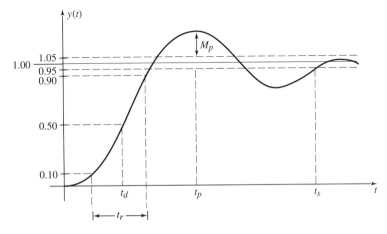

● FIGURE 4.5
A "Typical" Step Response

- The **rise time** t_r is defined as the time required for the response to increase or rise from 10% to 90% of its steady-state or *final value*, $y_\infty = y_{ss}(t) = \lim_{t \to \infty} y(t)$.
- The **delay time** t_d is defined as the time required for the response to reach 50% of its final value y_∞.
- The **maximum overshoot** M_p, which is often expressed as a percentage, defines the largest or peak excursion of the response above its final value. The *time* when the maximum overshoot occurs is called the **peak time** t_p. Therefore,

$$M_p \stackrel{\text{def}}{=} \frac{y(t_p) - y_\infty}{y_\infty} \qquad (4.2.23)$$

so that if $y_\infty = 2$ and $y(t_p) = 2.6$, $M_p = 0.3$ or 30%.

- The **settling time** t_s is defined as the time required for the response to reach and remain within some boundary limits $y_\infty(1 \pm \epsilon)$ around its final value. The limit ϵ is often expressed as some fixed percentage, such as $\pm 2\%$ or $\pm 5\%$, depending on the particular application.

All five of these step response parameters are illustrated in Figure 4.5 in the case when $y_\infty = 1$, with $\epsilon = 0.05$ or 5%.

The Step Response of a Second-Order System

The step response of an unknown system often can tell us a great deal about the location of certain of its poles, even when the actual dynamic order of a system is greater than two. In particular, the step response of a dynamic system is often determined by the location of its so-called **dominant poles,** which are defined as the poles (usually two) closest to the $j\omega$ axis in the complex $s\,(=\sigma + j\omega)$-plane (recall that the closer the poles are to the $j\omega$ axis, the longer it takes for the transient response to decay to zero). Therefore, a thorough understanding of the step response characteristics of second-order linear systems is useful in analyzing the step response of higher order systems.

In view of these observations, let us consider any SISO system characterized by a stable, zeroless, second-order transfer function

$$G(s) = \frac{y(s)}{u(s)} = \frac{c}{a(s)} = \frac{c}{s^2 + a_1 s + a_0} \qquad (4.2.24)$$

so that $a_1 > 0$ and $a_0 > 0$. Note that such a $G(s)$ always can be written in the alternative form

$$G(s) = \frac{c}{a(s)} = \tilde{G} \frac{\omega_n^2}{s^2 + 2\zeta \omega_n s + \omega_n^2} \qquad (4.2.25)$$

by defining $\omega_n = \sqrt{a_0}$, $\zeta = \frac{a_1}{2\sqrt{a_0}}$, and a scalar "gain" $\tilde{G} = \frac{c}{a_0}$.

For reasons that will soon become apparent, we now define

- ω_n as the **undamped natural frequency,**
- ζ as the **damping ratio,** and
- $\alpha = \zeta\omega_n$, as the **damping constant.**

We next note that the two poles of $G(s)$ are given by the roots of $s^2 + 2\zeta\omega_n s + \omega_n^2$, or

$$p_{1,2} = -\zeta\omega_n \pm \omega_n\sqrt{\zeta^2 - 1} = \omega_n\left(-\zeta \pm \sqrt{\zeta^2 - 1}\right) \qquad (4.2.26)$$

Therefore, for any fixed value of ω_n, the poles of $G(s)$ are dependent only on the damping ratio ζ. As ζ increases continuously from 0 to ∞, the location of the ($n = 2$) poles trace continuous paths in the complex s-plane, as depicted in Figure 4.6. These continuous paths, which might be termed a *root locus* of $a(s)$, begin at $\pm j\omega_n$, when $\zeta = 0$. They subsequently define circular arcs of radius ω_n in the left-half s-plane as ζ increases from 0 to 1, coming together as a repeated pair of poles at $s = -\omega_n$ when $\zeta = 1$. They then separate and move in opposite directions along the negative real axis as ζ increases beyond 1.

In view of the preceding discussion, three distinct cases are delineated:

- the **underdamped case,** when $0 < \zeta < 1$,
- the **critically damped case,** when $\zeta = 1$, and
- the **overdamped case,** when $1 < \zeta < \infty$.

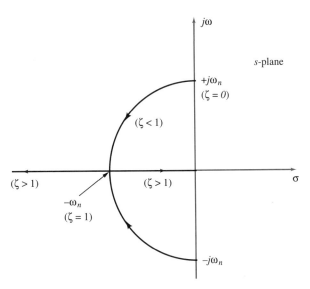

- **FIGURE 4.6**
  The ζ Dependent Root Locus of $a(s) = s^2 + 2\zeta\omega_n s + \omega_n^2$

In all three cases, the unit step response of the system can be obtained using Eq. (4.2.6), with $u(s) = \frac{1}{s}$. In particular, the unit step response of the system defined by Eq. (4.2.25), with $\tilde{G} = 1$, for convenience, is given by the relationship:

$$y(t) = \mathcal{L}^{-1}[G(s)u(s)] = \mathcal{L}^{-1}\left[\frac{\omega_n^2}{s(s^2 + 2\zeta\omega_n s + \omega_n^2)}\right] \qquad (4.2.27)$$

In the *underdamped case* when $0 < \zeta < 1$,

$$p_{1,2} = \underbrace{-\zeta\omega_n}_{-\alpha} \pm j\omega_n\sqrt{1 - \zeta^2} \qquad (4.2.28)$$

a complex conjugate pair in the half-plane $Re(s) < 0$, which implies the $t \geq 0$ unit step response (see Problem 4-9)

$$y(t) = 1 - \frac{e^{-\zeta\omega_n t}}{\sqrt{1 - \zeta^2}} \sin\left[\omega_n\sqrt{1 - \zeta^2}\, t + \tan^{-1}\frac{\sqrt{1 - \zeta^2}}{-\zeta}\right] \qquad (4.2.29)$$

In the *critically damped case* when $\zeta = 1$,

$$p_{1,2} = -\omega_n = -\alpha \qquad (4.2.30)$$

a repeated pair at the same real location in the half-plane $Re(s) < 0$, thus implying the unit step response

$$y(t) = 1 - e^{-\omega_n t} - \omega_n t e^{-\omega_n t} = 1 - e^{-\omega_n t}[1 + \omega_n t] \quad \text{for} \quad t \geq 0 \quad (4.2.31)$$

In the *overdamped case* when $\zeta > 1$,

$$p_{1,2} = -\zeta\omega_n \pm \omega_n\sqrt{\zeta^2 - 1} \qquad (4.2.32)$$

a real, disjoint pair in the stable half-plane $Re(s) < 0$, which implies the unit step response

$$y(t) = 1 + \frac{\omega_n}{2\sqrt{\zeta^2 - 1}}\left[\frac{e^{p_1 t}}{p_1} - \frac{e^{p_2 t}}{p_2}\right] \quad \text{for} \quad t \geq 0 \qquad (4.2.33)$$

In all three cases, $y(t)$ is defined by a steady-state value of 1 ($= y_{ss}(t) = y_\infty$), which represents the forced response $y_f(t)$, and a transient term that decays to zero in the limit as $t \to \infty$, which represents the natural response $y_n(t)$. Representative unit step responses for various values of ζ are given in Figure 4.7 as a function of the "normalized" time $\omega_n t$.

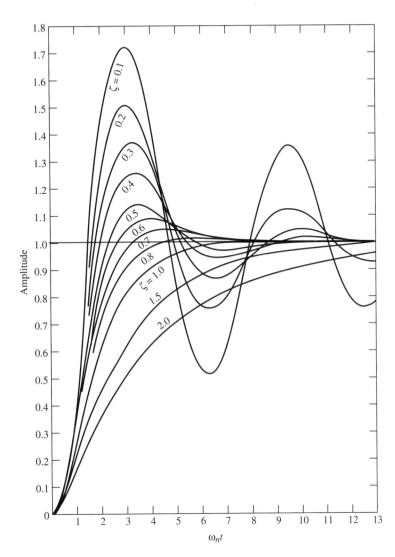

• **FIGURE 4.7**

The Step Response of $G(s) = \dfrac{\omega_n^2}{s^2 + 2\zeta\omega_n s + \omega_n^2}$

In view of the preceding discussion, and Figure 4.7 in particular, it may be noted that ζ is called the damping[3] ratio because it directly affects the amount of oscillation present in the response; that is, larger values of

---

[3]Damping is caused by friction in mechanical systems and by resistance in electrical circuits.

ζ imply more damping, hence less oscillation. Furthermore, when $\zeta \geq 1$, the step response exhibits no oscillation at all, with increasing values of ζ producing an increasingly "sluggish" response that takes longer to reach its steady-state value. In these cases, the peak time t_p is undefined and the maximum overshoot $M_p = 0\%$.

The undamped natural frequency ω_n is so named because it represents the frequency of oscillation in the undamped case (when $\zeta = 0$). In view of Eq. (4.2.29), ω_n also affects the **damped natural frequency**

$$\omega_d = \omega_n \sqrt{1 - \zeta^2} \tag{4.2.34}$$

which represents the frequency of the exponentially damped sinusoid in the underdamped case. Larger values of ω_n imply a faster response characterized by lower values of both the delay time t_d and the rise time t_r.

The Underdamped Case

In the underdamped case, analytical expressions for t_p and M_p can be obtained in terms of ζ and ω_n. In particular, M_p can be determined by equating the time derivative of Eq. (4.2.29) to zero, that is, since (see Problem 4-10)

$$\frac{dy(t)}{dt} = \frac{\omega_n e^{-\zeta \omega_n t}}{\sqrt{1 - \zeta^2}} \sin \underbrace{\omega_n \sqrt{1 - \zeta^2}}_{\omega_d} t \tag{4.2.35}$$

the unit step response reaches its peak value of $1 + M_p$ when $\sin \omega_n \sqrt{1 - \zeta^2}\, t = \sin \omega_d t = 0$, or when $\omega_d t = \pi$, so that the **peak time**

$$t_p = \frac{\pi}{\omega_d} = \frac{\pi}{\omega_n \sqrt{1 - \zeta^2}} \tag{4.2.36}$$

If we now substitute this value for t_p into Eq. (4.2.29), it follows that

$$y(t_p) = 1 + M_p = 1 + \frac{e^{\frac{-\zeta \pi}{\sqrt{1-\zeta^2}}}}{\sqrt{1 - \zeta^2}} \sin \left[\pi - \tan^{-1} \frac{\sqrt{1 - \zeta^2}}{-\zeta} \right]$$

and since (see Problem 4-3)

$$\tan^{-1} \frac{\sqrt{1 - \zeta^2}}{-\zeta} = \pi - \sin^{-1} \sqrt{1 - \zeta^2}$$

the **maximum overshoot**

$$M_p = e^{-\frac{\zeta \pi}{\sqrt{1-\zeta^2}}} \tag{4.2.37}$$

Explicit knowledge of ω_n and ζ implies explicit knowledge of the step response of the system, hence all five step response parameters for any second-order, zeroless, stable system. Conversely, the unit step response of an underdamped stable system, characterized by an otherwise unknown

$G(s)$ defined by Eq. (4.2.25), implies both M_p and t_p, hence corresponding values for both ζ and ω_n that, together with the scalar $\tilde{G}$, define $G(s)$.

In particular, Eq. (4.2.37) implies that

$$\left|\frac{\ln(M_p)}{\pi}\right| \overset{\text{def}}{=} \gamma = \frac{\zeta}{\sqrt{1-\zeta^2}} \qquad (4.2.38)$$

so that

$$\zeta = \frac{\gamma}{\sqrt{1+\gamma^2}} \qquad (4.2.39)$$

and

$$\omega_n = \frac{\pi}{t_p\sqrt{1-\zeta^2}} = \frac{\pi\sqrt{1+\gamma^2}}{t_p} \qquad (4.2.40)$$

in light of Eq. (4.2.36) and (4.2.39).

Moreover, the scalar $\tilde{G} = G(s=0)$ is equal to y_∞, the final value of the unit step response, using the final value theorem, that is, since $y(s) = G(s)u(s)$, with $G(s)$ given by Eq. (4.2.25) and $u(s) = \frac{1}{s}$, it follows that

$$y_\infty = \lim_{s\to 0} sy(s) = \lim_{s\to 0}\left[sG(s)\frac{1}{s}\right] = G(s=0) = \tilde{G}\frac{\omega_n^2}{\omega_n^2} = \tilde{G} \quad (4.2.41)$$

....

EXAMPLE 4.2.42 Suppose the unit step response of an otherwise unknown, zeroless, second-order system resembles that depicted in Figure 4.5, reaching a maximum value of 2.04 at $t_p = 1.22$, with a steady-state or final value of $1.5 = y_\infty = \tilde{G}$.

The maximum overshoot would then be given by Eq. (4.2.23)

$$M_p = \frac{y(t_p) - y_\infty}{y_\infty} = \frac{2.04 - 1.5}{1.5} = 0.36$$

so that $\ln(M_p) = -1.022$. Therefore,

$$\left|\frac{-1.022}{\pi}\right| = \gamma = 0.325, \qquad \zeta = \frac{0.325}{\sqrt{1.106}} = 0.31 \quad \text{and} \quad \omega_n = \frac{\pi\sqrt{1.106}}{1.22} = 2.71$$

using Eqs. (4.2.38), (4.2.39), and (4.2.40), respectively. As a consequence,

$$G(s) = \frac{\tilde{G}\omega_n^2}{s^2 + 2\zeta\omega_n s + \omega_n^2} = \frac{1.5(2.71)^2}{s^2 + 2(0.31)(2.71)s + (2.71)^2} = \frac{11.02}{s^2 + 1.68s + 7.34}$$

with poles at $s = -0.84 \pm j2.58$.

....

Even in those cases when a given dominant poles system is of order greater than two, a unit step response, such as that depicted in Figure 4.5,

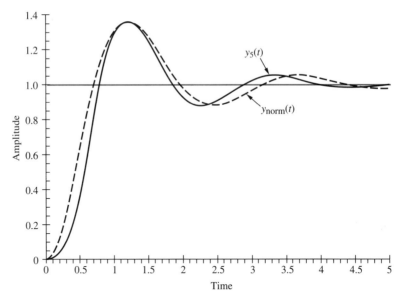

- **FIGURE 4.8**
  The Step Responses of $G_{\text{norm}}(s) = \dfrac{7.34}{s^2 + 1.68s + 7.34}$ and $G_5(s)$

implies a second-order approximation of the given system whose pole locations approximate the dominant pole locations of the given system.

EXAMPLE 4.2.43 For example, consider a fifth-order system defined by the transfer function

$$G_5(s) = \frac{1020s + 4080}{s^5 + 24s^4 + 218s^3 + 956s^2 + 2456s + 4080}$$

with poles at $s = -1 \pm j3$, -6, and $-8 \pm j2$, and a (single) zero at $s = -4$. The unit step response of this underdamped $G_5(s)$ is characterized by an $M_p = 0.36$ at a $t_p = 1.22$, the same two step response parameters that characterize the second-order $G(s)$ of Example 4.2.42. Therefore, the $G(s)$ of Example 4.2.42 represents a second-order approximation to $G_5(s)$. Note that the pole locations of the second-order system at $s = -0.84 \pm j2.58$ approximate the dominant pole locations of the fifth-order system at $s = -1 \pm j3$. The normalized (so that $y_\infty = 1$ in both cases) unit step responses of these two systems are depicted in Figure 4.8 for comparison purposes.

The Effect of Zeros

In the underdamped case, plant zeros can have a significant effect on the transient response of a system, especially if they are "close to" the origin. In particular, in light of Eq. (4.2.25), consider a SISO system defined by the stable, second-order transfer function

$$G(s) = \frac{y(s)}{u(s)} = \frac{\alpha \omega_n^2 s + \omega_n^2}{s^2 + 2\zeta \omega_n s + \omega_n^2} \tag{4.2.44}$$

which has a real zero at $s = -1/\alpha$, so that the larger the scalar α, the closer this zero is to the origin.

We now express $G(s)$ as the sum of two terms, namely,

$$G(s) = \frac{\omega_n^2}{s^2 + 2\zeta \omega_n s + \omega_n^2} + \alpha s \frac{\omega_n^2}{s^2 + 2\zeta \omega_n s + \omega_n^2} \tag{4.2.45}$$

where

$$G_0(s) \stackrel{\text{def}}{=} \frac{\omega_n^2}{s^2 + 2\zeta \omega_n s + \omega_n^2}$$

is a zeroless transfer function, analogous to that defined by Eq. (4.2.25). Since the Laplace transform of $Dy(t)$ is $sy(s)$, it follows that

$$y(t) = \underbrace{\mathcal{L}^{-1}[G_0(s)u(s)]}_{y_0(t)} + \underbrace{\mathcal{L}^{-1}[\alpha s G_0(s)u(s)]}_{\alpha D y_0(t)} = y_0(t) + \alpha \frac{dy_0(t)}{dt} \tag{4.2.46}$$

Therefore, if $y_0(t)$ is oscillatory (if $\zeta < 1$), a $G(s)$ zero close to the origin at $s = -1/\alpha$ could increase the overshoot of the step response significantly, due to the addition of the derivative term $\alpha D y_0(t)$. This phenomenon is illustrated in Figure 4.9, which depicts $D y_0(t)$ and the unit step response plots of the $G(s)$ defined by Eq. (4.2.44) when $\omega_n = 1$ and $\zeta = 0.7$, for $\alpha = 0$, 1, and 2.

The Overdamped Case and First-Order Systems

In the case of an unknown, overdamped, second-order system, it is nearly impossible to determine ζ and ω_n from a step response plot because any number of second-order systems, with different values of ζ and ω_n, can be characterized by the same, single, *dominant pole*. The step responses of all such systems will be nearly identical, and will "closely resemble" the step response of a first-order system that is defined, in general, by the transfer function

$$G(s) = \tilde{G} \frac{a}{s + a} \tag{4.2.47}$$

and the corresponding unit step response

$$y(t) = \mathcal{L}^{-1}\left[G(s)u(s) = \tilde{G} \frac{a}{s(s + a)} = \tilde{G}\left(\frac{1}{s} - \frac{1}{s + a} \right) \right] = \tilde{G}(1 - e^{-at}) \tag{4.2.48}$$

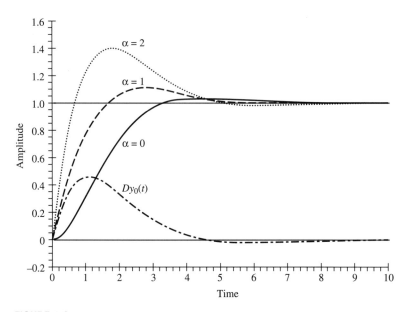

• **FIGURE 4.9**
The Effect of a Zero on the Step Response of an Underdamped System

The step response of a first-order system defined by the $G(s)$ of Eq. (4.2.47) is characterized by the single natural mode e^{-at} with **time constant**

$$t_c \overset{\text{def}}{=} \frac{1}{a} \tag{4.2.49}$$

In general, t_c is the time required for e^{-at} to decay from an initial value of 1, at $t = 0$, to $e^{-1} = 0.368$. In light of Eq. (4.2.48), t_c is also the time required for the step response of a first-order system to reach $(1 - e^{-1})\tilde{G} = 0.632\tilde{G}$, or 63.2% of its final value of $\tilde{G}$.

....

EXAMPLE 4.2.50 To illustrate the preceding discussion, consider two second-order systems that are defined by the transfer functions

$$G_1(s) = \frac{7}{(s + 1)(s + 7)} = \frac{7}{s^2 + 8s + 7}$$

with $\omega_n = 2.65$ and $\zeta = 1.51$, and

$$G_2(s) = \frac{14}{(s + 1)(s + 14)} = \frac{14}{s^2 + 15s + 14}$$

with $\omega_n = 3.74$ and $\zeta = 2$, both of which have a dominant pole at $s = -1$. As a consequence, their unit step responses, as depicted in Figure 4.10, are nearly identical. Moreover, they also are similar

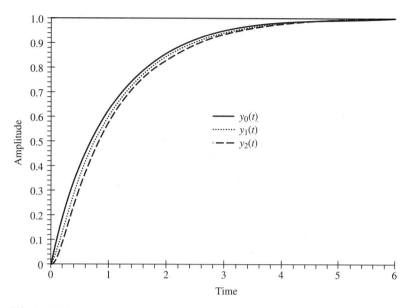

- **FIGURE 4.10**
  The Step Responses of Dominant Pole Systems (at $s = -1$)

to the step response of the first-order or single pole (at $s = -1$) system, which is defined by the transfer function

$$G_0(s) = \frac{1}{s+1}$$

The step response of the $G_0(s)$ system is depicted by the solid line in Figure 4.10 for comparison purposes.

4.3   THE FREQUENCY RESPONSE

Another important class of "test signals" that are employed in control system applications are sinusoidal inputs of the form

$$u(t) = K \sin \omega t \quad \text{for} \quad t \geq 0 \tag{4.3.1}$$

where the frequency ω can assume any positive value between 0 and ∞. Such inputs usually are applied to a stable system, which is initially at rest, in order to determine the resulting steady-state output, which is called the **frequency response** of the system. Generally speaking, frequency response data is simple and accurate to obtain experimentally using a variety of widely available signal generators and spectrum analyzers.

Frequency domain methods, which refer to those procedures that are based on knowledge of the frequency response characteristics of a system, were among the first control system analysis and design techniques to be developed. The initial motivation for this work was provided, in large part, by the desire for improved accuracy in automatically controlled weaponry during World War II. Prior to that time, the pioneering work of Nyquist [51] and Bode [7], which had focused on improved communication systems, provided many of the initial control system terms and techniques that remain very much in evidence to this day. Fundamental control concepts such as "feedback," "bandwidth," "gain and phase margins," and the "return difference" are due, in large part, to earlier work in communications, including the design of the first "feedback amplifiers" by Black [9].

In a communication system, transmitted information generally is restricted to a narrow band of frequencies around the "carrier" frequency, and a receiver is therefore designed to be sensitive to a narrow band of frequencies. A primary function of a control system, on the other hand, is to enable a controlled output to "follow" a command input as closely as possible at all times. The frequency content of a command input is dependent on its time variation; in other words, rapidly changing inputs are characterized by a wider range of frequencies than slowly varying signals.

Therefore, to follow a rapidly changing input, a control system must be responsive to a broad band of frequencies. Unfortunately, this is not always possible because of the associated power requirements. For example, the output of most physical systems usually can "follow" a sinusoidal input command more accurately at lower frequencies than at higher frequencies; for example, it requires less power to drive the output position of a radar antenna in a sinusoidal manner between $\pm 90°$ over a one-minute period rather than a one-second period. Therefore, if a sinusoidal input signal of fixed amplitude is applied to the antenna at various frequencies, the resulting sinusoidal output response will generally exhibit a decreasing amplitude as the input frequency is increased.

In many physical situations, the steady-state output response of a physical system to a sinusoidal input occurs at the same frequency as the that of the input $u(t)$. This always is the case when the system is both stable and linear, although the output amplitude and phase generally differ from that of $u(t)$, normally characterized by smaller output amplitudes at the higher frequencies. In certain cases, there is also the possibility of a maximum or peak output amplitude A_p occurring at some interim *resonant frequency* ω_p.[4] Figure 4.11 depicts a "typical" output amplitude frequency response of a dynamic system in response to input sinusoids of constant amplitude.

---

[4] Although such a peak often is desirable in communication systems, it usually is undesirable, but unavoidable, in control systems.

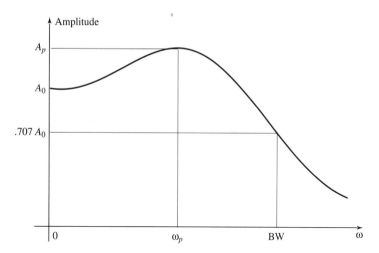

- **FIGURE 4.11**
A "Typical" Amplitude Frequency Response

The **bandwidth (BW)** of the frequency response is defined as the frequency at which the output amplitude drops to below 70.7% of its zero-frequency level A_0. This term has been adopted from communication systems, wherein the bandwidth of such systems usually is defined as the distance between the "half power" frequencies associated with a resonant response [7]. The bandwidth of a control system reflects a number of its characteristics; for example, lower bandwidths imply better noise filtering properties, while higher bandwidths reflect faster response times, and the resultant ability to track higher frequency sinusoidal signals with smaller errors. Larger values of A_p generally indicate poorer stability properties, as we will later show.

A thorough analysis of the frequency response of a system not only can tell us a great deal about its dynamic characteristics but also can lead to the design of compensators that "improve" its performance, without explicit knowledge of a state-space, differential operator, or transfer function description of the system. For these reasons, frequency response methods have retained their popularity in both control system analysis and design, although the primary emphasis in the remainder of this chapter will be on the analysis and the identification of systems by means of their frequency response.

In the next section we will show how a transfer function description of a system can be obtained from its experimentally determined frequency response. However, as in the previous section, we will begin with the "converse result." In particular, we will first show how the frequency response of a linear system can be obtained directly from its transfer function $G(s)$.

The Sinusoidal Transfer Function

We first define $G(s = j\omega)$ as the **sinusoidal transfer function** of a system defined by a transfer function $G(s)$, noting that for any positive real value of ω, the sinusoidal transfer function $G(j\omega)$ is a complex number that can be expressed in terms of its real (Re) and imaginary (Im) parts. In particular,

$$G(j\omega) = Re[G(j\omega)] + jIm[G(j\omega)] = |G(j\omega)|e^{j\phi} \qquad (4.3.2)$$

with **amplitude**

$$|G| \overset{\text{def}}{=} |G(j\omega)| = \sqrt{\{Re[G(j\omega)]\}^2 + \{Im[G(j\omega)]\}^2} \qquad (4.3.3)$$

and **phase angle**

$$\phi \overset{\text{def}}{=} \angle G(j\omega) = \tan^{-1}\left\{ \frac{Im[G(j\omega)]}{Re[G(j\omega)]} \right\} \qquad (4.3.4)$$

We note further, using Eqs. (4.2.10) and (4.2.16), that if a linear system is driven by the $u(t)$ of Eq. (4.3.1), the forced response $y_f(t)$ reflects that part of the total response that is both sinusoidal and of the same frequency ω as that of the input. In other words, the forced response $y_f(t)$ of a linear system driven by a $u(t) = K \sin \omega t$ represents its frequency response. It may be noted that this observation holds regardless of the stability properties of the system: An unstable linear system also has a well-defined frequency response, which is equal to its forced response $y_f(t)$. However, the unstable modes associated with its natural response $y_n(t)$ make it nearly impossible to physically measure the sinusoidal $y_f(t)$ unless the system is stabilized in some manner (see Problem 4-13).

In light of the preceding discussion, we will now show that if $u(t) = K \sin \omega t$ is applied to a system defined by a rational transfer function

$$G(s) = \frac{m(s)}{a(s)} = \frac{y(s)}{u(s)}$$

at $t = 0$, then its frequency response

$$y_f(t) = K|G| \sin(\omega t + \phi) \qquad (4.3.5)$$

with $|G| = |G(j\omega)|$ and $\phi = \angle G(j\omega)$ defined by Eqs. (4.3.3) and (4.3.4), respectively.

In particular, since

$$u(s) = \mathcal{L}[K \sin \omega t] = K\frac{\omega}{s^2 + \omega^2} = \frac{m_u(s)}{p_u(s)} \qquad (4.3.6)$$

(see Appendix B), Eq. (4.2.13) implies that

$$
\begin{aligned}
y(s) = G(s)u(s) &= \frac{m(s)m_u(s)}{a(s)p_u(s)} \\
&= \frac{KG(s)\omega}{(s + j\omega)(s - j\omega)} = \underbrace{\frac{\hat{m}(s)}{a(s)}}_{y_n(s)} + \underbrace{\frac{r_{n+1}}{s + j\omega} + \frac{r_{n+2}}{s - j\omega}}_{y_f(s)} \qquad (4.3.7)
\end{aligned}
$$

Equation (3.3.5) now implies that

$$r_{n+1} = \left[\frac{KG(s)\omega}{s - j\omega}\right]\Bigg|_{s=-j\omega} = \frac{KG(-j\omega)}{-j2} \qquad (4.3.8)$$

and

$$r_{n+2} = \left[\frac{KG(s)\omega}{s + j\omega}\right]\Bigg|_{s=+j\omega} = \frac{KG(j\omega)}{j2} \qquad (4.3.9)$$

a complex-conjugate pair. Therefore, using Eq. (4.3.7),

$$y_f(t) = \mathcal{L}^{-1}[y_f(s)] = \frac{KG(-j\omega)}{-j2}e^{-j\omega t} + \frac{KG(j\omega)}{j2}e^{j\omega t} \qquad (4.3.10)$$

Since $G(j\omega) = |G(j\omega)|e^{j\phi}$, in light of Eq. (4.3.2), and since $G(-j\omega) = |G(j\omega)|e^{-j\phi}$, Eq. (4.3.10) implies that

$$y_f(t) = K|G(j\omega)| \underbrace{\frac{e^{j(\omega t+\phi)} - e^{-j(\omega t+\phi)}}{j2}}_{\sin(\omega t + \phi)} \qquad (4.3.11)$$

thus establishing Eq. (4.3.5).

   In most situations, including all of those in which the frequency response is obtained experimentally, the defined system is stable, so that $\lim_{t\to\infty} y_n(t) = 0$. As a consequence,

$$y_{ss}(t) = \lim_{t\to\infty}\left[y(t) = y_n(t) + y_f(t)\right] = y_f(t) = K|G|\sin(\omega t + \phi) \qquad (4.3.12)$$

so that the steady-state output response represents the frequency response.

   We note finally that at any real frequency $\omega > 0$, the amplitude of the sinusoidal transfer function $|G(j\omega)| = |G|$ represents the ratio of the output amplitude to the input amplitude, while the phase of the sinusoidal transfer function $\phi = \angle G(j\omega)$ represents the amount by which the phase of the output $y_f(t)$ **leads** (if $\phi > 0°$) or **lags** (if $\phi < 0°$) that of the input $u(t)$. Figure 4.12 depicts "typical" sinusoidal signals that may characterize the input $u(t)$ and the corresponding forced output response $y_f(t)$.

4.4   BODE DIAGRAMS

Regardless of whether the frequency response of a system is obtained experimentally or through knowledge of its sinusoidal transfer function, it is useful to graphically depict the response in some appropriate manner, as in Figure 4.11. One of the most useful techniques for depicting the

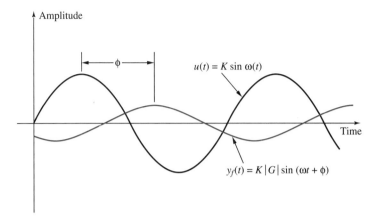

- **FIGURE 4.12**
"Typical" Input and Output Sinusoidal Signals

frequency response of a system is the one that was developed at Bell Laboratories by H. W. Bode [7]. Bode's procedure has retained its popularity over the years, and it has been enhanced further by the development of a variety of computer programs that have automated its implementation.

The basic idea behind the so-called **Bode diagrams** is to represent the frequency response of a system as two separate graphs, one displaying the logarithm of the amplitude of the response and the other the phase angle, both as functions of the input frequency ω, which is depicted on a log scale in *radians/second*. A major advantage of Bode's diagrams is that of converting nonlinear amplitude multiplication into linear amplitude addition, as we will now show.

The Corner Frequency Factored Form

To begin, a rational transfer function

$$G(s) = \frac{m(s)}{a(s)}$$

is defined to be **minimum phase** if all of its poles and zeros lie in the half-plane $Re(s) \leq 0$. Conversely, $G(s)$ is said to be **nonminimum phase** if one (or more) of its zeros lies in the unstable half-plane $Re(s) > 0$. Strictly speaking, an unstable $G(s)$ reflects "nonminimum phase behavior" as well, and may also be so designated. For traditional reasons, however, nonminimum phase behavior is usually attributed to a system with unstable, or nonminimum phase zeros, independent of its modal stability, and we will maintain this tradition here.

If $G(s)$ is a minimum phase transfer function, it can be expressed in a **corner frequency factored form**, namely,

$$G(s) = \frac{m(s)}{a(s)} = \tilde{G}s^k \frac{m_1(s)m_2(s)\ldots}{a_1(s)a_2(s)\ldots} \tag{4.4.1}$$

where $\tilde{G}$ is some positive real gain, and k is an integer that represents the number of system zeros (if $k > 0$) or poles (if $k < 0$) at $s = 0$. Moreover, each zero factor $m_i(s)$ and pole factor $a_q(s)$ can be represented as either a **first-order factor** (*corner frequency ω_τ*), namely,

$$m_i(s) \quad \text{or} \quad a_q(s) = (\tau s + 1) = \left(\frac{s}{\omega_\tau} + 1\right) \tag{4.4.2}$$

with real $\tau = \frac{1}{\omega_\tau} > 0$, which would correspond to a real zero or pole at $s = -\omega_\tau = -\frac{1}{\tau}$, or a **quadratic factor** (*corner frequency ω_n*), namely,

$$m_i(s) \quad \text{or} \quad a_q(s) = \left(\frac{s^2}{\omega_n^2} + \frac{2\zeta s}{\omega_n} + 1\right) \tag{4.4.3}$$

in light of Eq. (4.2.25), with $\omega_n > 0$ and $0 < \zeta < 1$, which would correspond to a complex-conjugate pair of zeros or poles at $s = -\zeta\omega_n \pm j\omega_n\sqrt{1-\zeta^2}$. Therefore, we can associate with each $m_i(s)$ and $a_q(s)$ term in a corner frequency factored form of $G(s)$ its particular *factor type*, namely, a first-order zero or pole factor or a quadratic zero or pole factor.

Although the corner frequency factored form of a rational transfer function is an important and useful Bode diagram concept, it is not necessary to actually convert a given $G(s)$ to this form in order to determine $\tilde{G}s^k$ and the corner frequencies and damping ratios associated with its $m_i(s)$ and $a_q(s)$ factors. As we will now show, knowledge of the zeros and poles of a given rational $G(s)$ directly implies all of this information.

In particular, if $G(s)$ is known, k is immediately apparent by inspection. Consequently, the gain $\tilde{G}$ can be determined by the relationship:

$$\tilde{G} = \lim_{s \to 0} \frac{G(s)}{s^k} \tag{4.4.4}$$

in light of Eq. (4.4.1), since $m_i(0) = a_q(0) = 1$. Moreover, any real zero or pole of $G(s)$ at $s = -\alpha$ implies a corresponding first-order zero or pole factor

$$m_i(s) \quad \text{or} \quad a_q(s) = \left(\frac{s}{\alpha} + 1\right) = \left(\frac{s}{\omega_\tau} + 1\right) \tag{4.4.5}$$

with a corner frequency

$$\omega_\tau = \alpha \tag{4.4.6}$$

Finally, if

$$(s + \beta + j\gamma)(s + \beta - j\gamma) = s^2 + 2\beta s + \beta^2 + \gamma^2 \tag{4.4.7}$$

is equated to $s^2 + 2\zeta\omega_n s + \omega_n^2$, it follows that a complex-conjugate pair of zeros or poles at $s = -\beta \pm j\gamma$ implies a corresponding quadratic zero or pole factor

$$m_i(s) \quad \text{or} \quad a_q(s) = \frac{s^2}{\beta^2 + \gamma^2} + \frac{2\beta s}{\beta^2 + \gamma^2} + 1 = \frac{s^2}{\omega_n^2} + \frac{2\zeta s}{\omega_n} + 1 \quad (4.4.8)$$

with a corner frequency

$$\omega_n = \sqrt{\beta^2 + \gamma^2} \quad (4.4.9)$$

and a damping ratio

$$\zeta = \frac{\beta}{\sqrt{\beta^2 + \gamma^2}} = \frac{\beta}{\omega_n} \quad (4.4.10)$$

· ·

EXAMPLE 4.4.11 To illustrate the preceding discussion, let us recall the particular $G(s)$ of Examples 4.1.3 and 4.2.17

$$G(s) = \frac{7s^2 + 18s + 15}{s^3 + 5s^2 + 11s + 15}$$

which has two zeros at $s = -1.286 \pm j.7$ and three poles at $s = -3$ and $-1 \pm j2$. Since $G(s)$ has no poles or zeros at $s = 0$, $k = 0$ in Eq. (4.4.1). Equation (4.4.4) therefore implies that

$$\tilde{G} = G(s = 0) = \frac{15}{15} = 1$$

Using Eqs. (4.4.5) and (4.4.6), the real pole at $s = -3 = -\alpha$ implies a first-order pole factor $a_1(s)$ with a corner frequency $\omega_\tau = 3$. Using Eqs. (4.4.9) and (4.4.10), the zeros at $s = -1.286 \pm j.7$ $= -\beta \pm j\gamma$ imply a quadratic zero factor $m_1(s)$ with a corner frequency

$$\omega_n = \sqrt{(1.286)^2 + (0.7)^2} = 1.464$$

and a damping ratio

$$\zeta = \frac{1.286}{1.464} = 0.878$$

Furthermore, the poles at $s = -1 \pm j2$ represent a quadratic pole factor $a_2(s)$ with a corner frequency

$$\omega_n = \sqrt{(1)^2 + (2)^2} = 2.236$$

and a damping ratio

$$\zeta = \frac{1}{2.236} = 0.447$$

In light of Eq. (4.4.1), $G(s)$ can therefore be written in corner frequency factored form as

$$G(s) = \tilde{G}s^k \frac{m_1(s)}{a_1(s)a_2(s)} = 1s^0 \frac{\left(\dfrac{s^2}{(1.464)^2} + \dfrac{2(0.878)s}{1.464} + 1\right)}{\left(\dfrac{s}{3} + 1\right)\left(\dfrac{s^2}{(2.236)^2} + \dfrac{2(0.447)s}{2.236} + 1\right)}$$

Note that the original rational expression for $G(s)$ can be recovered from its corner frequency factored form by scalar multiplication,

$$G(s) = \frac{\left(\dfrac{s^2}{(1.464)^2} + \dfrac{2(0.878)s}{1.464} + 1\right)}{\left(\dfrac{s}{3} + 1\right)\left(\dfrac{s^2}{(2.236)^2} + \dfrac{2(0.447)s}{2.236} + 1\right)}$$

$$= \frac{3(2.236)^2 \left[s^2 + 2(0.878)(1.464)s + (1.464)^2\right]}{(1.464)^2(s+3)\left[s^2 + 2(0.447)(2.236)s + (2.236)^2\right]} = \frac{7s^2 + 18s + 15}{s^3 + 5s^2 + 11s + 15}$$

The Sinusoidal Transfer Function

A $G(s)$ in the corner frequency factored form defined by Eq. (4.4.1) implies a corresponding sinusoidal transfer function given by

$$G(s = j\omega) = G(j\omega) = \tilde{G}(j\omega)^k \frac{m_1(j\omega)m_2(j\omega)\ldots}{a_1(j\omega)a_2(j\omega)\ldots} \qquad (4.4.12)$$

with

$$\tilde{G}(j\omega)^k = |\tilde{G}||\omega|^k(j)^k = \tilde{G}\omega^k e^{j\frac{k\pi}{2}} \qquad (4.4.13)$$

since both $\tilde{G}$ and ω are positive and $j = e^{j\frac{\pi}{2}}$. Moreover, each complex-valued $m_i(j\omega)$ and $a_q(j\omega)$ term can be expressed in exponential form as $|m_i|e^{j\alpha_i}$ and $|a_q|e^{j\beta_q}$, respectively, with real $|m_i| = |m_i(j\omega)| > 0$ and real $|a_q| = |a_q(j\omega)| > 0$. Therefore,

$$G(j\omega) = \underbrace{\tilde{G}\omega^k \frac{|m_1||m_2|\cdots}{|a_1||a_2|\cdots}}_{|G(j\omega)| = |G|} e^{j\left(\frac{k\pi}{2}+\alpha_1+\alpha_2+\cdots-\beta_1-\beta_2-\cdots\right)} \qquad (4.4.14)$$

so that[5]

$$\log|G(j\omega)| = \log\tilde{G} + k\log\omega + \log|m_1| + \log|m_2|$$
$$+ \cdots - \log|a_1| - \log|a_2| - \cdots \qquad (4.4.15)$$

and

$$\phi = \angle G(j\omega) = \frac{k\pi}{2} + \alpha_1 + \alpha_2 + \cdots - \beta_1 - \beta_2 - \cdots \qquad (4.4.16)$$

Traditionally, according to Bode [7], *power gain (PG) ratios* are measured in **decibels** (or dB), as defined by the relation:

$$(PG)_{\mathrm{dB}} \overset{\text{def}}{=} 10\log\frac{P_{\text{out}}}{P_{\text{in}}}$$

---

[5] All log functions will be to the base 10 unless stated otherwise.

Since power is proportional to the square of voltage, and voltage is a common measurement for transfer function amplitudes, the log amplitude relationship of Eq. (4.4.15) is often multiplied by 20 (or 2×10) and subsequently expressed in decibels as $|G|_{dB}$. Otherwise stated, the dB amplitude of $G(j\omega)$, namely,

$$|G|_{dB} \stackrel{\text{def}}{=} 20 \log |G(j\omega)| = 20 \log \tilde{G} + 20k \log \omega + 20 \log |m_1|$$

$$+ 20 \log |m_2| + \cdots - 20 \log |a_1| - 20 \log |a_2| - \cdots \quad (4.4.17)$$

Equations (4.4.17) and (4.4.16) express the amplitude and the phase of the frequency response of a system defined by a rational, minimum phase $G(s)$ as linear combinations of relatively simple terms that are mutually independent.

The Three Possible Factors

We now illustrate the Bode graphs for each of the three factor terms that can appear in a corner frequency factored form of a minimum phase $G(s)$, as given by Eq. (4.4.1):

- the *initial* or *low frequency factor*

$$G_0(s) = \tilde{G}s^k \quad (4.4.18)$$

- a *first-order zero* or *pole factor* defined by

$$G_1(s) = (\tau s + 1)^{\pm 1} = \left(\frac{s}{\omega_\tau} + 1\right)^{\pm 1} \quad (4.4.19)$$

- a *quadratic zero* or *pole factor* defined by

$$G_2(s) = \left(\frac{s^2}{\omega_n^2} + \frac{2\zeta s}{\omega_n} + 1\right)^{\pm 1} \quad (4.4.20)$$

The Bode Graphs of $\tilde{G}s^k$ Factors

For $G_0(s) = \tilde{G}s^k$, the amplitude of the sinusoidal transfer function

$$|G_0|_{dB} = 20 \log \tilde{G}\omega^k = 20 \log \tilde{G} + 20k \log \omega \quad (4.4.21)$$

assumes the values given in Table 4.1. Therefore, on a log ω scale, $|G_0|_{dB}$ plots as a straight line with a slope of $20k$ dB/decade[6] that passes through $|G|_{dB} = 20 \log \tilde{G}$ at $\omega = 1$. If $k = 0$, as is often the case, $|G_0|_{dB}$ plots as a horizontal (zero slope) line at $|G|_{dB} = 20 \log \tilde{G}$, which lies above the 0 dB line when $\tilde{G} > 1$ and below the 0 dB line when $\tilde{G} < 1$.

---

[6]A **decade** is defined as the distance between any ω and 10ω.

TABLE 4.1 $|G_0|_{dB}$ versus ω

ω	$|G_0|_{dB}$
0.01	$20 \log \tilde{G} - 40k$
0.1	$20 \log \tilde{G} - 20k$
1	$20 \log \tilde{G}$
10	$20 \log \tilde{G} + 20k$
100	$20 \log \tilde{G} + 40k$

In light of Eq. (4.4.13), the phase angle

$$\phi_0 = \angle G_0(j\omega) = \angle \tilde{G}(j\omega)^k = \frac{k\pi}{2} = k \times 90° \qquad (4.4.22)$$

which plots as a horizontal line at $\phi = k \times 90°$. Figure 4.13 depicts the Bode diagrams of $G_0(s) = \tilde{G}s^k$ when $\tilde{G} = 0.1$ and 10, for both $k = +1$ and -2.

The Bode Graphs of First-Order Factors

For $G_1(s) = (\tau s + 1)^{+1}$, which corresponds to a first-order zero factor of $G(s)$, the sinusoidal transfer function

$$G_1(s = j\omega) = (j\omega\tau + 1) = \left(j\frac{\omega}{\omega_\tau} + 1 \right) \qquad (4.4.23)$$

so that

$$|G_1(j\omega)| = |G_1| = \sqrt{\frac{\omega^2}{\omega_\tau^2} + 1} \qquad (4.4.24)$$

and

$$\phi_1 = \angle G_1(j\omega) = \tan^{-1}\left(\frac{\omega}{\omega_\tau} \right) \qquad (4.4.25)$$

In this case, $|G_1|_{dB}$ can be approximated by two straight line segments that are called the **asymptotes** of the actual amplitude graph. In particular, when $\omega << \omega_\tau$, $|G_1(j\omega)| \approx 1$, so that

$$|G_1|_{dB} \approx 20 \log 1 = 0 \, dB \qquad (4.4.26)$$

Therefore, the low frequency asymptote of $|G_1(j\omega)|$ plots as a horizontal line at 0 dB.

When $\omega >> \omega_\tau$, $|G_1(j\omega)| \approx \frac{\omega}{\omega_\tau}$, so that

$$|G_1|_{dB} \approx 20 \log \omega - 20 \log \omega_\tau \qquad (4.4.27)$$

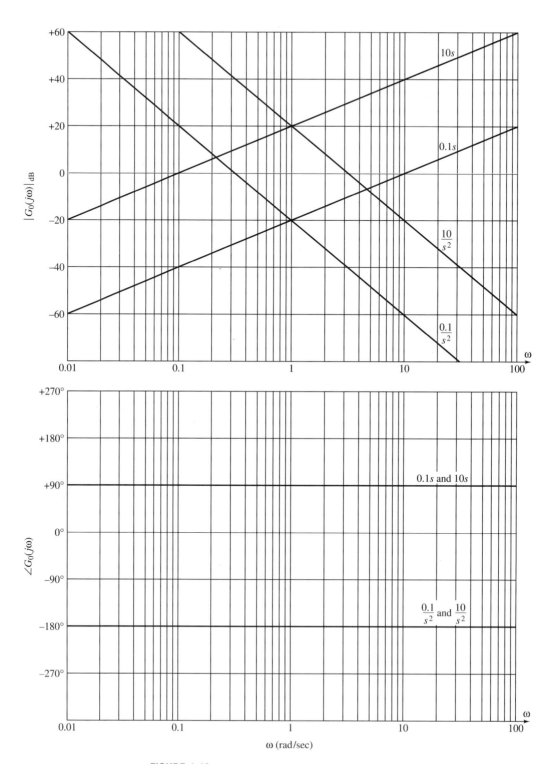

● **FIGURE 4.13**
Some Bode Diagrams of $G_0(s) = \tilde{G}s^k$

Therefore, the high frequency asymptote of $G_1(s)$ plots as a straight line with a positive slope of +20 dB/decade that passes through 0 dB at $\omega = \omega_\tau$. The intersection of these two asymptotes occurs on the 0 dB line at $\omega = \omega_\tau$, which is called the **corner frequency**[7] of the first-order factor. In light of Eq. (4.4.24), the actual amplitude of $G_1(j\omega)$ (in dB) at the corner frequency ω_τ is given by

$$|G_1(j\omega_\tau)|_{dB} = 20 \log \sqrt{2} = 10 \log 2 = 3.01 \text{ dB} \qquad (4.4.28)$$

which represents the maximum error between the actual and the asymptotic amplitude graphs. Using Eq. (4.4.25), the phase angle ϕ_1 increases continuously from 0° at $\omega = 0$, to +45° when $\omega = \omega_\tau$, to +90° as $\omega \to \infty$.

Since $\left(Ae^{j\theta}\right)^{-1} = \frac{1}{A}e^{-j\theta}$ and $20 \log \frac{1}{A} = -20 \log A$, *the Bode graphs of the inverse of a complex quantity plot as "mirror images" of the original graphs*, the amplitude mirror image axis being the 0 dB line and the phase angle mirror image axis being the 0° line. In light of this observation, the amplitude and phase Bode graphs of

$$G_1(s) = \frac{1}{\tau s + 1} = \frac{\omega_\tau}{s + \omega_\tau} \qquad (4.4.29)$$

which corresponds to a first-order pole factor of $G(s)$, plot as mirror images of those of $(\tau s + 1)$. The actual amplitude Bode graph of the system defined by Eq. (4.4.29) at the corner frequency $\omega_\tau = \frac{1}{\tau}$, namely,

$$\left|G_1\left(s = \frac{j}{\tau}\right)\right|_{dB} = 20 \log \left|\frac{1}{j+1}\right| = 20 \log \frac{1}{\sqrt{2}} = -3.01 \text{ dB} \quad (4.4.30)$$

which represents a "mirror image" value of that given by Eq. (4.4.28). The actual and the asymptotic amplitude Bode graphs of both zero and pole first-order factors are depicted in Figure 4.14 when $\tau = 2$, so that $\omega_\tau = 0.5$, while the corresponding phase plots are shown in Figure 4.15.

Since $\frac{1}{\sqrt{2}} = 0.707$, it may be noted (see Figure 4.11) that the bandwidth of a single pole system defined by Eq. (4.4.29) is equal to its corner frequency, ω_τ. Moreover, in light of Eq. (4.2.49), $\frac{1}{\omega_\tau} = \tau$ represents the time constant of the single pole system, so that larger values of ω_τ imply smaller time constants. Otherwise stated, *larger bandwidths imply faster response times*, as noted earlier.

---

[7]It is also known as the **break frequency** or the **breakpoint**.

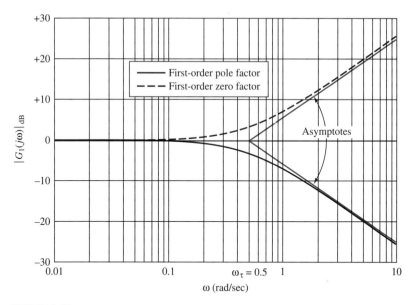

- **FIGURE 4.14**
First-Order Pole and Zero Magnitude Plots

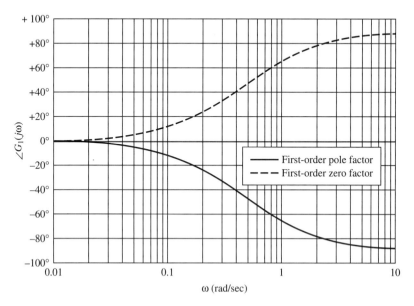

- **FIGURE 4.15**
First-Order Pole and Zero Phase Plots

The Bode Graphs of Quadratic Factors

For

$$G_2(s) = \left(\frac{s^2}{\omega_n^2} + \frac{2\zeta s}{\omega_n} + 1\right)^{+1} \qquad (4.4.31)$$

which corresponds to a quadratic zero factor of $G(s)$, the sinusoidal transfer function

$$G_2(s = j\omega) = 1 - \frac{\omega^2}{\omega_n^2} + j2\zeta\frac{\omega}{\omega_n} \qquad (4.4.32)$$

so that

$$|G_2(j\omega)| = |G_2| = \sqrt{\left(1 - \frac{\omega^2}{\omega_n^2}\right)^2 + \left(2\zeta\frac{\omega}{\omega_n}\right)^2} \qquad (4.4.33)$$

and

$$\phi_2 = \angle G_2(j\omega) = \tan^{-1}\left(\frac{2\zeta\omega\omega_n}{\omega_n^2 - \omega^2}\right) \qquad (4.4.34)$$

As in the case of first-order factors, $|G_2|_{dB} = 20 \log |G_2(j\omega)|$ can be approximated by two straight-line segments or asymptotes. When $\omega << \omega_n$, $|G_2(j\omega)| \approx 1$, so that

$$|G_2|_{dB} \approx 20 \log 1 = 0 \text{ dB} \qquad (4.4.35)$$

Therefore, the low frequency asymptote of $|G_2(j\omega)|$ is a horizontal line at 0 dB, as in the first-order case.

When $\omega >> \omega_n$, $|G_2(j\omega)| \approx \frac{\omega^2}{\omega_n^2}$, so that

$$|G_2|_{dB} \approx 40 \log \frac{\omega}{\omega_n} = 40 \log \omega - 40 \log \omega_n \qquad (4.4.36)$$

Therefore, the high frequency asymptote of $|G_2(j\omega)|$ is a straight line with a positive slope of +40 dB/decade that passes through 0 dB at $\omega = \omega_n$. The intersection of these two asymptotes occurs on the 0 dB line at $\omega = \omega_n$, the corner frequency of the quadratic factor.

The two asymptotes are independent of the damping ratio ζ. However, ζ determines the error between the Bode graphs of the actual and the asymptotic amplitudes at and near $\omega = \omega_n$. In particular,

$$|G_2(j\omega = j\omega_n)| = 2\zeta \qquad (4.4.37)$$

in light of Eq. (4.4.33), so that 2ζ represents the amplitude ratio of the actual graph to the asymptotic graph at the corner frequency ω_n. Therefore,

$$|G_2(j\omega_n)|_{dB} = 20 \log 2\zeta \qquad (4.4.38)$$

represents the dB error between the actual amplitude graph and the asymptotic graph when $\omega = \omega_n$. If $2\zeta > 1$ (when $0.5 < \zeta < 1$), the actual amplitude graph lies above the asymptotes at $\omega = \omega_n$, and when $2\zeta < 1$ (when $0 < \zeta < 0.5$), the actual amplitude graph lies below the asymptotes at $\omega = \omega_n$. In the former case, the error at $\omega = \omega_n$ approaches $20 \log 2 = 6$ dB as $\zeta \to 1$, while in the latter case, the dB error is unbounded, approaching $-\infty$ dB as $\zeta \to 0$.

In light of Eq. (4.4.34), the phase angle ϕ_2 increases continuously from $0°$ at $\omega = 0$ to $+90°$ when $\omega = \omega_n$ to $+180°$ as $\omega \to \infty$. The damping ratio ζ determines the rate at which this phase transition from $0°$ to $+180°$ occurs, with smaller values of ζ implying more abrupt transitions. In the limiting case when $\zeta = 0$, Eq. (4.4.34) implies that $\phi_2 = 0°$ for $\omega < \omega_n$ and $\phi_2 = +180°$ for $\omega > \omega_n$, reflecting an instantaneous change of $180°$ in ϕ_2 at $\omega = \omega_n$.

As in the first-order case, the Bode graphs of the amplitude and phase of

$$G_2(s) = \left(\frac{s^2}{\omega_n^2} + \frac{2\zeta s}{\omega_n} + 1 \right)^{-1} = \frac{\omega_n^2}{s^2 + 2\zeta \omega_n s + \omega_n^2} \qquad (4.4.39)$$

which corresponds to a quadratic pole factor of $G(s)$, plot as mirror images of those defined by Eq. (4.4.31). In these cases,

$$|G_2(j\omega = j\omega_n)| = \frac{1}{2\zeta} \qquad (4.4.40)$$

so that for quadratic pole factors, 2ζ represents the amplitude ratio of the asymptotic graph to the actual graph at the corner frequency ω_n. Figure 4.16 displays some Bode diagrams of the quadratic pole factor $G_2(s)$, as defined by Eq. (4.4.39), for different values of ζ.

The Amplitude Bode Graph of $G(s)$

A systematic procedure for plotting the asymptotic amplitude Bode graph of any minimum phase $G(s)$ in corner frequency factored form now follows. In particular, using Eqs. (4.4.1), (4.4.2), and (4.4.3), we associate with each

$$m_i(s) = \left(\frac{s}{\omega_{\tau i}} + 1 \right) \quad \text{or} \quad \left(\frac{s^2}{\omega_{ni}^2} + \frac{2\zeta_{mi} s}{\omega_{ni}} + 1 \right) \qquad (4.4.41)$$

and

$$a_q(s) = \left(\frac{s}{\omega_{\tau q}} + 1 \right) \quad \text{or} \quad \left(\frac{s^2}{\omega_{nq}^2} + \frac{2\zeta_{aq} s}{\omega_{nq}} + 1 \right) \qquad (4.4.42)$$

its corner frequency, namely, $\omega_{\tau i}$, ω_{ni}, $\omega_{\tau q}$ or ω_{nq}, and its corresponding factor type, namely, a first-order zero, a quadratic zero, a first-order pole, or a quadratic pole, respectively. Since the asymptotic amplitude of all $m_i(s = j\omega)$ and $a_q(s = j\omega)$ factors is 0 dB at frequencies below their respective corner frequencies, Eq. (4.4.17) implies that

$$|G(s = j\omega)|_{\mathrm{dB}} = |G|_{\mathrm{dB}} \approx 20 \log \tilde{G} + 20k \log \omega = |G_0|_{\mathrm{dB}} \qquad (4.4.43)$$

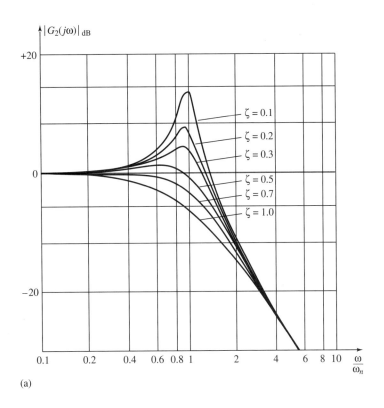

(a)

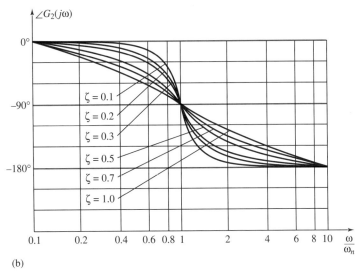

(b)

• **FIGURE 4.16**

Some Bode Diagrams of $G_2(s) = \left(\dfrac{s^2}{\omega_n^2} + \dfrac{2\zeta s}{\omega_n} + 1 \right)^{-1}$

as in Eq. (4.4.21), at all frequencies below the lowest corner frequency. Hence, Eq. (4.4.43) *represents the low frequency asymptotic Bode graph for G(s)*. As noted earlier, $|G_0|_{dB}$ of Eq. (4.4.43) plots as a straight line with a slope of $20k$ dB/decade that passes through $|G|_{dB} = 20 \log \tilde{G}$ at $\omega = 1$.

As each subsequent corner frequency is encountered, the slope of the asymptotic amplitude Bode graph changes to reflect the addition of a new amplitude term. The resultant slope changes are a direct function of the factor type involved. In particular, the change in slope is +20 dB/decade at each $\omega_{\tau i}$ that corresponds to a first-order zero factor of $G(s)$, and +40 dB/decade at each ω_{ni} that corresponds to a quadratic zero factor of $G(s)$. The change in slope is -20 dB/decade at each $\omega_{\tau q}$ that corresponds to a first-order pole factor of $G(s)$, and -40 dB/decade at each ω_{nq} that corresponds to a quadratic pole factor of $G(s)$. The straight line asymptotic amplitude graph continues (as $\omega \to \infty$) with a final slope equal to that determined at the highest corner frequency.

The Phase Angle Bode Graph of *G(s)*

The phase angles associated with each $m_i(s = j\omega)$ and $a_q(s = j\omega)$ term in Eq. (4.4.12) are approximately $0°$ at frequencies "sufficiently below" their corner frequencies, as noted earlier. Therefore, using Eq. (4.4.16),

$$\phi \angle G(j\omega) = \frac{k\pi}{2} = k \times 90° \qquad (4.4.44)$$

represents the low frequency value of ϕ.

The determination of the phase angle of a $G(s = j\omega)$ in corner frequency factored form also is a direct function of the factor type involved. In light of Eqs. (4.4.41) and (4.4.42), the phase angles contributed to $\phi \angle G(j\omega)$ by each $m_i(j\omega)$ and $a_q(j\omega)$ term at their corner frequencies, namely, $\omega_{\tau i}$, ω_{ni}, $\omega_{\tau q}$ and ω_{nq} equal $+45°$, $+90°$, $-45°$, and $-90°$, respectively. Moreover, as ω increases beyond each corner frequency, these angles approach $+90°$, $+180°$, $-90°$, and $-180°$, respectively. Therefore, an approximate value for ϕ can be obtained if, at each successive corner frequency, we add (to the previous value of ϕ) the phase angle contribution associated with the corner frequency term plus the remaining phase angle contribution associated with the previous corner frequency term.

The final value of ϕ (as $\omega \to \infty$) equals the value of ϕ determined at the highest corner frequency plus the remaining phase angle contribution associated with the highest corner frequency term. Once the values of ϕ have been determined, a "smooth" curve connecting them can be drawn to yield a phase angle Bode graph of $G(s = j\omega)$.

If there is a "large" distance between successive corner frequencies, the phase angle at their midpoint ω_{mp}, namely $\phi \angle G(j\omega_{mp})$, corresponds to the slope of the amplitude graph at ω_{mp} in the sense that an asymptotic amplitude slope of $20q$ dB/decade at ω_{mp} implies a corresponding phase angle of approximately $q \times 90°$ at $\omega = \omega_{mp}$. This fact is a characteristic of minimum phase systems and is actually why they are so named. The *phase angle of a minimum phase system represents the minimum phase angle possible consistent with the slope of its asymptotic amplitude graph.*

EXAMPLE 4.4.45 To illustrate the preceding discussion, let us plot the asymptotic amplitude Bode diagram of a system defined by the minimum phase transfer function

$$G(s) = \frac{2.4s^3 + 15.36s^2 + 19.2s + 12}{s^4 + 30.08s^3 + 2.44s^2 + 1.2s}$$

which is characterized by three zeros at $s = -0.7 \pm j.714$ and -5, and four poles at $s = 0, -0.04 \pm j.196$, and -30, so that $k = -1$ and

$$\tilde{G} = \lim_{s \to 0} sG(s) = \frac{12}{1.2} = 10$$

Using Eqs. (4.4.5) and (4.4.6), the zero at $s = -5$ represents a first-order zero factor of $G(s)$ with a corner frequency at $\omega = 5$, while the pole at $s = -30$ represents a first-order pole factor with a corner frequency at $\omega = 30$.

In view of Eqs. (4.4.9) and (4.4.10), the poles at $s = -0.04 \pm j.196$ represent a quadratic pole factor with a corner frequency at

$$\omega = \sqrt{(0.04)^2 + (0.196)^2} = 0.2$$

and a damping ratio

$$\zeta = \frac{0.04}{0.2} = 0.2$$

Furthermore, the zeros at $s = -0.7 \pm j.714$ represent a quadratic zero factor with a corner frequency at

$$\omega = \sqrt{(0.7)^2 + (0.714)^2} = 1$$

and a damping ratio

$$\zeta = \frac{0.7}{1} = 0.7$$

Explicit knowledge of the corner frequencies of $G(s)$, including the associated factor types, now enables us to determine appropriate asymptotic amplitude slope changes as well as approximate phase angle values at the corner frequencies of the Bode diagram. Table 4.2 summarizes such a determination at the particular corner frequencies associated with this example (as well as at $\omega = 0$ and $\omega \to \infty$), and Figure 4.17 depicts a Bode diagram of the asymptotic amplitude. The damping ratio values associated with each quadratic (zero or pole) factor can be used to determine the error between the actual and asymptotic amplitude graphs using Eqs. (4.4.37) and (4.4.40). Finally, this example also illustrates our earlier observation that it is not necessary to express $G(s)$ in a corner frequency factored form to plot its Bode diagram.

TABLE 4.2 A Bode Diagram Table for $G(s) = \dfrac{2.4s^3 + 15.36s^2 + 19.2s + 12}{s^4 + 30.08s^3 + 2.44s^2 + 1.2s}$

Corner Frequency	Slope Change (dB/decade)	New Slope (dB/decade)	Phase ϕ Change	Approximate $\phi = \angle G(j\omega)$
0	$20k = -20$	-20	$k \times 90° = -90°$	$-90°$ (actual)
0.2 $(\zeta = 0.2)$	-40	-60	$-90°$	$-180°$
1 $(\zeta = 0.7)$	$+40$	-20	$+90° - 90°$	$-180°$
5	$+20$	0	$+45° + 90°$	$-45°$
30	-20	-20	$-45° + 45°$	$-45°$
$\to \infty$	No change	-20	$-45°$	$-90°$ (actual)

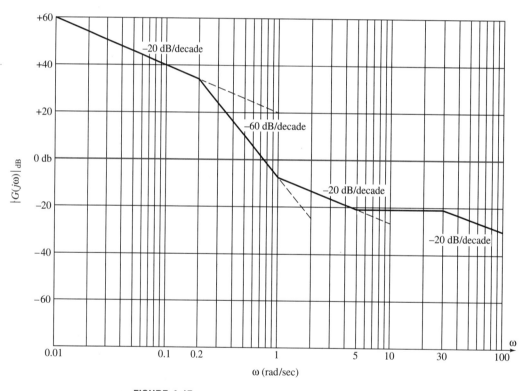

- **FIGURE 4.17**

  An Amplitude Bode Diagram of $G(s) = \dfrac{2.4s^3 + 15.36s^2 + 19.2s + 12}{s^4 + 30.08s^3 + 2.44s^2 + 1.2s}$

The Determination of $G(s)$ from a Bode Diagram

An approximation of the transfer function $G(s)$ of a minimum phase system can be obtained from a Bode diagram of its frequency response. To do
this, we first determine an asymptotic amplitude plot by approximating the
actual log amplitude graph by a sequence of connected straight lines, each
of which is characterized by a slope of $20q$ dB/decade for some positive
or negative integer q. Using Eq. (4.4.43), the initial low frequency line,
defined by a slope of $20k$ dB/decade, implies the integer k in Eq. (4.4.1).
Moreover, its amplitude value at $\omega = 1$ defines $20 \log \tilde{G}$, hence $\tilde{G}$.

A corner frequency is defined at any value of ω characterized by
the intersection of two consecutive lines, and its associated factor type
is determined by the change in slope at the corner frequency; that is,
slope changes of +20 dB/decade, +40 dB/decade, −20 dB/decade, and
−40 dB/decade imply first-order zero, quadratic-zero, first-order pole, and
quadratic-pole factors, respectively. Greater slope changes, if encountered,
imply multiple factor types. Approximate ζ values for the quadratic factors
are determined from the corner frequency errors between the actual and
asymptotic amplitude graphs, in light of Eqs. (4.4.37) and (4.4.40). In this
manner, an $m_i(s)$ or $a_q(s)$ factor of $G(s)$, as given by Eqs. (4.4.41) or
(4.4.42), can be determined at each corner frequency.

Knowledge of $\tilde{G}s^k$ and all of the $m_i(s)$ and $a_q(s)$ factors subsequently
implies a corner frequency factored form for $G(s)$, as given by Eq. (4.4.1).
The minimum phase assumption eliminates the need to include any phase
information in this approximation of $G(s)$, because of the direct correspondence between the slope of the asymptotic amplitude graph and the
corresponding phase angle. However, the phase angle information can be
used to verify the minimum phase assumption.

EXAMPLE 4.4.46 To illustrate this procedure, consider the actual Bode diagram of an otherwise
unknown minimum phase system, as depicted in Figure 4.18, which could have been obtained by
experimental frequency response measurements. Figure 4.19 displays a magnified amplitude graph
of the system with five superimposed approximation lines, labeled L1 through L5. Lines L1 and L5,
with slopes of −20 dB/decade, were the first to be drawn. The amplitude segment just below $\omega = 1$
was then determined to be characterized by a slope of −60 dB/decade, and L2 was drawn to "best
match" this segment. The transition between L2 and L5 was subsequently achieved by two straight
line segments, L3 and L4, with slopes of −20 dB/decade and 0 dB/decade, respectively.

Since the initial line (L1) has a slope of −20 (or $20k$) dB/decade, and an approximate amplitude
of +20 dB (or $20 \log \tilde{G}$) at $\omega = 1$, Eq. (4.4.43) implies that

$$\tilde{G}s^k \approx 10s^{-1} = \frac{10}{s}$$

the low frequency factor of $G(s)$.

The intersection of L1 and L2 defines the first corner frequency at $\omega \approx 0.2$, and the slope change
of −40 dB/decade at this frequency identifies the corresponding $G(s)$ factor as a quadratic pole term

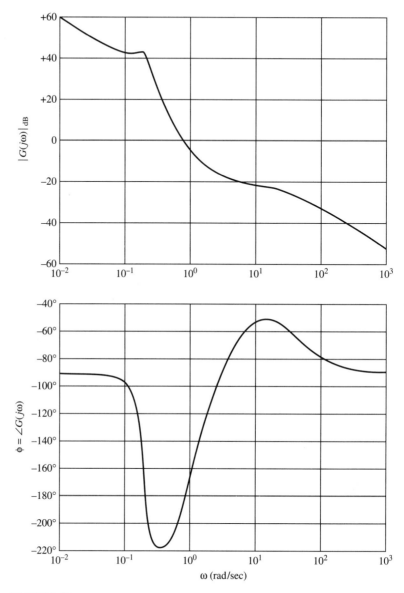

● **FIGURE 4.18**
The Actual Bode Diagram of an Unknown System

with a damping ratio

$$\zeta \approx \frac{45}{2(90)} = 0.25$$

according to Eq. (4.4.40), since 2ζ equals the amplitude ratio ($\approx 45/90$) of the asymptotic graph to the actual graph at the corner frequency. The intersection of L2 and L3 defines the next corner

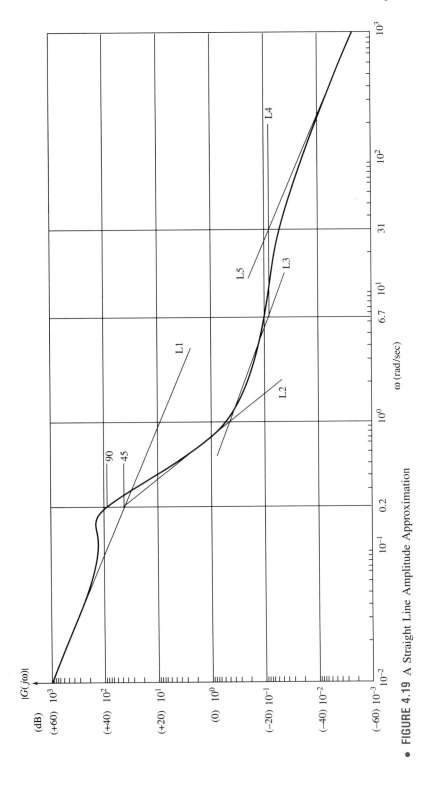

• **FIGURE 4.19** A Straight Line Amplitude Approximation

frequency at $\omega \approx 1$, and the slope change of $+40$ dB/decade identifies the corresponding $G(s)$ factor as a quadratic zero with a damping ratio

$$\zeta \approx \frac{0.6}{2(0.48)} = 0.625$$

since 2ζ equals the amplitude ratio ($\approx 0.6/0.48$) of the actual graph to the asymptotic graph at the corner frequency.

The intersection of L3 and L4 defines the next corner frequency at $\omega \approx 6.7$, and the corresponding slope change of $+20$ dB/decade identifies the corresponding $G(s)$ factor as a first-order zero. The intersection of L4 and L5 defines the final corner frequency at $\omega \approx 31$, and the corresponding slope change of -20 dB/decade identifies the corresponding $G(s)$ factor as a first-order pole.

All of this information is summarized in Table 4.3, which using Eq. (4.4.1) implies that

$$G(s) \approx \frac{10}{s} \frac{(s^2 + 2(0.625)s + 1)\left(\frac{s}{6.7} + 1\right)}{\left(\frac{s^2}{(0.2)^2} + \frac{2(0.25)s}{0.2} + 1\right)\left(\frac{s}{31} + 1\right)} = \frac{10(0.2)^2(31)}{s(6.7)} \frac{(s^2 + 1.25s + 1)(s + 6.7)}{(s^2 + 0.1s + 0.04)(s + 31)}$$

$$= \frac{1.85s^3 + 14.71s^2 + 17.34s + 12.4}{s^4 + 31.1s^3 + 3.14s^2 + 1.24s}$$

which is characterized by three zeros at $s = -0.625 \pm j.78$ and -6.7, and four poles at $s = 0, -0.05 \pm j.194$, and -31.

The astute reader has undoubtedly recognized the similarities between this system and the one defined in the previous example. This is not a coincidence, however, because the Bode diagram depicted in Figure 4.18 actually represents that of the Example 4.4.45 system, whose asymptotic Bode diagram is depicted in Figure 4.17. Note that the s-plane locations of the poles and zeros of the two systems are very close, which illustrates that a good approximation of the transfer function of a minimum phase system often can be obtained from a Bode diagram of its frequency response.

··

Nonminimum Phase Rational Transfer Functions

If a rational transfer function $G(s)$ is nonminimum phase or unstable, that is, if one or more of its zeros or poles lies in the unstable half-plane $Re(s) > 0$, then $G(s)$ can still be expressed in a corner frequency factored form "similar" to that given by Eq. (4.4.1). In particular, the stable factors would not change, and each unstable first-order factor is expressed as

$$m_i(s) \quad \text{or} \quad a_q(s) = (\tau s - 1) = \left(\frac{s}{\omega_\tau} - 1\right) \qquad (4.4.47)$$

with real $\tau = \frac{1}{\omega_\tau} > 0$, which corresponds to a real zero or pole at $s = +\omega_\tau = +\frac{1}{\tau}$. Furthermore, each unstable quadratic factor is expressed

TABLE 4.3 A Corner Frequency Table for a $G(s)$ Computation

Corner Frequency	New Slope (dB/decade)	Slope Change (dB/decade)	Factor Type	Approximate $\tilde{G}s^k$, $m_i(s)$, or $a_q(s)$
0	-20	no change	$k = 1$	$\tilde{G}s^k = 10s^{-1}$
0.2 ($\zeta = 0.25$)	-60	-40	Quadratic pole	$\dfrac{s^2}{(0.2)^2} + \dfrac{2(0.25)s}{0.2} + 1$
1 ($\zeta = 0.625$)	-20	$+40$	Quadratic zero	$s^2 + 2(0.625)s + 1$
6.7	0	$+20$	First-order zero	$\dfrac{s}{6.7} + 1$
31	-20	-20	First-order pole	$\dfrac{s}{31} + 1$

as

$$m_i(s) \quad \text{or} \quad a_q(s) = \left(\frac{s^2}{\omega_n^2} - \frac{2\zeta s}{\omega_n} + 1 \right) \tag{4.4.48}$$

with $\omega_n > 0$ and $0 < \zeta < 1$, which corresponds to a complex-conjugate pair of zeros or poles at $s = +\zeta\omega_n \pm j\omega_n\sqrt{1 - \zeta^2}$. Knowledge of the unstable pole and zero locations of $G(s)$ also implies these unstable factors using Eqs. (4.4.6), (4.4.9), and (4.4.10), as in the stable case.

The amplitudes associated with the corresponding first-order zero or pole terms defined by

$$G_1(s) = (\tau s - 1)^{\pm 1} = \left(\frac{s}{\omega_\tau} - 1 \right)^{\pm 1} \tag{4.4.49}$$

and the corresponding quadratic zero or pole terms defined by

$$G_2(s) = \left(\frac{s^2}{\omega_n^2} - \frac{2\zeta s}{\omega_n} + 1 \right)^{\pm 1} \tag{4.4.50}$$

are given by Eqs. (4.4.24) and (4.4.33), respectively, as in the stable cases.

The phase angle graphs, however, differ significantly from those defined by Eqs. (4.4.25) and (4.4.34), thereby reflecting the nonminimum phase angle properties of the system. In particular,

$$\phi_1 = \angle G_1(j\omega) = 180° - \tan^{-1}\left(\frac{\omega}{\omega_\tau} \right) \tag{4.4.51}$$

for the $G_1(s)$ of Eq. (4.4.49), and

$$\phi_2 = \angle G_2(j\omega) = \tan^{-1}\left(\frac{-2\zeta\omega\omega_n}{\omega_n^2 - \omega^2} \right) \tag{4.4.52}$$

for the $G_2(s)$ of Eq. (4.4.50). In the former case, ϕ_1 increases continuously from $180°$ at $\omega = 0$ to $135°$ when $\omega = \omega_\tau$ to $90°$ as $\omega \to \infty$, while

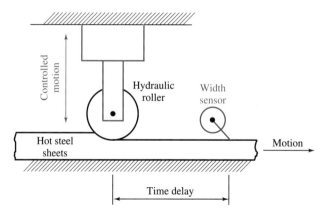

- **FIGURE 4.20**
Transportation Lag in Steel Processing

in the latter case, the phase angle ϕ_2 increases continuously from $0°$ at $\omega = 0$ to $-90°$ when $\omega = \omega_n$ to $-180°$ as $\omega \to \infty$. As in the case of stable factors, the damping ratio ζ determines the rate at which this phase transition from $0°$ to $-180°$ occurs, with smaller values of ζ implying more abrupt transitions.

Transportation Lag

Another common cause of nonminimum phase behavior is "transportation lag." In particular, many physical systems are characterized by a time delay between a control input action and the resultant measured or observed output response. This is often the case in process control systems when an output sensor might be placed some distance away from the source of a physical reaction to prevent its destruction. The time τ required for the effect of the reaction to travel to the measurement point is called the **transportation lag**.[8]

Figure 4.20 depicts a steel processing plant, in which a hydraulically actuated roller is employed to compress sheets of hot steel as they move through the process. A width sensing potentiometer must be placed some distance away from the roller because of the heat, thus implying a time delay between the measured output (the width of the steel) and the control input (the height of the roller).

In general, the relationship between the input $u(t)$ and the output $y(t)$ of a system characterized solely by transportation lag is defined by

---

[8]It is also known as the **transport lag**, the **dead time**, or the **time delay**.

the equation:

$$y(t) = u(t - \tau) \tag{4.4.53}$$

and since $\mathcal{L}[u(t-\tau)] = e^{-s\tau}u(s)$ (see Appendix B), the Laplace transform of Eq. (4.4.53) implies that

$$y(s) = e^{-s\tau}u(s) \tag{4.4.54}$$

In view of Eq. (3.1.2), it follows, therefore, that the nonrational transfer function

$$G_\tau(s) = \frac{y(s)}{u(s)} = e^{-s\tau} \tag{4.4.55}$$

represents the transportation lag relationship defined by Eq. (4.4.53).
    Since the sinusoidal transfer function

$$G_\tau(s = j\omega) = e^{-j\omega\tau} = \cos \omega\tau - j \sin \omega\tau \tag{4.4.56}$$

it follows that

$$|G_\tau| = |G_\tau(j\omega)| = 1 \tag{4.4.57}$$

and

$$\phi_\tau = \angle G_\tau(j\omega) = -\omega\tau \tag{4.4.58}$$

Hence, as in the case of a rational, nonminimum phase transfer function, transportation lag has no effect whatsoever on the amplitude graph of a Bode diagram. However, it does contribute additional *nonminimum phase lag* to the phase plot, which is directly proportional to the frequency ω.

4.5   POLAR PLOTS

We have now shown how Bode diagrams can be employed to depict the frequency response of dynamic systems. A major advantage of Bode diagrams is that they essentially convert nonlinear amplitude multiplication into linear amplitude addition, so that the log of the amplitude associated with each term in a corner frequency factored form of $G(s)$ can be summed separately. One disadvantage of Bode diagrams, however, is that different graphs must be utilized for the magnitude and the phase. This "deficiency" can be eliminated if we employ a single plot of both the magnitude and the phase of the frequency response in polar coordinates as the frequency ω varies from zero to infinity.

In particular, a **polar plot** of the frequency response of a system defined by a transfer function $G(s)$, hence a sinusoidal transfer function

$$G(j\omega) = Re[G(j\omega)] + jIm[G(j\omega)] = |G(j\omega)|e^{j\phi} \qquad (4.5.1)$$

is a plot of $Re[G(j\omega)]$ on the horizontal axis versus $Im[G(j\omega)]$ on the vertical axis in the complex $G(j\omega)$-plane as ω varies from 0 to ∞. Therefore, for each value of ω, a polar plot of $G(j\omega)$ is defined by a vector of length

$$|G(j\omega)| = \sqrt{\{Re[G(j\omega)]\}^2 + \{Im[G(j\omega)]\}^2} \qquad (4.5.2)$$

at an angle

$$\phi = \angle G(j\omega) = \tan^{-1}\left\{\frac{Im[G(j\omega)]}{Re[G(j\omega)]}\right\} \qquad (4.5.3)$$

in the $G(j\omega)$-plane.

Mathematically, this process may be regarded as a continuous "mapping" of the positive imaginary axis of the complex s(or $\sigma + j\omega$)-plane into the complex $G(j\omega)$-plane. Polar plots of the "loop gain" of a system can be very useful in determining its closed loop stability properties, as we will later show.

The Polar Plots of Some Simple Systems

In certain cases, polar plots are used in conjunction with Bode diagrams, the former providing more qualitative information and the latter more quantitative information relative to frequency response characteristics of a system. If the Bode diagram of a particular system is known, it is relatively straightforward to obtain its polar plot, and polar plots are sometimes determined in this manner. In simple cases, however, they often can be obtained directly, as we now illustrate.

For example, if

$$G(s) = G_0(s) = \tilde{G}s^k \qquad (4.5.4)$$

as in Eq. (4.4.18), and if $k > 0$, a polar plot of $G_0(s = j\omega) = \tilde{G}(j\omega)^k$ plots as a straight line in the $G(j\omega)$-plane, starting at the origin (when $\omega = 0$) and extending to ∞ (as $\omega \to \infty$) at a fixed angle $\phi = k \times 90°$. When $k < 0$, a polar plot of $G_0(j\omega)$ plots as a straight line starting at ∞ (when $\omega = 0$) and terminating at the origin (when $\omega = \infty$) at a fixed angle $\phi = |k| \times -90°$.

If $G(s)$ is a first-order zero factor, as in Eq. (4.4.2), so that

$$G(s = j\omega) = 1 + j\omega\tau \qquad (4.5.5)$$

its polar plot is a straight line starting at +1 in the complex $G(j\omega)$-plane (when $\omega = 0$) and extending to infinity (as $\omega \to \infty$) along a straight line that parallels the positive imaginary axis.

TABLE 4.4 $Re[G(j\omega)]$ and $Im[G(j\omega)]$ versus ω

ω	$Re[G(j\omega)]$	$Im[G(j\omega)]$
0	1	0
$0.1/\tau$	0.99	-0.099
$0.5/\tau$	0.8	-0.4
$1/\tau$	0.5	-0.5
$2/\tau$	0.2	-0.4
$10/\tau$	0.0099	-0.099
∞	0	0

If $G(s)$ is a first-order pole factor, so that

$$G(s = j\omega) = \frac{1}{1 + j\omega\tau} \tag{4.5.6}$$

then the data derived in Table 4.4 can be used to obtain a polar plot. In this case, it can be shown (see Problem 4-16) that the polar plot traces a semi-circular arc of radius 0.5 centered at +0.5 in the complex $G(j\omega)$-plane, starting at +1 when $\omega = 0$ and terminating at 0 when $\omega = \infty$. Figure 4.21 depicts the polar plots associated with these and other relatively simple transfer functions.

Polar Plots from Bode Diagrams

For more complex transfer functions, a table similar to Table 4.4 can be employed to determine as many points as may be necessary to obtain a particular polar plot with desired accuracy. In general, this process can be tedious because the effect of individual pole and zero terms cannot be determined independently, as with Bode diagrams. Indeed, in many cases a Bode diagram is determined first, and its magnitude and phase graphs are used to obtain a corresponding polar plot by converting decibels into ordinary magnitude. For example, if $G(s)$ is a quadratic-pole factor as in Eq. (4.4.39), so that

$$G(j\omega) = \frac{1}{1 - \dfrac{\omega^2}{\omega_n^2} + j2\zeta\dfrac{\omega}{\omega_n}} \tag{4.5.7}$$

the Bode diagrams of Figure 4.16 can be employed to obtain some corresponding polar plots for different values of ζ between 0.2 and 0.7, as depicted in Figure 4.22.

Alternatively, computer programs, such as Xmath [64] and Matlab [46], can be used to directly obtain Bode diagrams or polar plots from knowledge of either a transfer function or a state-space representation of a given system.

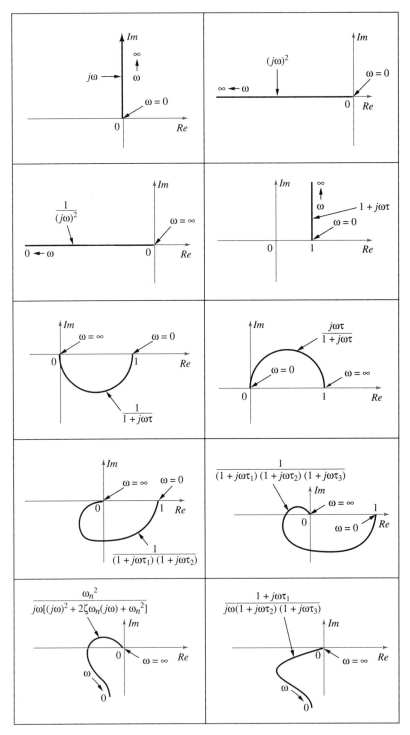

- **FIGURE 4.21**
  Some Simple Polar Plots

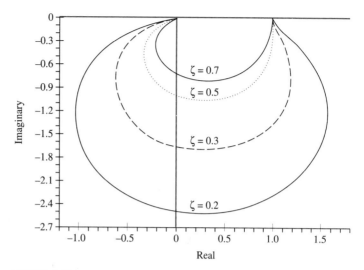

• **FIGURE 4.22**
Polar Plots of Quadratic Pole Factors

EXAMPLE 4.5.8 To further illustrate the determination of polar plots from Bode diagrams, consider the system defined in Example 4.4.45 by

$$G(s) = \frac{2.4s^3 + 15.36s^2 + 19.2s + 12}{s^4 + 30.08s^3 + 2.44s^2 + 1.2s}$$

If we translate the magnitude and phase information displayed in the Bode diagram of Figure 4.17 to polar coordinates, at frequencies between $\omega = 0.1$ and $\omega = 100$, we obtain a polar plot of the frequency response similar to that depicted in Figure 4.23.[9]

Transportation Lag

In light of Eq. (4.4.57), the frequency response amplitude of a system is unaffected by transportation lag, which is characterized by an $e^{-s\tau}$ multiplier of its transfer function. However, additional phase lag proportional to $\omega\tau$ is introduced in light of Eq. (4.4.58). The effect of such transportation lag essentially is to rotate every point defined by the frequency ω on a polar plot clockwise by an angular amount equal to $\omega\tau$. Figure 4.24 displays the effect that transportation lag equal to 0.4 seconds, so that $e^{-s\tau} = e^{-0.4j\omega}$, has on the polar plot of a system otherwise defined by a quadratic-pole factor $G(s)$, as given by Eq. (4.4.39), with $\zeta = 0.3$, that

---

[9]The actual polar plot depicted in Figure 4.23 was obtained using Xmath [64].

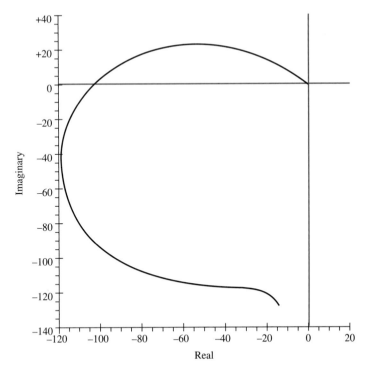

A Polar Plot of $G(s) = \dfrac{2.4s^3 + 15.36s^2 + 19.2s + 12}{s^4 + 30.08s^3 + 2.44s^2 + 1.2s}$ for $0.1 \leq \omega \leq 100$

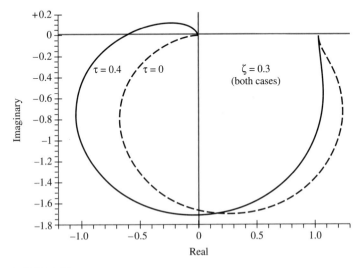

The Effect of Transportation Lag

is, when

$$G(s) = \frac{e^{-0.4s}\omega_n^2}{s^2 + 0.6\omega_n s + \omega_n^2} \qquad (4.5.9)$$

Generally speaking, the effect of transportation lag on the frequency response of a system is most easily seen by means of a polar plot.

Log-Magnitude versus Phase Plots

Another way of depicting the frequency response of a system by a single graph is to plot its logarithmic magnitude, $20 \log |G(j\omega)| = |G|_{dB}$ (on a vertical scale), versus its phase angle, $\angle G(j\omega)$ (on a horizontal scale), over an appropriate frequency range. The resulting curve is a function of the frequency ω, as in the case of polar plots. Some relatively simple **log-magnitude versus phase plots** are depicted in Figure 4.25, while Figure 4.26 displays all three frequency response representations of an underdamped, zeroless, second-order system defined by the transfer function

$$G(s) = \frac{\omega_n^2}{s^2 + 2\zeta\omega_n s + \omega_n^2}$$

By superimposing closed-loop magnitude and phase contours on such graphs, N. B. Nichols [36] was able to use them to design closed-loop (unity feedback) compensators. Unlike the linear scale associated with polar plots, which makes it difficult to capture some of the pertinent characteristics of $G(j\omega)$, Nichols showed how the logarithmic scale for magnitude allows these graphs to be used for feedback design. For this reason, these graphs are often called **Nichols charts**. The interested reader is referred to references [33], [16], and [52] for additional details regarding these graphs and their use in design.

4.6   SUMMARY

We have now shown how the dynamic behavior of a physical system can be described implicitly by its dynamic response. In particular, a common control system design goal is to have the output response track one or more of the three common reference inputs—the step, the ramp, or the parabolic—with as small an error as possible. Therefore, these three inputs (especially the step) often are used as test signals to analyze an otherwise unknown system. In such cases, the step response parameters that define stable, underdamped, zeroless systems of order two often can be used to obtain useful approximations to higher order systems, especially those that are characterized by a pair of dominant poles.

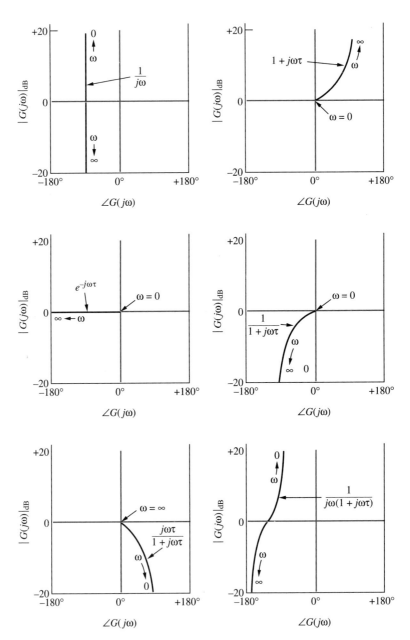

● **FIGURE 4.25**
Some Simple Log-Magnitude versus Phase Plots

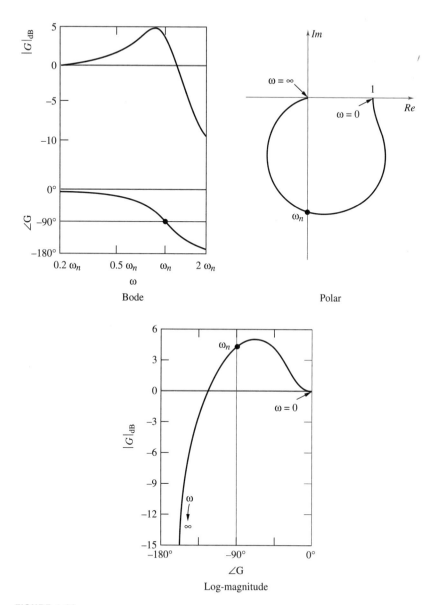

• **FIGURE 4.26**

Three Frequency Response Representations of $G(s) = \dfrac{\omega_n^2}{s^2 + 2\zeta\omega_n s + \omega_n^2}$

Sinusoidal signals were then shown to represent another important class of test signals that can be employed to obtain the frequency response of a dynamic system. A useful observation made here was that the substitution of $j\omega$ for s in the transfer function $G(s)$ of a system directly implies its frequency response, whether the system is stable or not.

Bode diagrams, polar plots, and Nichols charts were then shown to represent three different ways of graphically displaying this information. The corner frequency factored form of a given $G(s)$ was employed to simplify the actual construction of Bode diagrams, although such forms need not be explicitly determined if the zeros and poles of the system are known. The converse question of determining an approximation of $G(s)$ from Bode diagram plots of its frequency response also was addressed and illustrated. Polar plots were shown to be particularly illuminating in representing the detrimental effects of transportation lag.

PROBLEMS

4-1. Verify Eq. (4.1.2) using these relationships:

$$e^{j\theta} = \cos\theta + j\sin\theta \quad \text{and} \quad \cos(\alpha + \beta) = \cos\alpha\cos\beta - \sin\alpha\sin\beta$$

4-2. In view of Eq. (4.2.32), it follows that for an overdamped, second-order system, a ratio of the poles p_1 and p_2

$$r_2^1 \stackrel{\text{def}}{=} \frac{p_1}{p_2} = \frac{\zeta + \sqrt{\zeta^2 - 1}}{\zeta - \sqrt{\zeta^2 - 1}}$$

completely specifies the damping ratio ζ, independent of ω_n. Using this observation, determine ζ when $r_2^1 = 10$.

4-3. Show that in the underdamped case, when $0 < \zeta < 1$,

$$\tan^{-1}\frac{\sqrt{1 - \zeta^2}}{-\zeta} = \pi - \sin^{-1}\sqrt{1 - \zeta^2}$$

4-4. Consider a stable, second-order system defined by the rational transfer function

$$G(s) = \frac{m(s)}{a(s)} = \frac{4}{s^2 + 3s + 2} = \frac{4}{(s + 1)(s + 2)}$$

which is initially at rest. If the system is subjected to a unit ramp input $u(t) = r_r(t)$, express the complete output response as the sum of the natural and forced responses.

4-5. Consider a system defined by the rational transfer function

$$G(s) = \frac{m(s)}{a(s)} = \frac{s + 1}{s(s + 2)}$$

which is initially at rest. If the system is subjected to the exponentially decaying input $u(t) = e^{-t}$ for $t \geq 0$, so that

$$u(s) = \mathcal{L}[u(t)] = \frac{1}{s+1} = \frac{m_u(s)}{p_u(s)}$$

express the complete output response as the sum of the natural and forced response. Note, in particular, that the input "mode" e^{-t} will not appear at the output $y(t)$ because of the so-called *blocking zero* of $G(s)$ at $s = -1$. Explain this phenomenon using the controllability and observability properties of a series connection of a system defined by the transfer function $H(s) = u(s)$, followed by a system defined by $G(s)$, as in Problem 3-10.

4-6. Consider the

$$G(s) = \frac{7s^2 + 18s + 15}{s^3 + 5s^2 + 11s + 15} = \frac{m(s)}{a(s)}$$

of Example 4.2.17. If

$$u(s) = r_r(s) = \frac{1}{s^2} = \frac{m_u(s)}{p_u(s)}$$

determine the $\hat{m}(s)$ and $\hat{m}_u(s)$ of Eq. (4.2.13) using Eq. (4.2.15) by equating polynomial coefficients of equal degree.

4-7. Consider the closed loop, unity feedback system depicted in Figure 4.27. Note that although the forward loop system defined by $G_o(s) = \frac{2}{s-1}$ is unstable, the closed-loop system defined by

$$G_c(s) = \frac{y(s)}{r(s)} = \frac{G_o(s)}{1 + G_o(s)} = \frac{2}{s+1}$$

is stable, since its single pole lies at -1 in the complex s-plane.

    (a) Determine both $y(t)$ and $u(t) = r(t) - y(t)$ for $t > 0$ in response to a unit step external input $r(t) = r_s(t)$, assuming that the system is initially at rest.

    (b) Using your answer to part (a), define $K_u = \lim_{t \to \infty} u(t) = u_\infty$ and $K_y = \lim_{t \to \infty} y(t) = y_\infty$, and show that although $G_o(s)$ is

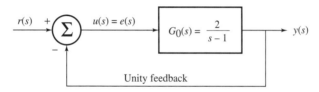

• **FIGURE 4.27**
The Feedback System of Problem 4-7

unstable, the final value theorem can be used to verify that

$$K_y = \lim_{s \to 0} s G_o(s) \frac{K_u}{s} = -2K_u$$

4-8. Use the initial value theorem to verify that the initial slope (at $t = 0$) of the unit step responses $y_1(t)$ and $y_2(t)$ of both of the second-order systems defined in Example 4.2.50 is 0, while the initial slope of the unit-step response $y_0(t)$ of the first-order system is 1, as depicted in Figure 4.10.

4-9. Show that in the underdamped case, when $0 < \zeta < 1$, $y(t)$ is given by Eq. (4.2.29).

4-10. Show that if $y(t)$ is given by Eq. (4.2.29), then its time derivative is given by Eq. (4.2.35).

4-11. Consider any stable, underdamped, second-order system defined by the $G(s)$ of Eq. (4.2.25). Let θ denote the positive angle between the negative real axis and a line drawn from the origin to either pole of $G(s)$ in the complex s-plane. Express θ as a function of the damping ratio ζ.

4-12. Use Eq. (4.2.6) to determine the unit step response of the system defined by the differential operator representation of Example 4.1.3. Verify that the step response also can be obtained by differentiating, with respect to time, the unit ramp response given in Example 4.2.17

$$y(t) = y_n(t) + y_f(t) = 0.333e^{-3t} - 0.894e^{-t}\cos(2t - 0.464) + 0.467 + t$$

Verify that the time derivative of the step response is equal to the impulse response given in Example 4.1.3, and discuss why this observation is true in general.

4-13. Consider the stable closed-loop system defined in Problem 4-7, which is depicted in Figure 4.27. If the external input $r(t) = K_r \sin \omega t = 0.5 \sin 2t$, determine the forced frequency response $y_f(t) = K_y \sin(2t + \phi_y)$ and $u_f(t) \stackrel{\text{def}}{=} r(t) - y_f(t) = K_u \sin(2t + \phi_u)$. Show that the unstable system defined by $G_o(s)$ does have a physically measurable frequency response in this case, because the residue $\hat{m}(s) = 0$ in Eq. (4.2.13). Also, show that the sinusoidal transfer function of the unstable system $G_o(j\omega = j2)$ can be used to determine the frequency response between $y(t)$ and $u(t)$ in this case; that is, verify that

$$|G_o(j2)| = \frac{K_y}{K_u} \quad \text{and} \quad \angle G_o(j2) = \phi_y - \phi_u$$

4-14. Show that when $0 < \zeta < 0.707$, the frequency response of a system defined by the transfer function $G_2(s)$ of Eq. (4.4.39) has a peak amplitude given by

$$A_p = \frac{1}{2\zeta\sqrt{1 - \zeta^2}}$$

which occurs at the resonant frequency

$$\omega_p = \omega_n\sqrt{1 - \zeta^2}$$

4-15. Consider a system defined by the transfer function

$$G(s) = \frac{K}{(\tau_1 s - 1)(\tau_2 s + 1)}$$

with $K > 0$ and $\tau_1 > \tau_2 > 0$.

(a) Sketch a Bode diagram (both the magnitude and the phase plots) of this system.

(b) Sketch a polar plot of this system.

4-16. Show that a polar plot of a first-order pole factor

$$G(s = j\omega) = \frac{1}{1 + j\omega\tau}$$

as defined by Eq. (4.5.6) traces a semicircular arc of radius 0.5 centered at $+0.5$ in the complex $G(j\omega)$-plane, starting at $+1$ when $\omega = 0$ and terminating at 0 as $\omega \to \infty$.

4-17. Consider the series RLC electrical circuit depicted in Figure 2.2. For what relative values of R, L, and C will the system be characterized by a quadratic-pole factor? In such cases, express ζ and ω_n as functions of R, L, and C.

4-18. Explain why the Bode diagram of the third-order system defined by the $G(s)$ of Example 4.4.11 is similar to that associated with the first-order differential operator system:

$$Dy(t) + 3y(t) = 3u(t)$$

4-19. Consider a minimum phase system whose asymptotic amplitude frequency response is depicted in Figure 4.28.

(a) Determine the transfer function $G(s)$ of this system.

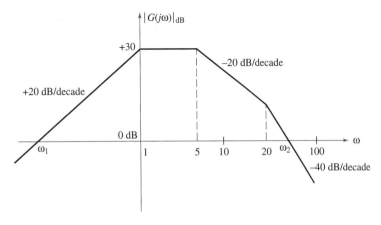

● **FIGURE 4.28**
An Asymptotic Amplitude Bode Plot for Problem 4-19

(b) Determine the two 0 dB "crossover frequencies," ω_1 and ω_2.

4-20.* Consider a system defined by the transfer function

$$G(s) = \frac{s + 100}{s^3 + 11.1s^2 + 11.1s + 1}$$

Express $G(s)$ in corner frequency factored form, and determine the *approximate* phase angles $\angle G(j\omega)$ at $\omega = 0$, 0.1, 1, 10, 100, and ∞. Compare your approximate values to the actual phase angle values.

4-21.* Sketch a polar plot of a system defined by the nonminimum phase transfer function

$$G(s) = \frac{0.1s - 1}{s(s + 1)(0.01s + 1)}$$

and determine the value of ω at which it crosses the positive real axis in the $G(j\omega)$-plane.

4-22. If the input $u(t) = 3 \sin(t + 1.57) + \cos 2t$ is applied to a system defined by the transfer function

$$G(s) = \frac{2e^{-s\tau}}{2s + 1}$$

determine the steady-state forced output response $y_f(t)$ when the transport lag $\tau = 2$.

4-23. Consider a state-space system defined by the state matrix

$$A = \begin{bmatrix} 0 & 1 & 0 & 0 \\ 0 & 0 & -\alpha & 0 \\ 0 & 0 & 0 & 1 \\ 0 & 0 & \beta & 0 \end{bmatrix}$$

with both $\alpha > 0$ and $\beta > 0$, as in the case of the linearized equations of motion of the inverted pendulum of Problem 2-5. Show that the system is unstable, since one of the eigenvalues of A (one of its poles) lies in the unstable half-plane $Re(s) > 0$. Determine the explicit locations of all of the system poles as functions of the parameters α and β.

PART II

PERFORMANCE
GOALS AND TESTS

Courtesy of Chrysler Corporation, Jefferson North Assembly Plant—producer of Jeep® Grand Cherokee

5

GENERAL CONTROL
AND NOMINAL
STABILITY

5.1 THE GENERAL CONTROL PROBLEM

In Part I of the text we focused on various ways the behavior of dynamic systems can be described or approximated by linear models. In many cases, such systems must be controlled if they are to produce acceptable "performance." We begin the second part of the text by presenting some of the reasons for control or compensation, as well as by describing certain of the difficulties that can be encountered when attempting to design appropriate controllers.

To begin, consider the "general control problem" represented in Figure 5.1. The system, whose defined output is to be controlled, is referred to as the **plant.** A plant can represent a variety of physical systems, such as an electronic circuit, an orbiting satellite, a motor driven cart that balances an inverted pendulum, or a diesel fuel injection system. In such cases, the defined output $y(t)$ may represent the charge on a capacitor or the current through an inductor, the altitude of the satellite, the angular offset of the pendulum, or the flow rate of diesel fuel. Moreover, a physical plant may be subjected to a variety of external disturbances, such as wind, waves, gravity, and temperature fluctuations, that can adversely affect $y(t)$.

Generally speaking, the overall **performance** of a controlled system or plant is based on the ability of its output $y(t)$ to robustly respond to the external, reference input $r(t)$, which is identically zero in the case of *regulation* and nonzero in the case of *tracking,* despite possible changes in the plant parameters or unmodeled dynamics, as well as the presence

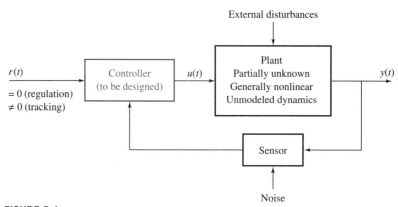

● **FIGURE 5.1**
The General Control System

of external disturbances or noisy measurements. Regulation implies the ability to maintain a constant output value, for example, to balance an inverted pendulum, while tracking implies the ability of the output $y(t)$ to equal or track a time-varying reference input $r(t)$, such as the constant acceleration lift-off of a missile.

In general, we will assume that the plant (output) to be controlled can be measured by some physical *sensor,* which may be corrupted by an external *noise disturbance,* $\eta(t)$. The output measurement is subsequently utilized in a *feedback* configuration in which both it and the reference input "drive" an appropriately designed controller whose output $u(t)$ then represents the *actuating signal,* or the physical control input to the plant.

In our subsequent discussions, we will assume that the dynamic behavior of the plant can be represented by a known, linear time-invariant system. However, it should be noted that such a *nominal representation* only approximates reality and that the dynamic behavior of an actual physical plant generally is not known entirely. The actual plant is often characterized by various nonlinearities that restrict its assumed linear range of operation and, in many cases, certain of its parameters are either partially unknown or vary with time or operating conditions. Moreover, a physical system may be characterized by unmodeled dynamics that, if ignored, can have a detrimental effect on its controlled, closed-loop performance.

Some Simplifying Assumptions

To develop some general procedures for the design of controllers, we henceforth will assume that the dynamic behavior of the plant can be represented (or approximated with "sufficient" accuracy) by a linear time-invariant, nth order system, which is defined by a strictly proper, rational transfer function

$$G(s) = \frac{y(s)}{u(s)} = \frac{c(s)}{a(s)} \qquad (5.1.1)$$

called the **nominal plant**, with $a(s)$ a known polynomial of degree n and $c(s)$ a known polynomial of degree strictly less than n:

$$\deg[c(s)] < n = \deg[a(s)] \qquad (5.1.2)$$

We will assume that $a(s)$ and $c(s)$ are coprime, so that any nth order state-space or differential operator realization of $G(s)$ will be both controllable and observable.

We will also assume that any *external disturbances* that act on the plant can be represented by a single, output-additive signal $d(t)$, with known dynamic properties, or its Laplace transform $d(s)$, which may depend on the plant if, for example, the physical disturbance enters the plant prior to its output. Moreover, sensor dynamics will be ignored here,[1] and we will assume that the sensor produces a continuous measure of the potentially noisy output $y(t) - \eta(t)$.

We will also assume that the controller can be represented by a linear time-invariant system, whose single output $u(t)$ is produced by the two inputs $r(t)$ and $y(t) - \eta(t)$. Therefore, its dynamic behavior can be described, in the most general case, by a (1×2) transfer vector:

$$u(s) = \frac{[q(s), \ -h(s)]}{k(s)} \begin{bmatrix} r(s) \\ y(s) - \eta(s) \end{bmatrix} \qquad (5.1.3)$$

with

$$\deg[q(s), -h(s)] \le \deg[k(s)] \qquad (5.1.4)$$

where $k(s)$ is a monic polynomial whose degree defines the order of the controller.

The assumption Eq. (5.1.4) of a proper compensator is made to avoid the high-frequency noise amplification problems, as well as possible plant input saturation, that can be caused by the derivative terms associated with improper transfer functions (see Problem 5-6). Although the nominal plant polynomials $a(s)$ and $c(s)$ are known, those that define the compensator are not. Indeed, the primary goal of this text will be the determination of the compensator polynomials $h(s)$, $k(s)$, and $q(s)$ that produce the "best" overall controlled performance.

In light of Eq. (2.7.17), we now define a *partial state* $z(t)$ for the plant, whose dynamic behavior can be described by a Laplace transformed differential operator representation in controllable canonical form, namely,

$$a(s)z(s) = u(s); \qquad y(s) = c(s)z(s) + d(s) \qquad (5.1.5)$$

---

[1] Sensor dynamics can be included in the design process, as we will later show in Section 7.3.

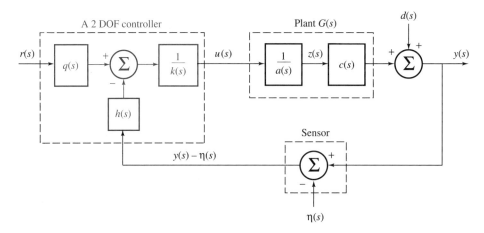

• **FIGURE 5.2**
The Nominal Control System

The general control problem depicted in Figure 5.1 can be represented by
the nominal closed-loop system depicted in Figure 5.2. The three external
signals that affect the dynamic behavior of the plant, $r(t)$, $d(t)$, and $\eta(t)$,
are called the **exogenous inputs.**

The Transfer Matrix

The dynamic behavior of the nominal control system of Figure 5.2 normally is characterized by the response of not only the defined output $y(t)$
but also the actuating signal $u(t)$, as well as the error $e(t)$ between the
reference input and the plant output, that is, $e(t) = r(t) - y(t)$, so that

$$e(s) = r(s) - y(s) \qquad (5.1.6)$$

The (3×3) transfer matrix that represents the exogenous input/response
behavior of the closed-loop system can now be obtained, in light of Figure 5.2, by the following Laplace transformed differential operator equations:

$$a(s)k(s)z(s) = q(s)r(s) - h(s)[y(s) - \eta(s)] \qquad (5.1.7)$$

$$y(s) = c(s)z(s) + d(s) \qquad (5.1.8)$$

If we substitute $c(s)z(s) + d(s)$ for $y(s)$ in Eq. (5.1.7), it follows that

$$[a(s)k(s) + c(s)h(s)]z(s) = [q(s), \ -h(s), \ h(s)] \begin{bmatrix} r(s) \\ d(s) \\ \eta(s) \end{bmatrix} \qquad (5.1.9)$$

with

$$\begin{bmatrix} y(s) \\ u(s) \\ e(s) \end{bmatrix} = \begin{bmatrix} c(s) \\ a(s) \\ -c(s) \end{bmatrix} z(s) + \begin{bmatrix} 0 & 1 & 0 \\ 0 & 0 & 0 \\ 1 & -1 & 0 \end{bmatrix} \begin{bmatrix} r(s) \\ d(s) \\ \eta(s) \end{bmatrix} \tag{5.1.10}$$

Therefore, looking back at Eq. (3.1.26),

$$\begin{bmatrix} y(s) \\ u(s) \\ e(s) \end{bmatrix} = \left\{ \begin{bmatrix} c(s) \\ a(s) \\ -c(s) \end{bmatrix} \frac{[q(s),\ -h(s),\ h(s)]}{a(s)k(s) + c(s)h(s)} + \begin{bmatrix} 0 & 1 & 0 \\ 0 & 0 & 0 \\ 1 & -1 & 0 \end{bmatrix} \right\} \begin{bmatrix} r(s) \\ d(s) \\ \eta(s) \end{bmatrix}$$

$$\tag{5.1.11}$$

or

$$\begin{bmatrix} y(s) \\ u(s) \\ e(s) \end{bmatrix} = \frac{\begin{bmatrix} c(s)q(s) & a(s)k(s) & c(s)h(s) \\ a(s)q(s) & -a(s)h(s) & a(s)h(s) \\ a(s)k(s)+c(s)[h(s)-q(s)] & -a(s)k(s) & -c(s)h(s) \end{bmatrix} \begin{bmatrix} r(s) \\ d(s) \\ \eta(s) \end{bmatrix}}{a(s)k(s) + c(s)h(s)}$$

$$\tag{5.1.12}$$

which defines the (3×3) *transfer matrix* associated with the nominal control system of Figure 5.2.

The poles of the closed-loop system or the *closed-loop poles* of the system are given by the roots of the denominator polynomial of Eq. (5.1.12), namely, $a(s)k(s) + c(s)h(s)$, which is defined as $\delta(s)$, that is,

$$\delta(s) \stackrel{\text{def}}{=} a(s)k(s) + c(s)h(s) \tag{5.1.13}$$

Using Eqs. (5.1.2) and (5.1.4), $\deg[c(s)h(s)] < \deg[a(s)k(s)]$, so that $\delta(s)$ is a well-defined monic polynomial with

$$\deg[\delta(s)] = \deg[a(s)k(s)] = n + \deg[k(s)] \tag{5.1.14}$$

This relationship, together with those in Eqs. (5.1.2) and (5.1.4), ensure that the closed-loop system depicted in Figure 5.2 is **well-posed**—all nine transfer function elements of Eq. (5.1.12) are both nonzero and proper.

The Primary Transfer Functions

The overall performance of the system depicted in Figure 5.2 depends on the ability of its output $y(t)$ to accurately track the reference input $r(t)$ while simultaneously minimizing the effect of both a potential disturbance signal $d(t)$ and possible sensor noise $\eta(t)$ on its behavior.[2] Therefore, the transfer function relationships between $y(s)$ and the exogenous inputs $r(s)$, $d(s)$, and $\eta(s)$, which constitute the first row of Eq. (5.1.12), are of

---

[2]External disturbances or noisy measurements are not present or relevant in all situations, and their effect often can be ignored.

primary concern in the design of an appropriate controller. These three transfer functions are defined as the **output response transfer function**,

$$T(s) \overset{\text{def}}{=} \frac{y(s)}{r(s)} = \frac{c(s)q(s)}{\delta(s)} \tag{5.1.15}$$

the **sensitivity (transfer) function**,

$$S(s) \overset{\text{def}}{=} \frac{a(s)k(s)}{\delta(s)} = \frac{y(s)}{d(s)} \tag{5.1.16}$$

and the **complementary sensitivity (transfer) function**

$$C(s) \overset{\text{def}}{=} \frac{c(s)h(s)}{\delta(s)} = \frac{y(s)}{\eta(s)} \tag{5.1.17}$$

respectively. We ideally would like $T(s) = 1$ and $S(s) = C(s) = 0$. However, this is both mathematically and physically impossible.

The **input response transfer function**, which reflects the effect of the external input $r(t)$ on the plant input $u(t)$, namely,

$$\frac{u(s)}{r(s)} = \frac{a(s)q(s)}{\delta(s)} \tag{5.1.18}$$

is also important, since it determines the amplitude response of $u(t)$ that results from an applied input reference signal $r(t)$.

Finally, the **error response transfer function**, namely,

$$\frac{e(s)}{r(s)} = \frac{a(s)k(s) + c(s)[h(s) - q(s)]}{\delta(s)} \tag{5.1.19}$$

which reflects the effect of the reference input on the error, also is of concern, since a zero steady-state error

$$e_{ss}(t) = \lim_{t \to \infty} \{e(t) = r(t) - y(t)\} = 0 \tag{5.1.20}$$

is often desired. The five transfer functions defined by the relationships in Eqs. (5.1.15) through (5.1.19) will be referred to as the **primary transfer functions**.

Note that only two of these transfer functions, Eqs. (5.1.16) and (5.1.17), are independent of the polynomial $q(s)$, which is loop independent. These two transfer functions have been given specific designations, $S(s)$, the so-called sensitivity function, and $C(s)$, the so-called complementary sensitivity function, for reasons that will be discussed in the next chapter. At this point, however, it can be noted that $C(s)$ is called the "complementary" sensitivity function because

$$S(s) + C(s) = \frac{a(s)k(s)}{\delta(s)} + \frac{c(s)h(s)}{\delta(s)} = \frac{\delta(s)}{\delta(s)} = 1 \tag{5.1.21}$$

As a consequence of Eq. (5.1.21), the choice of a particular sensitivity function $S(s)$ directly implies a corresponding unique choice for the complementary sensitivity function

$$C(s) = 1 - S(s) \tag{5.1.22}$$

and vice versa. This observation underlies a most important factor in loop transfer function design, namely the "tradeoffs" that must be made when selecting these two mutually dependent transfer functions.

One- and Two-Degree-of-Freedom Configurations

The particular controller design depicted in Figure 5.2 is called a **two-degree-of-freedom** (2 DOF) configuration because $k(s)u(s)$ is an independent function of the two compensator inputs, $r(s)$ and $y(s) - \eta(s)$. It may be noted that if the reference input $r(s)$ and $y(s) - \eta(s)$ are the only two compensator inputs, and the plant input $u(s)$ is its only output, then a *three polynomial, 2 DOF configuration represents the most general possible linear controller* (see Problem 5-4).

In many cases involving more traditional, or so-called classical control techniques, $q(s)$ is assumed to be equal to $h(s)$. This implies the more restrictive unity feedback configuration depicted in Figure 5.3, which is characterized by a one-degree-of-freedom (1 DOF) forward loop compensator

$$H(s) \stackrel{\text{def}}{=} \frac{h(s)}{k(s)} \qquad (5.1.23)$$

Note that the selection of $q(s) = k(s)$ would result in another 1 DOF configuration, with $H(s)$ in the feedback path. Also, the choice of $q(s) = 1$ would imply yet another 1 DOF configuration. In the vast majority of 1 DOF cases, however, the unity feedback configuration is preferred, because the error $e(s) = r(s) - y(s)$ is explicitly present both to drive the controller and to be zeroed via feedback. Therefore, unless stated otherwise, any future reference to 1 DOF compensation will imply the explicit error, unity feedback configuration of Figure 5.3. In both the

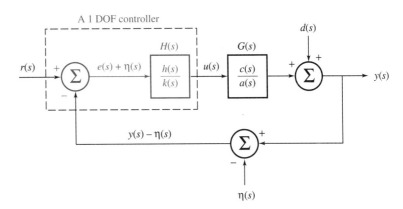

● **FIGURE 5.3**
The 1 DOF, Unity Feedback Configuration

1 DOF and 2 DOF cases, the *compensator order* is equal to the degree of $k(s)$, so that a physical implementation of the 2 DOF configuration of Figure 5.2 requires the same number of integrators as the 1 DOF design.

The 1 DOF Restriction

Since $q(s) = h(s)$ in the 1 DOF cases, all five of the primary transfer functions can be expressed in terms of the loop transfer functions $S(s)$ and $C(s)$, as defined by Eqs. (5.1.16) and (5.1.17). In view of Eq. (5.1.19), the error response transfer function is

$$\frac{e(s)}{r(s)} = \frac{a(s)k(s)}{\delta(s)} = S(s) \tag{5.1.24}$$

and in view of Eq. (5.1.15), the output response transfer function is

$$T(s) = \frac{y(s)}{r(s)} = \frac{c(s)h(s)}{\delta(s)} = C(s) \tag{5.1.25}$$

so that the complementary sensitivity function $C(s) = T(s)$ in the unity feedback cases.

Moreover, since

$$\frac{y(s)}{u(s)} = G(s) \implies \frac{u(s)}{y(s)} = \frac{1}{G(s)} = \frac{a(s)}{c(s)}$$

Eq. (5.1.18) implies that the input response transfer function

$$\frac{u(s)}{r(s)} = \frac{y(s)}{r(s)}\frac{u(s)}{y(s)} = \frac{C(s)}{G(s)} = \frac{a(s)h(s)}{\delta(s)} \tag{5.1.26}$$

Therefore, in the 1 DOF case, the choice of either $S(s) = 1 - C(s)$ or $C(s) = 1 - S(s)$ directly implies all five primary transfer functions, an observation that will be referred to as the **1 DOF restriction**.

Although the 1 DOF configuration is more restrictive than the 2 DOF configuration, there is a "simplicity-of-design" benefit associated with the 1 DOF case, because only a single transfer function compensator, $H(s) = \frac{h(s)}{k(s)}$, need be determined. Many of the more traditional, classical control techniques assume a unity feedback (1 DOF) configuration, because such designs often result in acceptable system performance, and they are simpler to design. Furthermore, certain 1 DOF controller designs can be implemented directly from the frequency response characteristics of the plant. Chapter 8 will focus on some of these more traditional control system designs.

The Loop Gain and the Return Difference

In both the 1 DOF and the 2 DOF cases, the **loop gain $L(s)$** is defined as the transfer function product (the net "gain") around the entire feedback loop; that is, in light of Figure 5.2 and Figure 5.3,

$$L(s) \stackrel{\text{def}}{=} \frac{c(s)h(s)}{a(s)k(s)} = G(s)H(s) \tag{5.1.27}$$

is a transfer function whose zeros [the roots of $c(s)h(s)$] represent the *loop zeros* of the system and whose poles [the roots of $a(s)k(s)$] represent the *open-loop poles* of the system.

Using Eqs. (5.1.16) and (5.1.17), both the sensitivity function and the complementary sensitivity function can be expressed directly in terms of the loop gain $L(s)$:

$$S(s) = \frac{1}{1 + L(s)} = \frac{a(s)k(s)}{\delta(s)} \tag{5.1.28}$$

and

$$C(s) = \frac{L(s)}{1 + L(s)} = \frac{c(s)h(s)}{\delta(s)} \tag{5.1.29}$$

Also, observe that the zeros of $1 + L(s)$, which Bode [7] termed the **return difference**, are equal to the roots of $\delta(s)$, the closed-loop poles of the system, while the poles of the return difference correspond to the open-loop poles of the system; that is,

$$1 + L(s) = 1 + \frac{c(s)h(s)}{a(s)k(s)} = \frac{a(s)k(s) + c(s)h(s)}{a(s)k(s)} = \frac{\delta(s)}{a(s)k(s)} \tag{5.1.30}$$

is the ratio of the closed-loop poles of the system to its open-loop poles. Since the sensitivity function $S(s)$ is equal to the reciprocal of the return difference, $S(s)$ represents the ratio of the open-loop poles of the system to its closed-loop poles.

Nominal Closed-Loop Stability

The polynomial $\delta(s) = a(s)k(s) + c(s)h(s)$ represents the denominator of all nine transfer functions associated with the nominal control system depicted in Figure 5.2. Therefore, the s-plane locations of the roots of $\delta(s)$, which correspond to the closed-loop poles of the system, are of primary importance in both control system analysis and design, and various procedures have been developed for determining their locations not only when $G(s)$ and $H(s)$ are fixed but also when one or more parameters of these transfer functions vary.

The remaining sections of this chapter will outline three well-known techniques that can be employed to determine the s-plane locations of the roots of a polynomial, such as $\delta(s)$, in both the fixed and the variable parameter cases. A primary goal of all three of these procedures will be to establish **nominal closed-loop stability**, which implies that $\delta(s)$ is a stable polynomial or, equivalently, that all of the roots of $\delta(s)$ lie in the stable half-plane $Re(s) < 0$. However, since the specific closed-loop pole locations of a system also affect its transient response, an alternative goal will be to determine certain response characteristics of the system as well.

5.2   THE ROUTH-HURWITZ STABILITY CRITERION

As we noted in the previous section, a primary goal associated with any control system design is nominal stability, which implies that all of the poles of a system lie in the stable half-plane $Re(s) < 0$. Determining the locations of all of the system poles or, without loss of generality, all of the roots of an nth degree polynomial such as

$$a(s) = a_n s^n + a_{n-1} s^{n-1} + \cdots + a_1 s + a_0 \qquad (5.2.1)$$

usually requires a digital computer if $n > 3$. Note, however, that it is not necessary to determine the exact pole locations in order to ensure stability if it can be established "in some manner" that none them lie in the unstable half-plane $Re(s) > 0$. This observation serves as a basis for the remainder of this section.

A Necessary Condition for Stability

We note first that it is possible to make certain general statements about the root locations of $a(s)$. In particular, if $a_0 \neq 0$ in Eq. (5.2.1), $a(s)$ can be factored as the product of a_0 times a number of first-order and quadratic factors, as defined by Eqs. (4.4.2) and (4.4.3), respectively. Moreover, ω_τ, ω_n, and ζ in each of these factored expressions will be positive whenever the roots of the factor lie in the stable half-plane $Re(s) < 0$. It therefore follows that *a necessary (but not sufficient) condition for $a(s)$ to be a stable polynomial is that all of its coefficients $a_i \neq 0$ and have the same sign.* Note that this necessary condition for *absolute* (yes or no) *stability* can be determined directly by inspection of $a(s)$.

A Necessary and Sufficient Condition for Stability

In the late 1800s, E. J. Routh [55] and A. Hurwitz [35] published independent results relative to the determination of the "absolute locations" of the roots of a given polynomial from knowledge of its real coefficients. Their findings will now be presented without formal proof.

First we employ the coefficients of

$$a(s) = a_n s^n + a_{n-1} s^{n-1} + \cdots + a_1 s + a_0$$

to define the following **Routh-Hurwitz array**:

$$
\begin{array}{ll}
s^n & \text{row}: \quad a_n \;\; a_{n-2} \;\; a_{n-4} \;\; a_{n-6} \;\cdots \\
s^{n-1} & \text{row}: \; a_{n-1} \;\; a_{n-3} \;\; a_{n-5} \;\; a_{n-7} \;\cdots \\
s^{n-2} & \text{row}: \; b_{n-2} \;\; b_{n-4} \;\; b_{n-6} \;\;\cdots \\
s^{n-3} & \text{row}: \; c_{n-3} \;\; c_{n-5} \;\; c_{n-7} \;\;\cdots \\
& \quad\quad \vdots \quad\quad \vdots \quad\quad \vdots \\
s^1 & \text{row}: \quad d_1 \\
s^0 & \text{row}: \quad e_0
\end{array}
$$

Note that the elements that comprise the s^n and the s^{n-1} rows of the array are appropriately ordered coefficients of $a(s)$. The third (s^{n-2}) row elements are then formed from its preceding two rows as follows:

$$
b_{n-2} = -\det\begin{bmatrix} a_n & a_{n-2} \\ a_{n-1} & a_{n-3} \end{bmatrix} \div a_{n-1} = \frac{a_{n-1}a_{n-2} - a_n a_{n-3}}{a_{n-1}}
$$

$$
b_{n-4} = -\det\begin{bmatrix} a_n & a_{n-4} \\ a_{n-1} & a_{n-5} \end{bmatrix} \div a_{n-1} = \frac{a_{n-1}a_{n-4} - a_n a_{n-5}}{a_{n-1}}
$$

$$
b_{n-6} = -\det\begin{bmatrix} a_n & a_{n-6} \\ a_{n-1} & a_{n-7} \end{bmatrix} \div a_{n-1} = \frac{a_{n-1}a_{n-6} - a_n a_{n-7}}{a_{n-1}}, \dots \quad (5.2.2)
$$

The fourth (s^{n-3}) row elements are then formed from its preceding two rows as in Eq. (5.2.2). In particular,

$$
c_{n-3} = -\det\begin{bmatrix} a_{n-1} & a_{n-3} \\ b_{n-2} & b_{n-4} \end{bmatrix} \div b_{n-2} = \frac{b_{n-2}a_{n-3} - a_{n-1}b_{n-4}}{b_{n-2}}
$$

$$
c_{n-5} = -\det\begin{bmatrix} a_{n-1} & a_{n-5} \\ b_{n-2} & b_{n-6} \end{bmatrix} \div b_{n-2} = \frac{b_{n-2}a_{n-5} - a_{n-1}b_{n-6}}{b_{n-2}}, \dots (5.2.3)
$$

Each subsequent row of the $(n+1)$-row Routh-Hurwitz array is formed from its preceding two rows in an analogous fashion. Zeros can be inserted for undefined elements, as required, and any row can be multiplied or divided by a positive scalar to simplify the computation of a subsequent row. The completed array will be triangular in form, with the final two (s^1 and s^0) rows having only one term each, which have been designated here as d_1 and e_0.

Once the Routh-Hurwitz array has been determined, the **Routh-Hurwitz Stability Criterion** can be stated as follows:

> The number of roots of $a(s)$ with positive real parts is equal to the number of sign changes in the first column of the Routh-Hurwitz array.

For example, a pattern of $+$, $-$, $+$ represents two sign changes, hence two unstable roots. It therefore follows that $a(s)$ will be a stable polynomial if and only if a_n, a_{n-1}, b_{n-2}, c_{n-3}, ..., d_1, and e_0 all have the same sign and are nonzero.

. . . .

EXAMPLE 5.2.4 To illustrate the Routh-Hurwitz stability criterion, consider a sixth-order system whose ($n = 6$) poles correspond to the roots of

$$a(s) = s^6 + 5s^5 + 15s^4 + 55s^3 + 154s^2 + 210s + 100$$

Note that the necessary condition for stability is satisfied since all of the coefficients of $a(s)$ are present and have the same sign (positive).

We determine next that the first two rows of the Routh-Hurwitz array are given by

$$
\begin{array}{cccccc}
s^6 & \text{row}: & 1 & 15 & 154 & 100 \\
s^5 & \text{row}: & 5 & 55 & 210 & 0
\end{array}
$$

We now divide the s^5 row by 5 and replace it by the new expression

$$
\begin{array}{cccccc}
s^5 & \text{row}: & 1 & 11 & 42 & 0
\end{array}
$$

to simplify our subsequent computations, since the resulting $a_{n-1} = a_5 = 1$, rather than 5. In light of Eq. (5.2.2), the third, or s^4, row is then given by

$$
\begin{array}{cccc}
s^4 & \text{row}: & \underbrace{15 - 11}_{4} & \underbrace{154 - 42}_{112} & 100 & 0
\end{array}
$$

or if we divide by 4, the new expression is

$$
\begin{array}{cccccc}
s^4 & \text{row}: & 1 & 28 & 25 & 0
\end{array}
$$

The fourth, or s^3, row is now given by Eq. (5.2.3), that is,

$$
\begin{array}{cccc}
s^3 & \text{row}: & \underbrace{11 - 28}_{-17} & \underbrace{42 - 25}_{17} & 0
\end{array}
$$

or[3] if we divide by 17, the new expression is

$$
\begin{array}{ccccc}
s^3 & \text{row}: & -1 & 1 & 0
\end{array}
$$

The final three rows of the array are now determined to be given by

$$
\begin{array}{cccc}
s^2 & \text{row}: & \underbrace{(-28 - 1) \div -1}_{29} & 25
\end{array}
$$

$$
\begin{array}{ccc}
s^1 & \text{row}: & \underbrace{(29 + 25) \div 29}_{1.862}
\end{array}
$$

$$
\begin{array}{ccc}
s^0 & \text{row}: & 25
\end{array}
$$

---

[3]If absolute stability is the only concern, the Routh-Hurwitz procedure can be terminated at this time, because -17 represents a first-column sign change. However, by completing the entire array, the actual number of unstable poles can be determined.

Therefore, the first column of the resulting Routh-Hurwitz array is given by

$$
\begin{array}{ll}
s^6 & \text{row}: +1 \\
s^5 & \text{row}: +1 \\
s^4 & \text{row}: +1 \\
s^3 & \text{row}: -1 \\
s^2 & \text{row}: +29 \\
s^1 & \text{row}: +1.862 \\
s^0 & \text{row}: +25
\end{array}
$$

Since there are exactly two sign changes in the first-column elements, the first due to the sign change from +1 to −1 and the second due to the sign change from −1 to +29, it follows that exactly two roots of $a(s)$ lie in the unstable half-plane $Re(s) > 0$. We may finally verify that this is indeed the case, since $a(s)$ can be factored as

$$a(s) = (s - 1 \pm j3)(s + 1)(s + 2)(s + 2 \pm j)$$

so that its roots lie at $s = +1 \pm j3$, -1, -2, and $-2 \pm j$ in the complex s-plane.

· ·

A Solitary Zero First-Column Element

In applying the Routh-Hurwitz procedure, it is possible to obtain a zero first-column element, which implies one or more roots with nonnegative real parts. Two special cases associated with such an occurrence will now be discussed. In particular, if the first-column term in any row is zero, but there are nonzero remaining terms, then one or more roots of $a(s)$ will have positive real parts. In such cases, the zero can be replaced by an arbitrarily small positive or negative constant ϵ and the procedure continued. Alternatively, $a(s)$ can be multiplied by a simple factor, such as $(s + \alpha)$, and the resulting polynomial $(s + \alpha)a(s)$ subsequently tested using the Routh-Hurwitz procedure (see Problem 5-9).

· ·

EXAMPLE 5.2.5 To illustrate this situation, consider a fourth-order system whose ($n = 4$) poles correspond to the roots of

$$a(s) = -2s^4 - 4s^3 - 4s^2 - 8s - 1$$

Note again that the necessary condition for stability is satisfied since all of the coefficients of $a(s)$ are present and have the same sign (negative). We determine next that the first two rows of the Routh-Hurwitz array are given by

$$
\begin{array}{lccc}
s^4 & \text{row}: & -2 & -4 & -1 \\
s^3 & \text{row}: & -4 & -8 & 0
\end{array}
$$

In light of Eq. (5.2.2), the third, or s^2, row is given by

$$
\begin{array}{lcc}
s^2 & \text{row}: & \underbrace{(16 - 16) \div -4}_{0} & \quad -1
\end{array}
$$

The 0 first-column element is now replaced by ϵ, which implies the new expression

$$s^2 \quad \text{row}: \qquad \epsilon \qquad -1$$

The final two rows of the array are then determined to be given by

$$s^1 \quad \text{row}: \qquad \underbrace{(-8\epsilon - 4) \div \epsilon}_{\approx -4\epsilon^{-1}}$$

$$s^0 \quad \text{row}: \qquad\qquad -1$$

Therefore, the entire first column of the resulting Routh-Hurwitz array is given by

$$
\begin{array}{lll}
s^4 & \text{row}: & -2 \\
s^3 & \text{row}: & -4 \\
s^2 & \text{row}: & +\epsilon \\
s^1 & \text{row}: & -4\epsilon^{-1} \\
s^0 & \text{row}: & -1
\end{array}
$$

Since there are exactly two sign changes in the first column elements, *independent of whether ϵ is positive or negative,* it follows that exactly two roots of $a(s)$ lie in the unstable half-plane $Re(s) > 0$. We may finally verify that this is indeed the case, since $a(s)$ can be factored as

$$a(s) = -(s + 1.956)(s + 0.133)(s - 0.044 \pm j1.387)$$

so that its roots lie at $s = -1.956, -0.133$, and $+0.044 \pm j1.387$ in the complex s-plane.

An Entire Zero Row

A second special case associated with a zero first-column element is one in which all of the remaining row elements also are zero. This implies the presence of mirror-image roots relative to the imaginary axis or one or more pairs of imaginary-conjugate roots ($\pm j\omega$). In such cases, an *auxiliary polynomial $p_a(s)$* can be defined directly by the elements of the previous row. The zero-row elements can then be replaced by the coefficients of the derivative of the auxiliary polynomial, and the procedure continued. Alternatively, $p_a(s)$ can be factored out of $a(s)$, and the resulting, lower degree polynomial can be tested using the Routh-Hurwitz procedure.

EXAMPLE 5.2.6 To illustrate this situation, consider a sixth-order system whose ($n = 6$) poles correspond to the roots of

$$a(s) = s^6 + 2s^5 - 9s^4 - 12s^3 + 43s^2 + 50s - 75$$

which clearly is not a stable polynomial since it has both positive and negative coefficients. We next determine that the first two rows of the Routh-Hurwitz array are given by

$$
\begin{array}{lllll}
s^6 & \text{row}: & 1 & -9 & 43 & -75 \\
s^5 & \text{row}: & 2 & -12 & 50 & 0
\end{array}
$$

We next divide the s^5 row by 2 and replace it by the new expression

$$s^5 \quad \text{row}: \qquad 1 \quad -6 \quad 25 \quad 0$$

to simplify our subsequent computations. In light of Eq. (5.2.2), the s^4 row is now given by

$$s^4 \quad \text{row}: \qquad \underbrace{-9+6}_{-3} \quad \underbrace{43-25}_{18} \quad -75 \quad 0$$

or if we divide by 3, the new expression is

$$s^4 \quad \text{row}: \qquad -1 \quad 6 \quad -25 \quad 0$$

Since this s^4 row is the negative of the preceding s^5 row, the entire s^3 row will be zero in this case. At this point, we define the auxiliary polynomial

$$p_a(s) \overset{\text{def}}{=} -s^4 + 6s^2 - 25$$

using the s^4 row elements, noting that $p_a(s)$ divides $a(s)$:

$$a(s) = (-s^2 - 2s + 3)(-s^4 + 6s^2 - 25)$$

We can now replace the entire zero s^3 row by the coefficients of

$$\frac{d}{ds}\underbrace{(-s^4 + 6s^2 - 25)}_{p_a(s)} = -4s^3 + 12s, \,^4$$

which implies the resulting

$$s^3 \quad \text{row}: \qquad -4 \quad 12 \quad 0$$

or if we divide by 4, the new expression is

$$s^3 \quad \text{row}: \qquad -1 \quad 3 \quad 0$$

If we now continue the procedure, the final three rows of the array are given by

$$s^2 \quad \text{row}: \qquad \underbrace{(-6+3) \div -1}_{3} \quad -25$$

$$s^1 \quad \text{row}: \qquad \underbrace{(9-25) \div 3}_{-5.333}$$

$$s^0 \quad \text{row}: \qquad -25$$

---

[4]We can also determine the roots of $p_a(s) = -(s + 2 \pm j)(s - 2 \pm j)$, and apply the Routh-Hurwitz procedure to the resulting polynomial, $-s^2 - 2s + 3$.

Therefore, the entire first column of the resulting Routh-Hurwitz array is given by

$$
\begin{array}{lll}
s^6 & \text{row}: & +1 \\
s^5 & \text{row}: & +1 \\
s^4 & \text{row}: & -1 \\
s^3 & \text{row}: & -1 \\
s^2 & \text{row}: & +3 \\
s^1 & \text{row}: & -5.333 \\
s^0 & \text{row}: & -25
\end{array}
$$

Since there are three sign changes in the first column, it follows that three roots of $a(s)$ lie in the unstable half-plane $Re(s) > 0$. We may finally verify that this is indeed the case, since $a(s)$ can be factored as

$$
a(s) = \underbrace{(s + 2 \pm j)(s - 2 \pm j)}_{-p_a(s)}(s - 1)(s + 3)
$$

Therefore, the roots of $a(s)$ lie at the imaginary axis, mirror-image locations $s = \pm 2 + j$ and $\pm 2 - j$, which correspond to the roots of the auxiliary polynomial $p_a(s)$, as well as at $s = -3$ and $+1$.

If $a(s)$ has no roots in the unstable half-plane $Re(s) > 0$, then the subsequent determination of an entire zero row of the Routh-Hurwitz array implies one or more pairs of imaginary-conjugate roots ($\pm j\omega$). In such cases, the employment of the derivative of the auxiliary polynomial implies no sign changes in the first column of the resulting Routh-Hurwitz array, because there are no roots with positive real parts (see Problem 5-12). However, in such *marginally stable* situations, even slight changes in the nominal values of the coefficients of $a(s)$ can imply an unstable resulting polynomial.

We note finally that the Routh-Hurwitz test often is used to determine the range of variation of a parameter, such as a scalar loop gain K, for which a system will be stable. Example 5.3.30 and Problem 5-20 illustrate this observation.

5.3   THE ROOT LOCUS

In the previous section, we introduced the Routh-Hurwitz technique for determining the absolute (yes or no) stability of a given system from knowledge of a polynomial that defines its poles. In this section we will outline a procedure for determining the actual pole locations, or the so-called *relative stability* of a system. Many computer programs are now available that readily determine the exact locations of all n roots of a polynomial for any fixed set of its coefficients (see references [46] and [64]) and, as it happens, such a determination represents the starting point for the procedure that will be presented here.

In 1948, W. R. Evans [20] introduced the **root locus** as a graphic procedure for depicting the change in the complex s-plane locations (locus) of the roots of a polynomial as a function of a single variable parameter. The initial motivation for the root locus was to determine the variation in the closed-loop poles of a system characterized by a proper loop gain transfer function, such as[5]

$$L(s) = \frac{Km(s)}{a(s)} = K\frac{s^r + m_{r-1}s^{r-1} + \cdots + m_1 s + m_0}{s^n + a_{n-1}s^{n-1} + \cdots + a_1 s + a_0} \qquad (5.3.1)$$

as a function of an adjustable gain K. In view of Eq. (5.1.30), the closed-loop poles of such a system are given by the n zeros of the return difference

$$1 + L(s) = 1 + \frac{Km(s)}{a(s)} = \frac{a(s) + Km(s)}{a(s)} = \frac{\delta(s)}{a(s)}$$

or the n roots of the polynomial

$$\delta(s) = a(s) + Km(s) \qquad (5.3.2)$$

The root locus method can be used in other situations as well, for example, to plot the variation in the roots of $s^2 + 2\zeta\omega_n s + \omega_n^2$ as a function of a varying damping ratio ζ, as in Figure 4.6. In all root locus cases, an equation analogous to Eq. (5.3.2) will hold and will represent the base equation for a root locus plot. For example, note that a choice of $a(s) = s^2 + \omega_n^2$, $m(s) = 2\omega_n s$, and $K = \zeta$ will imply Eq. (5.3.2) when $\delta(s) = s^2 + 2\zeta\omega_n s + \omega_n^2$.

Some Root Locus Facts

In light of Eq. (5.3.2), we may first observe that since the coefficients of $a(s)$ and $m(s)$ are real, any complex roots of $\delta(s)$ will appear as symmetrical complex-conjugate pairs. Moreover, if $K = 0$ represents the starting point for an s-plane root locus plot, which usually is the case, then we have the first of our **root locus facts**.

Fact 1   A root locus plot of $\delta(s)$ starts at the n roots of $a(s)$ (the open-loop system poles) when $K = 0$, and it is symmetrical with respect to the real axis.

We note next that a root locus plot depicts those particular values s^* of s for which

$$\delta(s^*) = a(s^*) + Km(s^*) = 0 \qquad (5.3.3)$$

---

[5]Note that $L(s)$ is represented here as $\frac{Km(s)}{a(s)}$, rather than the $\frac{c(s)h(s)}{a(s)k(s)}$ given by Eq. (5.1.27). However, the results presented are independent of the particular transfer function description employed for $L(s)$.

or, those s^* for which

$$\frac{m(s^*)}{a(s^*)} = -\frac{1}{K} \tag{5.3.4}$$

Therefore, when $0 < K < \infty$, Eq. (5.3.4) implies both the *magnitude condition*:

$$\left|\frac{m(s^*)}{a(s^*)}\right| = \frac{1}{K} \tag{5.3.5}$$

and the **phase condition**:

$$\angle\frac{m(s^*)}{a(s^*)} = \pm(2q+1)180° \quad \text{for} \quad q = 0,\, 1,\, 2,\, \ldots \tag{5.3.6}$$

Clearly, the magnitude condition Eq. (5.3.5) can be satisfied for any value s^* of s by a $K = \frac{|a(s^*)|}{|m(s^*)|}$, so that *the phase condition defined by Eq. (5.3.6) represents the fundamental positive ($K > 0$) root locus relationship.*

In particular, if z_j for $j = 1, 2, \ldots, r$ represent the roots of $m(s)$, hence the r loop zeros of the system, and if p_i for $i = 1, 2, \ldots, n$ represent the roots of $a(s)$, hence the n poles of the open-loop system, then in view of Eq. (3.3.3),

$$\frac{m(s^*)}{a(s^*)} = \frac{(s^* - z_1)(s^* - z_2) \cdots (s^* - z_r)}{(s^* - p_1)(s^* - p_2) \cdots (s^* - p_n)} \tag{5.3.7}$$

Equations (5.3.7) and (4.4.16) then imply that Eq. (5.3.6) can be written as

$$\angle\frac{m(s^*)}{a(s^*)} = \angle(s^* - z_1) + \angle(s^* - z_2) + \cdots + \angle(s^* - z_r)$$

$$-\angle(s^*-p_1)-\angle(s^*-p_2)-\cdots-\angle(s^*-p_n) = \pm(2q+1)180° \tag{5.3.8}$$

We next note that for any complex quantity γ, $\angle(s^* - \gamma)$ is equal to the counterclockwise angle measured from a horizontal line drawn through and to the right of γ to the line drawn from γ to s^*. For example, if $\gamma = -2 - j$ and $s_1^* = -2 + j2$, then $\angle(s_1^* - \gamma) = \angle(j3) = +90°$, and if $s_2^* = 1 + j2$, then $\angle(s_2^* - \gamma) = \angle(3 + j3) = +45°$, as depicted in Figure 5.4.

If

$$\theta_{zj} = \angle(s^* - z_j) \quad \text{for} \quad j = 1, 2, \ldots, r \tag{5.3.9}$$

represent the angles measured from the loop zeros at $s = z_j$ to s^*, and if

$$\theta_{pi} = \angle(s^* - p_i) \quad \text{for} \quad i = 1, 2, \ldots, n \tag{5.3.10}$$

represent the angles measured from the open-loop poles at $s = p_i$ to s^*, then Eq. (5.3.8) implies that s^* will be a point on the positive ($K > 0$) root locus if and only if for some $q = 0, 1, 2, \ldots$

$$\theta_{z1} + \theta_{z2} + \cdots + \theta_{zr} - \theta_{p1} - \theta_{p2} - \cdots - \theta_{pn} = \pm(2q+1)180° \tag{5.3.11}$$

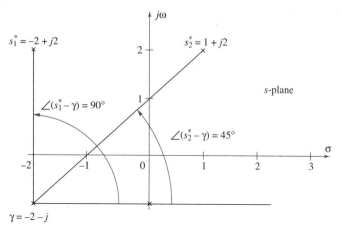

- **FIGURE 5.4**
A Graphic Angle Determination

The root locus usually begins with an s-plane plot of the open-loop system poles at $s = p_i$, whose locations are depicted by $\times$ symbols, and the loop zeros at $s = z_j$, whose locations are depicted by $\odot$ symbols, as in Figure 5.5. We can then determine those points on the real (σ) axis that belong to the positive root locus using Eq. (5.3.11). In particular, any complex-conjugate pairs of zeros or poles (such as the zeros at $s = z_1$ and z_2, or the poles at $s = p_3$ and p_4) will be irrelevant to this determination, since any such pairs contribute a net zero angle in Eq. (5.3.11) for any real values of s^*. For example, in the case depicted in Figure 5.5, $\theta_{z1} = -\theta_{z2}$ and $\theta_{p3} = -\theta_{p4}$ for all real values of s^*. Furthermore, any poles or zeros on the real axis contribute either $0°$ (if they lie to the left of a real s^*

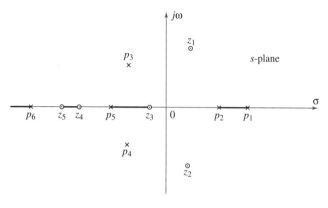

- **FIGURE 5.5**
A Real Axis (Positive) Root Locus Plot

point) or 180° (if they lie to the right of a real s^* point). In light of these observations, we now have the second of our root locus facts.

Fact 2   A real axis point s^* belongs to the positive root locus if the number of real poles plus the number of real zeros to the right of s^* is odd $[(2q + 1)$ for $q = 0, 1, 2, \ldots]$.

The appropriate regions of the real axis that belong to the positive root locus, as defined by Fact 2, are indicated by the black lines in Figure 5.5.

We consider next the angles at which the root locus leaves the individual open-loop poles at $s = p_i$ (when $K = 0+$), or the so-called **departure angles**. If s^* is a point on the positive root locus that is "very close" to the pole at $s = p_k$, then $\angle(s^* - p_k)$ defines the departure angle from p_k. To determine this angle, note that for such an s^*, each θ_{zj} in Eq. (5.3.11) will be approximately equal to the angle measured from the zero at $s = z_j$ to the pole at $s = p_k$. We will denote these r angles as θ_{zj}^{pk}. Moreover, each θ_{pi} in Eq. (5.3.11), other than θ_{pk}, will be approximately equal to the angle measured from the pole at $s = p_i$ to the pole at $s = p_k$. We will denote these $n - 1$ angles as θ_{pi}^{pk}. If we now denote $\angle(s^* - p_k)$, the departure angle from $s = p_k$, as θ_{pk}^{pk}, Eq. (5.3.11) implies the third of our root locus facts.

Fact 3   As $K \to 0+$, the departure angle from each open-loop system pole at $s = p_k$ is given by

$$\theta_{pk}^{pk} = \sum_{j=1}^{r} \theta_{zj}^{pk} - \sum_{i=1(\neq k)}^{n} \theta_{pi}^{pk} \mp (2q + 1)180° \qquad (5.3.12)$$

We consider next the root locus consequences as $K \to \infty$. If we write Eq. (5.3.3) as

$$\frac{a(s^*)}{K} + m(s^*) = 0 \qquad (5.3.13)$$

then as $K \to \infty$, those r values of s that zero $m(s)$ will also zero Eq. (5.3.13), so that r of the n closed-loop poles will terminate at the loop zeros as $K \to \infty$. If $n > r$, which is usually the case, the remaining $n - r$ poles will be defined by values of $s^* \to \infty$ along $n - r$ different angular directions or **asymptotes.**

For very large values of s, all of the open-loop poles and zeros will appear to be clustered on the real axis near the origin in the complex s-plane at some asymptotic starting point, which is called the **centroid** σ_c of the root locus. Since r zeros essentially "cancel" an equal number of poles near σ_c, it follows that for large values of s,

$$\frac{m(s)}{a(s)} \approx \frac{1}{(s - \sigma_c)^{n-r}} \qquad (5.3.14)$$

where

$$\frac{a(s)}{m(s)} \approx (s - \sigma_c)^{n-r} \approx s^{n-r} - (n - r)\sigma_c s^{n-r-1} \qquad (5.3.15)$$

using the first two terms of a binomial expansion.

For large values of s, Eq. (5.3.1) implies that

$$\frac{m(s)}{a(s)} \approx \frac{s^r + m_{r-1}s^{r-1}}{s^n + a_{n-1}s^{n-1}}$$

so that by long division,

$$\frac{a(s)}{m(s)} \approx \frac{s^n + a_{n-1}s^{n-1}}{s^r + m_{r-1}s^{r-1}} \approx s^{n-r} + (a_{n-1} - m_{r-1})s^{n-r-1} \qquad (5.3.16)$$

In light of Eq. (5.3.7), we next note that a negative sum of the roots of $m(s)$ and $a(s)$ define the s^{r-1} and s^{n-1} coefficients of $m(s)$ and $a(s)$, respectively, that is, $m_{r-1} = -\sum_1^r z_j$ and $a_{n-1} = -\sum_1^n p_i$. Therefore, for large values of s, Eq. (5.3.16) implies that

$$\frac{a(s)}{m(s)} \approx s^{n-r} + \left(\sum_1^r z_j - \sum_1^n p_i\right) s^{n-r-1} \qquad (5.3.17)$$

Equating the large-s approximations for $\frac{a(s)}{m(s)}$ given by Eqs. (5.3.15) and (5.3.17), we note finally that

$$-(n-r)\sigma_c = \left(\sum_1^r z_j - \sum_1^n p_i\right)$$

or that

$$\sigma_c = \frac{\sum_1^n p_i - \sum_1^r z_j}{n - r} \qquad (5.3.18)$$

Therefore, the *centroid* is given by the sum of the (open-loop) poles minus the sum of the zeros divided by the difference between the number of poles and zeros.

We note further that as $s^* \to \infty$, the phase angles associated with all of the θ_{zj} and all of the θ_{pi} are approximately equal to the same *asymptotic angles* θ_A, such as the one depicted in Figure 5.6; Eq. (5.3.11) implies that

$$r \times \theta_A - n \times \theta_A = (r - n)\theta_A = \pm(2q + 1)180° \quad \text{for} \quad q = 0, 1, 2, \ldots$$

Therefore, the $n - r$ asymptotic angles are given by the relationship

$$\theta_A = \frac{(2q + 1)180°}{n - r} \quad \text{for} \quad q = 0, 1, 2, \ldots, n - r - 1 \qquad (5.3.19)$$

which establishes the fourth of our root locus facts.

Fact 4   As $K \to \infty$, the n roots of $\delta(s)$ terminate at the r loop zeros and at ∞ along $n - r$ asymptotes that originate at σ_c on the real axis, as defined by Eq. (5.3.18), and $\to \infty$ at angular directions θ_A defined by Eq. (5.3.19).

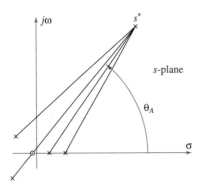

● **FIGURE 5.6**
A Large s^* Asymptotic Angle Determination

Our final root locus fact relates to the determination of **multiple root points** or, in light of Eq. (5.3.2), s^* values of s where, for certain values $\hat{K}$ of K,

$$\delta(s) = a(s) + \hat{K}m(s) = (s - s^*)^p \hat{\delta}(s), \qquad (5.3.20)$$

for some integer $p \geq 2$. If, for any function $f(s)$ of s,

$$f'(s) \overset{\text{def}}{=} \frac{df(s)}{ds}$$

then Eq. (5.3.20) implies that

$$\delta'(s) = a'(s) + \hat{K}m'(s) = p(s - s^*)^{p-1}\hat{\delta}(s) + (s - s^*)^p \hat{\delta}'(s) \quad (5.3.21)$$

Therefore, at a multiple point s^*, both

$$\delta(s^*) = a(s^*) + \hat{K}m(s^*) = 0 \qquad (5.3.22)$$

and

$$\delta'(s^*) = a'(s^*) + \hat{K}m'(s^*) = 0 \qquad (5.3.23)$$

for

$$\hat{K} = -\frac{a(s^*)}{m(s^*)} = -\frac{a'(s^*)}{m'(s^*)} \qquad (5.3.24)$$

which implies our fifth and final root locus fact.

Fact 5 If s^* is a multiple point on the positive root locus of $\delta(s)$, then

$$a(s^*)m'(s^*) - a'(s^*)m(s^*) = 0 \qquad (5.3.25)$$

Note that Eq. (5.3.25) represents a necessary (but not sufficient) condition for determining the multiple points on a root locus plot; that is, a root of $a(s)m'(s) - a'(s)m(s)$ need not represent a multiple point. However, all multiple points s^* must satisfy Eq. (5.3.25).

If the real axis region between two adjacent poles belongs to the positive root locus, then there will generally be a $p = 2$ multiple point between the poles. Such a point is called a **breakaway point** on the root locus, because the two poles usually will meet at the point (as $K \rightarrow \hat{K}$) and subsequently "break away" from one another as a complex-conjugate pair as K increases beyond $\hat{K}$. Conversely, if the real axis region between two adjacent zeros belongs to the positive root locus, then there will generally be a $p = 2$ multiple point between the zeros that is called a **break-in point** on the root locus, because two complex-conjugate poles will usually meet or "break in" to one another at the point (as $K \rightarrow \hat{K}$) and subsequently approach the zeros in opposite directions on the real axis as K increases beyond $\hat{K}$. Breakaway and break-in points are not restricted to the real axis (see Problem 5-16). Table 5.1 summarizes these positive root locus facts.

The Negative Root Locus

We will now extend our root locus results to include negative values of the gain K as well. In particular, if $K < 0$ in Eq. (5.3.4), then the $(K > 0)$ phase condition defined by Eq. (5.3.6) can be replaced by the $(K < 0)$ *phase condition*:

$$\angle \frac{m(s^*)}{a(s^*)} = \pm(2q)180° \quad \text{for} \quad q = 0, 1, 2, \ldots \quad (5.3.26)$$

If θ_{zj} and θ_{pi} remain as defined by Eqs. (5.3.9) and (5.3.10), respectively, it then follows that s^* will be a point on the negative $(K < 0)$ root locus if and only if for some $q = 0, 1, 2, \ldots$

$$\theta_{z1} + \theta_{z2} + \cdots + \theta_{zr} - \theta_{p1} - \theta_{p2} - \cdots - \theta_{pn} = \pm(2q)180° \quad (5.3.27)$$

As a consequence of the preceding, the positive root locus facts imply analogous negative root locus facts, once $(2q + 1)$ is replaced by $(2q)$.

More specifically, the positive root locus Facts 1 and 5 will hold, as stated, in the negative case as well. However, the positive root locus Facts 2, 3, and 4 will respectively imply the following in the negative case.

Fact 2N   A real axis point s^* belongs to the negative root locus if the number of real poles plus the number of real zeros to the right of s^* is even [$(2q)$ for $q = 0, 1, 2, \ldots$].

Fact 3N   As $K \rightarrow 0-$, the departure angle from each open-loop system pole at $s = p_k$ is given by

$$\theta_{pk}^{pk} = \sum_{j=1}^{r} \theta_{zj}^{pk} - \sum_{i=1(\neq k)}^{n} \theta_{pi}^{pk} \mp (2q)180° \quad (5.3.28)$$

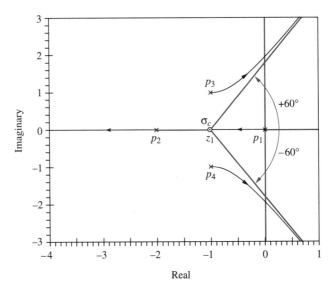

- **FIGURE 5.7**
  A $K > 0$ Root Locus Plot of $\delta(s) = s^4 + 4s^3 + 6s^2 + 4s + K(s + 1)$

Fact 4N   As $K \to -\infty$, the n roots of $\delta(s)$ terminate at the r loop zeros and at ∞ along $n - r$ asymptotes that originate at the centroid σ_c, as defined by Eq. (5.3.18), and $\to \infty$ at asymptotic angular directions defined by

$$\theta_A = \frac{(2q)180°}{n - r} \quad \text{for} \quad q = 0, 1, 2, \ldots, n - r - 1 \qquad (5.3.29)$$

Note that Facts 2 and 2N together imply that every real axis point belongs to either the positive or the negative root locus of a given $\delta(s)$ (see Problem 5-15).

....

EXAMPLE 5.3.30 To illustrate the root locus, consider a system defined by the loop gain transfer function

$$L(s) = \frac{Km(s)}{a(s)} = \frac{K(s + 1)}{s^4 + 4s^3 + 6s^2 + 4s} = \frac{K(s + 1)}{s(s + 2)(s + 1 \pm j)}$$

which is characterized by a single zero ($r = 1$) at $s = z_1 = -1$ and four ($= n$) poles at $s = p_1 = 0$, $p_2 = -2$, $p_3 = -1 + j$, and $p_4 = -1 - j$, which, in light of Fact 1, represent the starting points for the root locus plot. These s-plane locations are depicted by $\times$ symbols in Figure 5.7, while the zero at $s = -1$ is depicted by a $\odot$.

TABLE 5.1 Some Positive ($K > 0$) Root Locus Facts

The closed-loop poles of a system characterized by a proper loop gain transfer function

$$L(s) = \frac{Km(s)}{a(s)} = K\frac{s^r + m_{r-1}s^{r-1} + \cdots + m_1 s + m_0}{s^n + a_{n-1}s^{n-1} + \cdots + a_1 s + a_0}$$

with $r \le n$ and K adjustable, are given by the n roots of

$$\delta(s) = a(s) + Km(s)$$

An s-plane plot of the n roots of $\delta(s)$ as K varies continuously from 0 to $+\infty$ is called the **positive root locus**.

- **Fact 1:** A root locus plot of $\delta(s)$ starts at the n roots of $a(s)$ (the open-loop system poles) when $K = 0$, and it is symmetrical with respect to the real axis.

- **Fact 2:** A real axis point s^* belongs to the positive root locus if the number of real poles plus the number of real zeros to the right of s^* is odd [$(2q+1)$ for $q = 0, 1, 2, \ldots$].

- **Fact 3:** As $K \to 0+$, the departure angle from each open-loop system pole at $s = p_k$ is given by

$$\theta_{pk}^{pk} = \sum_{j=1}^{r} \theta_{zj}^{pk} - \sum_{i=1(\ne k)}^{n} \theta_{pi}^{pk} \mp (2q+1)180°$$

- **Fact 4:** As $K \to \infty$, the n roots of $\delta(s)$ terminate at the r loop zeros and at ∞ along $n - r$ asymptotes that originate at the centroid

$$\sigma_c = \frac{\sum_1^n p_i - \sum_1^r z_j}{n - r}$$

on the real axis and $\to \infty$ at asymptotic angular directions defined by

$$\theta_A = \frac{(2q+1)180°}{n - r} \quad \text{for} \quad q = 0, 1, 2, \ldots, n-r-1$$

- **Fact 5:** If s^* is a multiple point on the positive root locus of $\delta(s)$, then

$$a(s^*)m'(s^*) - a'(s^*)m(s^*) = 0$$

Fact 2 now implies that the real axis region between the pole at $p_1 = 0$ and the zero at $z_1 = -1$ as well as that between the pole at $p_2 = -2$ and $-\infty$ belong to the positive root locus. It is obvious from these real axis, root locus regions that the departure angle from both of the real poles (p_1 and p_2) is 180°.

Fact 3 can now be used to determine the departure angle from the pole at $s = p_3 = -1 + j$. In particular,

$$\theta_{p3}^{p3} = \underbrace{\theta_{z1}^{p3}}_{90°} - \underbrace{\theta_{p1}^{p3}}_{135°} - \underbrace{\theta_{p2}^{p3}}_{45°} - \underbrace{\theta_{p4}^{p3}}_{90°} + 180° = 0°$$

which is also the departure angle from the pole at $s = p_4 = -1 - j$; that is, by real axis symmetry, the sum of the departure angles from any two complex-conjugate poles equals $0°$.

Fact 4 now implies that the centroid of the root locus

$$\sigma_c = \frac{p_1 + p_2 + p_3 + p_4 - z_1}{n - r} = \frac{0 - 2 + (-1 + j) + (-1 - j) - (-1)}{4 - 1} = \frac{-3}{3} = -1$$

which represents the origin of the $n - r = 3$ asymptotes whose angular directions are equal to $60°$, $180°$, and $300°$, or $\pm 60°$ and $180°$. The positive root locus in this case can now be drawn, as in Figure 5.7. Note, in particular, that the pole at $p_1 = 0$ approaches the zero at $z_1 = -1$ as $K \to \infty$, while the remaining three poles $\to \infty$ in the defined asymptotic directions.

In light of Figure 5.7, it is clear that for lower values of the gain $K > 0$, the closed-loop system is stable. However, as K increases, the poles at p_3 and p_4 cross the imaginary axis and enter the unstable half-plane $Re(s) > 0$. The Routh-Hurwitz stability criterion can be employed to determine the value of K at which this occurs. In particular,

$$\delta(s) = s^4 + 4s^3 + 6s^2 + (4 + K)s + K$$

in this case, which implies the Routh-Hurwitz array:

s^4	row :	1	6	K
s^3	row :	4	$4 + K$	0
s^2	row :	$\dfrac{20 - K}{4}$	K	
s^1	row :	$4 + K - \dfrac{16K}{20 - K}$		
s^0	row :	K		

The first-column element in the s^2 row implies that $K < 20$ for stability. If we assume this to be the case, the first-column element in the s^1 row then implies that $(20 - K)(4 + K) - 16K = -K^2 + 80 > 0$, or that $K < \sqrt{80} = 8.944$ for stability. The final s^0 row implies that $K > 0$ as well. In summary, therefore, the closed-loop system will be stable provided $0 < K < 8.944$, and at $K = 8.944$ the root locus will cross the imaginary axis and enter the unstable half-plane.

We can verify this latter observation if we now solve for the particular value of $s^* = j\omega^*$ that zeros $\delta(s)$ when $K = 8.944$. In this case,

$$\delta(j\omega^*) = \omega^{*4} - 6\omega^{*2} + 8.944 - j(4\omega^{*3} - 12.944\omega^*) = 0$$

which implies that both the real and imaginary parts must equal zero, that is, both

$$\omega^{*4} - 6\omega^{*2} + 8.944 = 0 \quad \text{and} \quad 4\omega^{*3} - 12.944\omega^* = 0$$

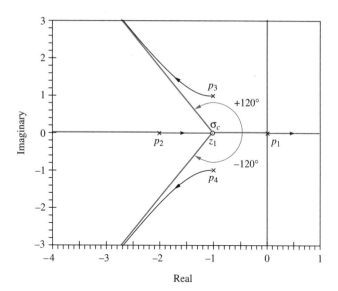

• **FIGURE 5.8**
  A $K < 0$ Root Locus Plot of $\delta(s) = s^4 + 4s^3 + 6s^2 + 4s + K(s+1)$

The latter imaginary relationship implies that $\omega^{*2} = 12.944 \div 4 = 3.236$, or that $\omega^* = 1.799$, a value of ω^* that zeros $\omega^{*4} - 6\omega^{*2} + 8.944$ as well. We thus conclude that the positive root locus crosses the imaginary axis at $\pm j1.799$ when $K = 8.944$.

The negative root locus for this example is depicted in Figure 5.8. Using Fact 2N, note that as $K \to -\infty$, the pole at $s = p_2 = -2$ approaches the zero at $z_1 = -1$, while the pole at $p_1 = 0 \to \infty$ along the positive real axis. As a consequence, every real axis point belongs to either the positive or the negative root locus, an observation that was noted earlier.

Fact 3N can now be used to determine the departure angle from the pole at $s = p_3 = -1 + j$. In particular,

$$\theta_{p3}^{p3} = \underbrace{\theta_{z1}^{p3}}_{90°} - \underbrace{\theta_{p1}^{p3}}_{135°} - \underbrace{\theta_{p2}^{p3}}_{45°} - \underbrace{\theta_{p4}^{p3}}_{90°} = -180°$$

which, by real axis symmetry, is also the departure angle from the pole at $s = p_4 = -1 - j$. This is to be expected, however, since the negative root locus represents a smooth continuation of the positive root locus. Therefore, as K decreases through zero from positive to negative values, the negative root locus departure angles from distinct poles will differ from the positive root locus departure angles by 180°.

Finally, Fact 4N implies that the $n - r = 3$ poles at p_1, p_3, and $p_4 \to \infty$ at asymptotic angular directions that are equal to 0°, 120°, and 240°, or 0° and ±120°, as depicted in Figure 5.8.

EXAMPLE 5.3.31 To further illustrate the root locus, consider a system similar to that of Example 5.3.30, but with the positions of z_1 and p_2 interchanged. In particular, consider a system defined by the loop gain transfer function

$$L(s) = \frac{Km(s)}{a(s)} = \frac{K(s+2)}{s^4 + 3s^3 + 4s^2 + 2s} = \frac{K(s+2)}{s(s+1)(s+1\pm j)}$$

which is characterized by a single zero ($r = 1$) at $s = z_1 = -2$ and four ($n = 4$) poles at $s = p_1 = 0$, $p_2 = -1$, $p_3 = -1 + j$, and $p_4 = -1 - j$, which represent the starting points for the root locus plot.

Fact 2 implies that the real axis region between the open-loop poles at $p_1 = 0$ and $p_2 = -1$ as well as that between the zero at $z_1 = -2$ and $-\infty$ belong to the positive root locus. It is obvious from these root locus regions that the departure angle from the pole at $p_1 = 0$ is $180°$, while the departure angle from the pole at $p_2 = -1$ is $0°$.

Fact 3 now can be used to determine the departure angle from the pole at $s = p_3 = -1 + j$. In particular,

$$\theta_{p3}^{p3} = \underbrace{\theta_{z1}^{p3}}_{45°} - \underbrace{\theta_{p1}^{p3}}_{135°} - \underbrace{\theta_{p2}^{p3}}_{90°} - \underbrace{\theta_{p4}^{p3}}_{90°} + 180° = -90°$$

By real axis symmetry, this implies a departure angle of $+90°$ from the pole at $s = p_4 = -1 - j$.

Fact 4 implies that the centroid

$$\sigma_c = \frac{p_1 + p_2 + p_3 + p_4 - z_1}{n - r} = \frac{0 - 1 + (-1+j) + (-1-j) - (-2)}{4 - 1} = -\frac{1}{3}$$

which represents the origin of the $n - r = 3$ asymptotes, whose angular directions are given by $\pm 60°$ and $180°$.

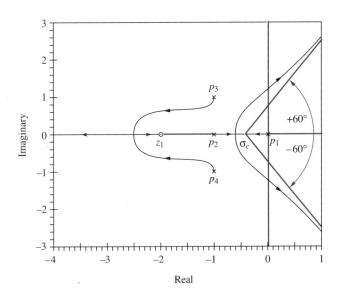

• **FIGURE 5.9**
A $K > 0$ Root Locus Plot of $\delta(s) = s^4 + 3s^3 + 4s^2 + 2s + K(s+2)$

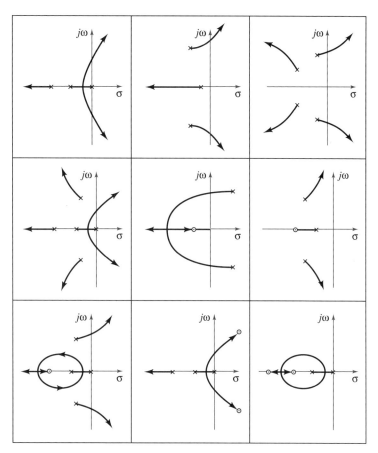

● **FIGURE 5.10**
Some Positive Root Locus Plots

The positive root locus in this case can now be drawn, as in Figure 5.9. Note, in particular, that the poles at p_1 and p_2 approach one another as K increases, eventually coming together at a real axis breakaway point that can be determined using Fact 5. In particular, since

$$a(s)m'(s) - a'(s)m(s) = s^4 + 3s^3 + 4s^2 + 2s - (4s^3 + 9s^2 + 8s + 2)(s + 2)$$

$$= -3s^4 - 14s^3 - 22s^2 - 16s - 4$$

$$= -(s + 0.48)(s + 2.503)(s + 0.842 \pm j.633)$$

it follows that the breakaway point between p_1 and p_2 occurs at $s = -0.48$. Furthermore, the point at $s = -2.503$ represents the break-in point where the poles at p_3 and p_4 meet and subsequently move in opposite directions along the real axis, one pole approaching the zero at z_1 and the other approaching $-\infty$ as K approaches $+\infty$.

Observing Figure 5.9, it is clear that for lower values of the gain $K > 0$, the closed-loop system is stable. However, as K increases, the poles at p_1 and p_2 will eventually cross the imaginary axis and enter the unstable half-plane $Re(s) > 0$. The Routh-Hurwitz stability criterion can be used to determine that this occurs when $K = 1.71$ (see Problem 5-20). Figure 5.10 displays a variety of other positive root locus plots that correspond to different initial loop gain pole and zero locations.

5.4   THE NYQUIST STABILITY CRITERION

As we have noted, stability is a primary goal of any control system design, and both the Routh-Hurwitz stability criterion and the root locus represent techniques for determining the stability properties of a system from explicit knowledge of its transfer function. In this section we will present an alternative stability test based on a polar plot of the loop gain frequency response of a system, as defined in Section 4.5.

The particular results that will be presented are due to Nyquist [51] who, in the early 1930s, was involved with the design of feedback amplifiers at the Bell Telephone Laboratories. At that time, it was generally observed that a feedback amplifier would go unstable if its gain was increased sufficiently. In certain situations, however, an amplifier would go unstable if its gain was decreased as well. Nyquist's investigations explained these observations by establishing a fundamental relationship between the open-loop frequency response of a system, and its closed-loop stability.

The Principle of the Argument

The basic concept that Nyquist employed to derive his now classical result was the so-called "principle of the argument," due to Cauchy, from the field of complex variables [15]. The principle of the argument pertains to the mapping of a known function $F(s)$, of the complex variable $s = \sigma + j\omega$, from the s-plane to the $F(s)$-plane.

Suppose $F(s)$ is a rational function of s with real coefficients that is analytic everywhere in the s-plane except at its poles, and Γ_s is any closed, continuous *contour* in the s-plane that does not intersect any poles or zeros of $F(s)$. If the values of s on Γ_s are substituted for s in $F(s)$, it follows that the resulting values of $F(s)$ will define another closed, continuous contour Γ_F in the $F(s)$-plane, which is called a *mapping of Γ_s into Γ_F*, that is, $\Gamma_F = F(\Gamma_s)$. The **principle of the argument** or **Cauchy's theorem** can now be stated as follows:

> If P represents the number of poles and Z the number of zeros of $F(s)$ encircled by a closed, clockwise contour Γ_s in the s-plane, then the net number N of times that Γ_F encircles the

TABLE 5.2 Some Mapped $F(s)$
Values when $r = 1.414$

Γ_s Values	Γ_F Values
1.414	0.414
$1 - j$	$0.4 + j.2$
$-j1.414$	$0.333 + j.47$
$-1 - j$	j
-1.414	-2.415
$-1 + j$	$-j$
$j1.414$	$0.333 - j.47$
$1 + j$	$0.4 - j.2$

origin of the $F(s)$-plane (positive if clockwise and negative if counterclockwise) is equal to $Z - P$, or

$$N = Z - P \qquad (5.4.1)$$

EXAMPLE 5.4.2 To illustrate the principle of the argument, consider a rational

$$F(s) = \frac{1}{s + 1}$$

which has a single pole at $s = -1$ and no zeros, so that $Z = 0$. If Γ_s is a circular s-plane contour of radius $r = \sqrt{2} = 1.414$, centered at $s = 0$, a clockwise path around Γ_s does encircle the $F(s)$ pole at $s = -1$, as in Figure 5.11, so that $P = 1$ in Eq. (5.4.1).

Table 5.2 lists eight different values of s on Γ_s, beginning at $s = 1.414$, and the corresponding mapped values of $F(s)$ that define Γ_F in the $F(s)$-plane, beginning at $F(s = 1.414) = 0.414$. A continuous, directed contour that connects these Γ_F values in the $F(s)$-plane also is depicted in Figure 5.11. Note that in this case, Γ_F encircles the $F(s)$-plane origin exactly once in the

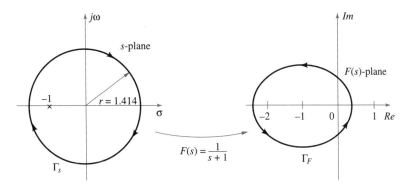

● **FIGURE 5.11**
An s-Plane Mapping of a Circular ($r = 1.414$) Γ_s into Γ_F

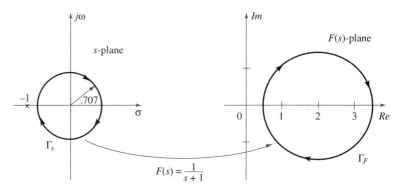

- **FIGURE 5.12**
An s-Plane Mapping of a Circular ($r = 0.707$) Γ_s into Γ_F

counterclockwise direction, so that $N = -1$,[6] thus verifying Eq. (5.4.1):

$$\underbrace{N}_{-1} = \underbrace{Z}_{0} - \underbrace{P}_{1}$$

We next note that if the radius of Γ_s is reduced to $r = \sqrt{2}/2 = 0.707$, as in Figure 5.12, a clockwise path around Γ_s does *not* encircle the $F(s)$ pole at $s = -1$, so that $P = 0$ in Eq. (5.4.1). Table 5.3 lists eight different values of s on Γ_s, beginning at $s = 0.707$, and the corresponding mapped values of $F(s)$ that define Γ_F in the $F(s)$-plane, beginning at $F(s = 0.707) = 0.586$. A continuous, directed contour that connects these Γ_F values in the $F(s)$-plane also is depicted in Figure 5.12. Note that in this particular case, Γ_F does not encircle the $F(s)$-plane origin, so that $N = 0$, again verifying Eq. (5.4.1):

$$\underbrace{N}_{0} = \underbrace{Z}_{0} - \underbrace{P}_{0}$$

....

The Nyquist Contour

Let us now consider a closed-loop system characterized by a proper loop gain transfer function

$$L(s) = \frac{Km(s)}{a(s)} = K\frac{s^r + m_{r-1}s^{r-1} + \cdots + m_1 s + m_0}{s^n + a_{n-1}s^{n-1} + \cdots + a_1 s + a_0} \tag{5.4.3}$$

as defined by Eq. (5.3.1). In light of Eq. (5.1.30), the closed-loop poles of the system are given by the zeros of the return difference

$$1 + L(s) = 1 + \frac{Km(s)}{a(s)} = \frac{a(s) + Km(s)}{a(s)} = \frac{\delta(s)}{a(s)} \tag{5.4.4}$$

---

[6]Note that the actual shape of Γ_F is unimportant relative to the determination of N.

TABLE 5.3 Some Mapped $F(s)$ Values
when $r = 0.707$

Γ_s Values	Γ_F Values
0.707	0.586
$0.5 - j.5$	$0.6 + j.2$
$-j.707$	$0.667 + j.47$
$-0.5 - j.5$	$1 + j$
-0.707	3.413
$-0.5 + j.5$	$1 - j$
$j.707$	$0.667 - j.47$
$0.5 + j.5$	$0.6 - j.2$

or the n roots of the polynomial $\delta(s) = a(s) + Km(s)$. Therefore, the closed-loop system will be stable if and only if all of the roots of $\delta(s)$ lie in the stable half-plane $Re(s) < 0$, that is, if and only if $\delta(s)$ is a stable polynomial.

Using the principle of the argument, let us now set

$$F(s) = 1 + L(s) = 1 + \frac{Km(s)}{a(s)} = \frac{\delta(s)}{a(s)} \qquad (5.4.5)$$

and assume that none of the roots of $a(s)$ or $\delta(s)$ lie on the imaginary ($j\omega$) axis in the complex s-plane. We will now define Γ_s to be the s-plane **Nyquist contour** that encloses the entire, unstable, right half-plane $Re(s) > 0$ in a clockwise direction, as depicted in Figure 5.13. In light of the principle of the argument, it then follows that N corresponds to the net number of encirclements of the origin of the $1 + L(s)$-plane by Γ_F, the contour mapped by Γ_s.

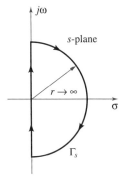

• **FIGURE 5.13**
The Nyquist Contour Γ_s

Note, however, that if Γ_F is shifted to the left by one unit, it will correspond to a Γ_L contour that is mapped by the Nyquist contour Γ_s into the $L(s)$-plane, whose -1 point corresponds to the $F(s)$-plane origin. In particular, the resulting Γ_L mapping represents an *extended polar plot* of the loop gain $L(s)$; that is, a normal polar plot of $L(j\omega)$ as ω varies from 0 to $+\infty$, followed by a plot of $L(s)$ as s traverses a clockwise semicircle of infinite radius from $+j\infty$ to $-j\infty$, followed by a plot of $L(j\omega)$ as $j\omega$ varies from $-j\infty$ to 0, as in Figure 5.13. An extended polar plot also is called a **Nyquist plot** or a **Nyquist diagram**, a name sometimes given normal polar plots as well.

Since $L(s)$ is a proper transfer function, it follows that

$$L(\infty) = \lim_{s\to\infty} \frac{K}{s^{n-r}}$$

which is a constant equal to K (if $n = r$) or 0 (if $n > r$). Therefore, the entire semicircular portion of Γ_s, of infinite radius, maps into the stationary point K or 0 in the $L(s)$-plane. Also note that the values of $L(-j\omega)$ are equal to the complex-conjugates of the $L(+j\omega)$ values, so that they map as the real axis mirror images of the normal polar plot values.

It thus follows that N is equal to the net number of encirclements of the -1 point in the $L(s)$-plane by an extended polar plot of $L(s)$. Furthermore, Eq. (5.4.5) implies that Z is equal to the number of unstable roots of $\delta(s)$ (the number of unstable closed-loop poles), and P corresponds to the number of unstable roots of $a(s)$ (the number of unstable open-loop poles).

To summarize these observations, let us now consider a system characterized by a nominal loop gain $L(s)$, as defined by Eq. (5.4.3), whose closed-loop poles are given by the roots of $\delta(s) = a(s) + Km(s)$. In light of the principle of the argument, the **Nyquist stability criterion** can now be stated as follows:

A necessary and sufficient condition for the nominal closed-loop stability of a system defined by $L(s)$ is that

$$Z = N + P \tag{5.4.6}$$

be equal to 0, or that $N = -P$, where N equals the net number of encirclements (positive if clockwise and negative if counterclockwise) of the -1 point in the $L(s)$-plane by an extended polar plot of the loop gain $L(s)$, and P corresponds to the number of unstable poles of $L(s)$.

Clearly, if the loop gain transfer function $L(s)$ is stable, so that $P = 0$, N also must equal 0 for closed-loop stability, so that an extended polar plot of $L(s)$ must encircle the -1 point the same number of times in a clockwise direction as in a counterclockwise direction. It may be

TABLE 5.4 Some Frequency Response Values of $L(j\omega)$

ω	$|L(j\omega)|$	$\angle L(j\omega)$
0	K	$0°$
0.1	$0.9937K$	$-9.15°$
0.5	$0.8666K$	$-43.46°$
2.5	$0.2251K$	$-133.58°$
5.657	$0.0505K$	$-180°$
10	$0.0138K$	$-208°$
100	$0.00002K$	$-262.57°$
∞	0	$-270°$

noted that in many of these closed-loop stable cases, both the number of clockwise and the number of counterclockwise encirclements of the -1 point will be zero.

EXAMPLE 5.4.7 To illustrate the Nyquist stability criterion, consider a system defined by the loop gain

$$L(s) = \frac{20K}{s^3 + 13s^2 + 32s + 20} = \frac{20K}{(s+1)(s+2)(s+10)}$$

whose three poles lie at $s = -1, -2$, and -10 in the stable half-plane $Re(s) < 0$, so that $P = 0$ in Eq. (5.4.6). For closed-loop stability, it follows that N must equal 0 or, in this case, that an extended polar plot of $L(s)$ must not encircle the -1 point.

Some frequency response values associated with $L(j\omega)$ are given in Table 5.4. Note that such frequency response values can be obtained experimentally using a variety of widely available signal generators and spectrum analyzers. They can then be used to plot Γ_L, thus determining the stability properties of the closed-loop system, *without explicit knowledge of the loop gain $L(s)$*. This is a major advantage of the Nyquist stability criterion when compared with both the Routh-Hurwitz and the root locus methods.

The values given in Table 5.4 imply a polar plot of $L(s = j\omega)$ that begins at $K > 0$ on the positive real axis in the $L(s)$-plane, as depicted in Figure 5.14. The magnitude then continues to decrease toward 0 as ω approaches $+\infty$, crossing the negative real axis at a magnitude of $0.0505K$ when $\omega = 5.657$. Since $L(s)$ is strictly proper, the entire $r = \infty$ semicircular portion of Γ_s then maps into the $L(s)$-plane origin; Γ_L then continues as the mirror image of the $+j\omega$ (or normal) polar plot as $j\omega$ increases from $-j\infty$ to 0.

The precise shape of the Γ_L contour in Figure 5.14 is unimportant, relative to the stability properties of the closed-loop system, and it is usually necessary to accurately plot Γ_L only near the critical -1 point. Also, the mirror image of the normal polar plot of $L(s = +j\omega)$ usually need not be explicitly depicted, as in Figure 5.14, since its effect relative to the determination of N is usually apparent from the normal polar plot.

The Nyquist stability criterion now implies that for closed-loop stability, the negative axis crossover point at $-0.0505K$ must lie to the right of the -1 point to prevent it from being encircled.

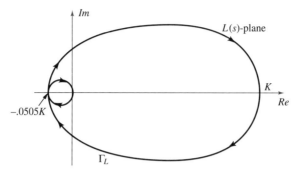

• **FIGURE 5.14**

An Extended Polar Plot (Γ_L) of $L(s) = \dfrac{20K}{s^3 + 13s^2 + 32s + 20}$

Therefore, $0.0505K < 1$, or $K < 19.8$ for closed-loop stability, a loop gain magnitude constraint that can be verified using the Routh-Hurwitz stability criterion (see Problem 5-22a). If $K > 19.8$, Γ_L will encircle the -1 point twice in a clockwise direction, so that $N = 2$ in Eq. (5.4.6). As a consequence, $Z = N + P = 2 + 0 = 2$, and two poles of the closed-loop system will lie in the unstable half-plane $Re(s) > 0$, an observation which can be verified by means of the root locus (see Problem 5-22b).

Imaginary Axis Poles

In certain cases, one or more poles p_i of $L(s)$ may lie on the s-plane imaginary ($j\omega$) axis. When this occurs, a Nyquist contour Γ_s can be modified so that it encircles each such p_i, without intersecting it, along a semicircular path of radius $\epsilon \approx 0$. Such a path usually is chosen so that p_i is not enclosed by the resultant Γ_s, although this need not be the case (see Problem 5-24).

A common cause of imaginary axis poles is a loop gain transfer function $L(s)$ of *type* $p > 0$, which implies that $L(s)$ has exactly p poles (and no zeros) at $s = 0$. In such cases,

$$L(s) \approx \frac{\hat{K}}{s^p} \tag{5.4.8}$$

along a semicircular Γ_s path of radius ϵ centered at $s = 0$. The resulting Γ_L contour will be a large circular path of radius $\hat{K}\epsilon^{-p}$ which, in view of Eq. (5.4.8), starts at an angle of $p \times 90°$ (when $s = -j\epsilon$ on Γ_s) and passes through $0°$ (when $s = +\epsilon$), eventually terminating at an angle of $p \times -90°$ (when $s = +j\epsilon$).

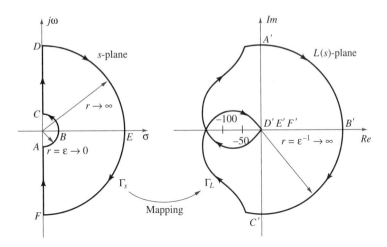

• **FIGURE 5.15**
A Nyquist Contour Γ_s and Its Mapping Γ_L for a Type 1 $L(s)$

. . . .

EXAMPLE 5.4.9 To illustrate the preceding discussion, let us determine the stability properties of a closed-loop system defined by a loop gain transfer function equal to K times the $G(s)$ of Example 4.4.45, that is, a minimum phase

$$L(s) = K\frac{m(s)}{a(s)} = K\frac{2.4s^3 + 15.36s^2 + 19.2s + 12}{s^4 + 30.08s^3 + 2.44s^2 + 1.2s}$$

which is a strictly proper, type 1 transfer function with three zeros at $s = -0.7 \pm j.714$ and -5, and four poles at $s = 0, -0.04 \pm j.196$, and -30.

A Γ_s contour that encircles the pole at $s = 0$, without enclosing it, is depicted in Figure 5.15, together with the contour Γ_L it maps into the $L(s)$-plane when $K = 1$. The points denoted by the letters A through F on the Γ_s contour will map into the corresponding Γ_L points A' through F', respectively. Note that the Γ_L contour that maps from $s = j\omega$ values between $j.1$ and $j100$ corresponds to the normal polar plot of $L(s) = G(s)$ depicted in Figure 4.23. Note further that it is virtually impossible to determine the number of encirclements of the -1 point in either Figure 4.23 or in Figure 5.15 because of the relatively large scale required to display the entire Γ_L contour.

To remedy this situation, we can restrict the polar plot to frequencies above $\omega = 0.6$, thereby obtaining the polar plot depicted in Figure 5.16. This $\omega > 0.6$ plot represents an amplification of the polar plot of Figure 4.23 near the critical -1 point, which is the region of primary interest relative to the stability properties of the closed-loop system. In particular, in view of Figures 5.15 and 5.16, it follows that when $K = 1$, an extended polar plot of $L(s) = G(s)$ does encircle the -1 point two times in a clockwise direction, so that

$$Z = N + P = 2 + 0 = 2$$

in Eq. (5.4.6), thus implying two closed-loop poles in the unstable half-plane $Re(s) > 0$.

If K is increased beyond 1.156, a polar plot of $L(s)$ will cross the negative real axis to the left of the -1 point (see Problem 5-24). In such a case, Γ_L will enclose the -1 point by both a

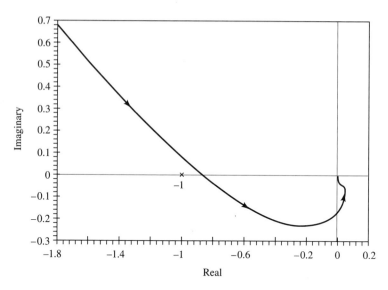

● **FIGURE 5.16**

A Polar Plot of $L(s) = G(s) = \dfrac{2.4s^3 + 15.36s^2 + 19.2s + 12}{s^4 + 30.08s^3 + 2.44s^2 + 1.2s}$ for $\omega > 0.6$

relatively large clockwise $+1$ path and a relatively small counterclockwise -1 path, so that $N = 0$. As a consequence,

$$Z = \underbrace{N}_{0} + \underbrace{P}_{0} = 0$$

in Eq. (5.4.6), thus implying closed-loop stability. Moreover, in view of Figure 5.15, if K is decreased below 0.0098 (see Problem 5-24), the -1 point will lie outside of the entire Γ_L contour. In such cases, $N = 0$ as well, also implying closed-loop stability.

A root locus plot of the closed-loop poles of the system as the gain K increases from 0 to $+\infty$ can be used to verify these observations. In particular, such a plot starts (when $K = 0$) at the open-loop pole locations at $s = 0, -0.04 \pm j.196$, and -30. As K increases, the poles at $s = 0$ and -30 remain stable, moving towards the zeros at $s = -5$ and $-\infty$, respectively. However, the complex-conjugate poles at $-0.04 \pm j.196$ cross into the unstable half-plane as K increases beyond 0.0098, and they remain there until $K = 1.156$, when they again cross the imaginary axis into the stable half-plane to approach the zeros at $s = -0.7 \pm j.714$ as K approaches $+\infty$, as depicted in Figure 5.17. A system such as this, which is stable for larger values of K but becomes unstable as K is reduced, is called **conditionally stable**.

··

We remark finally that a Γ_L contour that intersects the -1 point implies one or more closed-loop poles on the imaginary s-plane axis. Therefore, if Γ_L almost intersects the -1 point, it follows that one or more of the closed-loop poles will lie close to the $j\omega$ axis, which is not a

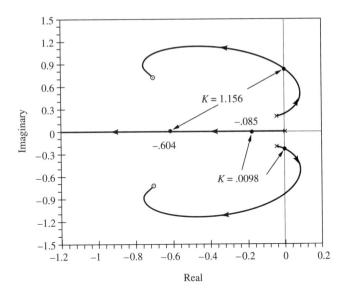

- **FIGURE 5.17**
A Root Locus Plot of the Complex-Conjugate Poles of $L(s)$

desirable situation relative to the stability properties of the system. Indeed, we will show in the next chapter that the minimum distance between a polar plot of $L(s = j\omega)$ and the critical -1 point represents a measure of the relative stability of a closed-loop system. Therefore, the polar plot of a nominally stable system should stay as far away from the -1 point as possible in order to insure a "robustly stable" closed-loop design.

Transportation Lag

The Nyquist stability criterion represents a rather illuminating way of displaying the destabilizing effect that transportation lag has on the closed-loop performance of a system. Consider a system defined by a loop gain $L(s) = KG(s)$, with $G(s)$ defined by Eq. (4.5.9):

$$L(s) = KG(s) = \frac{Km(s)}{a(s)} = \frac{Ke^{-0.4s}\omega_n^2}{s^2 + 0.6\omega_n s + \omega_n^2} \qquad (5.4.10)$$

Since $a(s)$ is stable, $P = 0$ in Eq. (5.4.6). Looking at Figure 4.24, the polar plot of $L(s) = KG(s)$ crosses the negative real axis at -0.6 when $K = 1$. Therefore, if K is increased beyond $1/0.6 = 1.667$, a polar plot of $L(s)$ will cross the negative real axis to the left of the -1 point. As a consequence, the resulting Γ_L contour will encircle the -1 point twice in a clockwise direction, so that

$$Z = N + P = N = 2$$

thus implying a closed-loop system with two unstable poles. It should be noted that if there were no transportation lag present in Eq. (5.4.10), then the gain K could be increased without bound and the closed-loop system would remain stable.

5.5   SUMMARY

This chapter has served to define the general control problem, as depicted in Figure 5.1. Acceptable performance of a controlled plant generally implies an output $y(t)$ that robustly responds to the reference input $r(t)$ despite possible changes in the plant parameters or unmodeled dynamics, as well as the presence of external disturbances or noisy measurements.

When the nominal behavior of a plant can be represented by a strictly proper, rational transfer function $G(s)$, a 2 DOF linear controller, as defined by a (1×2) proper transfer vector, was shown to imply five primary transfer functions, which define the closed-loop performance. Moreover, the 2 DOF controller presented represents the most general possible linear compensator. If a more restrictive, 1 DOF design is employed, the five primary transfer functions were shown to be mutually dependent.

In both cases, however, the closed-loop poles of the system are defined by the roots of the denominator polynomial $\delta(s) = a(s)k(s) + c(s)h(s)$, and the remainder of the chapter focused on three well-known techniques that can be employed to ensure nominal closed-loop stability, or that all of the roots of $\delta(s)$ lie in the stable half-plane $Re(s) < 0$.

The Routh-Hurwitz stability criterion of Section 5.2 was shown to represent a test for absolute (yes or no) stability, providing the number of unstable poles but not their relative locations. The root locus facts presented in Section 5.3 were then shown to represent a relative stability procedure, graphically displaying the actual s-plane variations of the closed-loop poles as a function of an adjustable gain.

The Nyquist stability criterion of Section 5.4 was shown to represent another test for relative stability. Unlike the Routh-Hurwitz and the root locus methods, however, the frequency response characteristics of a system directly imply the Nyquist criterion, without explicit knowledge of the loop gain $L(s)$. This represents an important advantage of the Nyquist test, relative to the design of appropriate compensators, as we will later show.

PROBLEMS

5-1. Consider the nominal control system depicted in Figure 5.2. If the additive disturbance signal $d(s)$ enters the loop at the plant input, rather than the plant output, so that $u(s)$ equals the sum of $d(s)$ and the output of the 2 DOF controller, determine the transfer function relationship between $y(s)$ and $d(s)$.

5-2. Consider the nominal control system depicted in Figure 5.2. If the controller polynomials $q(s)$, $h(s)$, and $k(s)$ can be chosen such that
- $s = 0$ is a root of $k(s)$,
- $\delta(s) = a(s)k(s) + c(s)h(s)$ is a stable polynomial, and
- $q(s = 0) = h(s = 0)$,

determine the steady-state error, $e_{ss}(t) = \lim_{t \to \infty}\{r(t) - y(t)\}$, if the reference input $r(t) = r_s(t)$, the unit step function.

5-3. Consider the 1 DOF, unity feedback configuration of Figure 5.3, when the proper, rational transfer function

$$H(s) = \frac{h(s)}{k(s)} = \frac{a(s)}{c(s)\tilde{k}(s)}$$

with the roots of $\tilde{k}(s)$ arbitrary but disjoint from those of $a(s)$.

(a) Discuss the controllability and observability properties of the closed-loop system.

(b) Determine a lower bound on the dynamic order of a minimal realization of the resulting closed-loop system.

(c) For what class of plants $G(s)$, if any, can the closed-loop system be made internally stable?

5-4. Consider a system defined by a rational transfer function $G(s) = \frac{c(s)}{a(s)}$, which is compensated by five rational transfer functions $H_i(s) = \frac{h_i(s)}{k_i(s)}$, as depicted in Figure 5.18(a).

(a) Show that the five-part compensator has an equivalent 2 DOF implementation, as depicted in Figure 5.18(b), by expressing the three controller polynomials, $q(s)$, $h(s)$, and $k(s)$, in terms of the ten polynomials, $h_i(s)$ and $k_i(s)$, which comprise the $H_i(s)$, for $i = 1, 2, \ldots, 5$.

(b) Determine the loop gain $L(s)$ and the sensitivity function $S(s)$ associated with this system.

(c) Determine the transfer function relationship between $y(s)$ and $r(s)$ using both a "brute force" approach and one of Mason's formulas (see Problem 3-22).

5-5. The controller configuration depicted in Figure 5.19, which consists of a reference input prefilter $H_1(s)$ and a series loop compensator $H_2(s)$, has been termed a "two-degree-of-freedom" feedback structure (see References [33] and [41]). Show that such a design actually represents a restrictive form of the most general 2 DOF controller configuration, which is depicted in Figure 5.18(b); that is, show that the Figure 5.19 con-

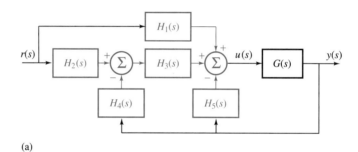

(a)

(b)

- **FIGURE 5.18**
  The Two Compensators of Problem 5-4

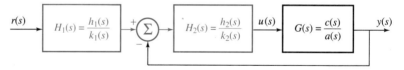

- **FIGURE 5.19**
  The Restrictive 2 DOF Configuration of Problem 5-5

figuration is equivalent to the Figure 5.18(b) configuration, but with constrained (partially dependent) compensator polynomials $k(s) = k_1(s)k_2(s)$, $h(s) = k_1(s)h_2(s)$, and $q(s) = h_1(s)h_2(s)$. Show that a similar constraint characterizes the five-part compensator of Problem 5-4.

5-6. Consider a proper, second-order, 2 DOF controller configuration, as depicted in Figure 5.18(b), with

$$k(s) = s^2 + k_1 s + k_0, \qquad q(s) = q_2 s^2 + q_1 s + q_0,$$

$$\text{and} \quad h(s) = h_2 s^2 + h_1 s + h_0$$

Verify that

$$\frac{h(s)}{k(s)} = \frac{(h_1 - h_2 k_1)s + h_0 - h_2 k_0}{s^2 + k_1 s + k_0} + h_2$$

and $\dfrac{q(s)}{k(s)} = \dfrac{(q_1 - q_2 k_1)s + q_0 - q_2 k_0}{s^2 + k_1 s + k_0} + q_2$

In light of the observable canonical representations of Table 3.1, show that the following differential operator representation defines the dynamic behavior of the 2 DOF configuration:

$$(D^2 + k_1 D + k_0)z(t) = [(q_1 - q_2 k_1)D + q_0 - q_2 k_0]r(t)$$
$$+[(h_1 - h_2 k_1)D + h_0 - h_2 k_0]y(t);$$
$$u(t) = z(t) + q_2 r(t) + h_2 y(t)$$

Draw an analog diagram of an equivalent state-space system in order to verify that such a proper 2 DOF compensator does not require the use of differentiators. Note that this observation holds irrespective of the compensator order.

5-7. Consider a unity feedback system defined by the loop gain transfer function

$$L(s) = G(s)H(s) = \dfrac{K_1 s + K_2}{s^2 + 2s - 1}$$

Determine all values of the $\{K_1, K_2\}$ pair for which the following conditions hold:

(i) The steady-state error due to a positional (step) change in the reference input is less than 10%.

(ii) The closed-loop response is characterized by an undamped natural frequency $\omega_n > 3$ and a damping ratio $\zeta = 0.8$.

5-8.* Consider the 1 DOF closed-loop system depicted in Figure 5.20.

(a) Determine the closed-loop transfer function $\dfrac{y(s)}{r(s)}$.

(b) Determine the values of K that imply closed-loop stability.

(c) Plot the positive root locus for this system, explicitly identifying the centroid, the asymptotes, the departure angles, and the $\pm j\omega^*$ crossover points.

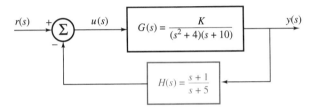

● FIGURE 5.20
The 1 DOF Feedback System of Problem 5-8

5-9. Multiply the fourth-degree polynomial

$$a(s) = -2s^4 - 4s^3 - 4s^2 - 8s - 1$$

$$= -(s + 1.956)(s + 0.133)(s - 0.044 \pm j1.387)$$

of Example 5.2.5 by $(s + 1)$, and then apply the Routh-Hurwitz stability criterion to the resulting fifth-degree polynomial, $\hat{a}(s) = (s+1)a(s)$. Note, in particular, that a first-column 0 element, such as that encountered in the s^2 row of Example 5.2.5, is eliminated by this procedure.

5-10. Consider a third-order system whose poles correspond to the roots of

$$\delta(s) = \delta_3 s^3 + \delta_2 s^2 + \delta_1 s + \delta_0$$

If all four coefficients $\delta_i > 0$, determine a necessary and sufficient condition for $\delta(s)$ to be a stable polynomial.

5-11.* Verify that the polynomial $p(s) = s^4 + 8s^3 + 32s^2 + 80s + 100$ is stable. Now determine the number of roots of $p(s)$ that are to the right of a vertical line drawn at $s = -2$ in the complex s-plane by substituting $\hat{s} - 2$ for s and applying the Routh-Hurwitz test to the resulting polynomial. Determine the actual locations of the roots of $p(s)$ to verify your answer. Note that this result can be used to determine a measure of the relative stability of a system.

5-12.* Use the Routh-Hurwitz criterion to determine the number of roots of

$$a(s) = s^5 + 6s^4 + 12s^3 + 12s^2 + 11s + 6$$

in the unstable half-plane $Re(s) > 0$, and explicitly determine the location of any roots that may be on the imaginary ($j\omega$) axis. Determine the actual locations of the roots of $a(s)$ in order to verify your answer.

5-13. Complete the entire positive and negative root locus plot of the system whose open-loop poles and zeros are depicted in Figure 5.5.

5-14. Let s^* be a point on the positive root locus that is very close to a loop zero at $s = z_q$, so that $\angle(s^* - z_q)$ defines the **arrival angle** of the root locus plot to z_q as $K \to \infty$. Determine an expression for

$$\theta_{zq}^{zq} \overset{\text{def}}{=} \lim_{s^* \to z_q} \angle(s^* - z_q)$$

which is analogous to Eq. (5.3.12) for determining the departure angle from an open-loop pole at $s = p_k$.

5-15. Plot the locus of the roots of the polynomial $\delta(s) = s^3 + (k+4)s^2 + 3s + 4k$ in the complex s-plane for all real values of k between $-\infty$ and $+\infty$. Explicitly identify the asymptotes of the locus as $k \to +\infty$, as well as any values of k (and the corresponding $\pm j\omega^*$) at which the locus crosses the imaginary axis. Note that for any real value σ of s, $\delta(\sigma) = 0$ for some corresponding, real value of k.

5-16.* Consider the state-space system: $\dot{\mathbf{x}}(t) = A\mathbf{x}(t) + Bu(t)$; $y(t) = C\mathbf{x}(t)$, with

$$A = \begin{bmatrix} 0 & 0 & -10 \\ 1 & 0 & -9 \\ 0 & 1 & -4 \end{bmatrix}, \qquad B = \begin{bmatrix} 1 \\ 0 \\ 0 \end{bmatrix}, \quad \text{and} \quad C = [0 \ \ 0 \ \ 1]$$

which is subjected to the integral of the error, unity feedback control law:

$$u(t) = K \int \{r(t) - y(t)\}\, dt, \quad \text{for} \ \ 0 < K < \infty$$

Plot a root locus for this system, explicitly identifying the centroid and the asymptotes, *all* of the breakaway points and any values of $\pm j\omega^*$ at which the locus crosses the imaginary axis.

5-17. Draw an extended polar plot of the type 1 system whose transfer function

$$G(s) = \frac{10s - 100}{s^3 + 101s + 100s} = L(s)$$

and use the Nyquist stability criterion to determine the stability of this system when closed under unity feedback.

5-18. Consider a system defined by the loop gain transfer function

$$G(s)H(s) = L(s) = \frac{10^4(s + 1)}{(s + 0.1)(s^2 + 90s - 1000)}$$

(a) Sketch a Bode diagram of the system.

(b) Use your answer to Part (a) to sketch an extended polar plot of the system.

(c) Determine the closed-loop stability of the system via the Nyquist stability criterion.

(d) Verify your answer to Part (c) via the root locus.

5-19.* Consider a state-space system: $\dot{\mathbf{x}}(t) = A\mathbf{x}(t) + Bu(t)$; $y(t) = C\mathbf{x}(t)$, with

$$A = \begin{bmatrix} 0 & 1 & 0 \\ 0 & 0 & 1 \\ -10 & -1 & 0 \end{bmatrix}, \qquad B = \begin{bmatrix} 0 \\ 0 \\ 1 \end{bmatrix}, \quad \text{and} \quad C = [0 \ \ 1 \ \ 1]$$

which is subjected to the error driven, unity feedback control law: $u(t) = Ke(t) = K\{r(t) - y(t)\}$. Sketch an entire root locus for this system, and determine the value of K at which the plot crosses the imaginary axis and the corresponding values of $\pm j\omega^*$.

5-20. Consider the system defined in Example 5.3.31 by the loop gain transfer function

$$L(s) = \frac{Km(s)}{a(s)} = \frac{K(s + 2)}{s^4 + 3s^3 + 4s^2 + 2s} = \frac{K(s + 2)}{s(s + 1)(s + 1 \pm j)}$$

Use the Routh-Hurwitz criterion to verify that the closed-loop system will be unstable if $K > 1.71$.

5-21. Consider a conditionally stable system defined by the loop gain transfer function

$$L(s) = \frac{K}{(\tau_1 s - 1)(\tau_2 s + 1)^2}$$

with $K > 0$ and $\tau_1 > \tau_2 > 0$.

(a) Sketch a Bode diagram for this system.

(b) Determine the closed-loop stability of this system using the Nyquist stability criterion for high, medium, and low values of the gain K.

(c) Verify your answer to Part (b) via the root locus.

5-22. Consider a closed-loop system defined by the loop gain transfer function of Example 5.4.7, namely,

$$L(s) = \frac{20K}{s^3 + 13s^2 + 32s + 20}$$

(a) Use the Routh-Hurwitz criterion to verify that the system will be unstable if $K > 19.8$.

(b) Sketch a positive root locus for this system, and explicitly identify the $\pm j\omega^*$ points where it crosses into the unstable half-plane.

5-23. Consider an open-loop system that is defined by the differential equation:

$$\ddot{y}(t) + 4\dot{y}(t) = Ku(t - 1)$$

If the system is closed under unity feedback, determine the values of $K > 0$ for which it is stable.

5-24.* Consider a closed-loop system defined by the loop gain transfer function

$$L(s) = K\frac{2.4s^3 + 15.36s^2 + 19.2s + 12}{s^4 + 30.08s^3 + 2.44s^2 + 1.2s}$$

as in Example 5.4.9. Apply the Nyquist stability criterion (when $K = 1$) by defining a Γ_s contour that *encloses* the pole at $s = 0$ with a semicircular path of radius $\epsilon \approx 0$ that passes through $s = -\epsilon$ (rather than $+\epsilon$, as in Example 5.4.9). Verify that this alternative Γ_s contour also implies the same number (two) of unstable closed-loop poles. Employ the Routh-Hurwitz stability criterion to verify that the closed-loop system will be stable if $0 < K < 0.0098$ or if $K > 1.156$

5-25. Consider the state-space system: $\dot{\mathbf{x}}(t) = A\mathbf{x}(t) + Bu(t)$; $y(t) = C\mathbf{x}(t)$, when

$$A = \begin{bmatrix} \gamma - \alpha & \alpha\gamma \\ 1 & 0 \end{bmatrix}, \qquad B = \begin{bmatrix} 1 \\ 0 \end{bmatrix}, \qquad \text{and} \quad C = [1 \quad \beta],$$

with $0 < \alpha < 1 < \beta < \gamma$ and $\alpha\gamma < \beta$.

(a) Draw an asymptotic amplitude Bode plot of this system, explicitly identifying all corner frequencies.

(b) Express both the low frequency gain K_0 (as $\omega \to 0$), and the frequency ω_g at which the plot crosses the 0 dB line as functions of the parameters α, β, and γ.

(c) Determine the *approximate* phase angles for the Bode diagram at $\omega = 0$, at all of the corner frequencies, and in the limit as $\omega \to +\infty$.

(d) Sketch an extended polar plot of the frequency response of this system, and discuss the closed-loop stability properties of this system if $u(t) = e(t) = r(t) - y(t)$.

LOOP GOALS

6.1 ROBUST STABILITY FOR PLANT PARAMETER VARIATIONS

Good performance in a controlled, closed-loop system usually requires the attainment of a variety of goals that, in general, can be associated with either the loop or the response characteristics of the system. This chapter will focus on the primary loop goals that should be obtained, and the next chapter will discuss the equally important response goals. The first loop goal that we will discuss is perhaps the most important, namely **robust stability**, which, in view of the general control problem outlined in Section 5.1, implies that the closed-loop poles of the system depicted in Figures 5.2 and 5.3 remain in the stable half-plane $Re(s) < 0$ despite any *uncertainty* in the nominal plant model $G(s)$.

Typical sources of uncertainty can include plant parameter variations due to such environmental factors as temperature, pressure, and age; a variety of potential nonlinearities, such as hysteresis, deadzone, and friction; and unmodeled, high frequency dynamics, caused by such factors as flexible "bending modes." If the model uncertainty is due to known parameters α, whose values are uncertain, so that $G(s, \alpha)$ has a known structure, then the uncertainty is termed **structured uncertainty.** If, on the other hand, the uncertainty has no known structure, and it only can be characterized by frequency response magnitude bounds, then it is termed **unstructured uncertainty.** This section will deal with the structured uncertainty caused by plant parameter variations, and Section 6.2 will focus on robust stability with respect to the unstructured uncertainty that generally characterizes unmodeled, high frequency plant dynamics.

The selection of the loop compensator polynomials $h(s)$ and $k(s)$ represents the primary means of achieving robust stability, and it is for this reason that stability is a loop goal. To obtain robust stability, $h(s)$

and $k(s)$ must first be chosen to ensure nominal closed-loop stability, that is, a denominator polynomial

$$\delta(s) = a(s)k(s) + c(s)h(s) \qquad (6.1.1)$$

as defined by Eq. (5.1.13), whose roots lie in the half-plane $Re(s) < 0$.

The Definition of Gain Margin and Phase Margin

To ensure robust stability, or that $\delta(s)$ remains a stable polynomial, $h(s)$ and $k(s)$ should be chosen to position the poles of the nominal closed-loop system "far enough" into the stable half-plane $Re(s) < 0$ to ensure that they remain there despite any potential changes in the plant. Two traditional measures of the effectiveness of such a choice for $h(s)$ and $k(s)$ are the gain margin and the phase margin.

Gain margin and phase margin are usually defined for stable, closed-loop systems that are characterized by a minimum phase, loop gain transfer function

$$L(s) = G(s)H(s) = \frac{c(s)h(s)}{a(s)k(s)} \qquad (6.1.2)$$

In such cases, the **gain margin** (GM) is defined as the multiplicative amount that the magnitude of $L(s)$ can be increased before the closed-loop system goes unstable, and the **phase margin** (ΦM) is defined as the amount of additional phase lag that can be associated with $L(s)$ before the closed-loop system goes unstable.

Gain and Phase Margins from Polar Plots

The Nyquist stability criteria of Section 5.4 can be employed to obtain quantitative values for these two measures of robust stability. Recall that when $L(s)$ is stable, so that $P = 0$ in Eq. (5.4.6), closed-loop stability is ensured if an extended polar plot of $L(s)$ does not encircle the critical -1 point in the complex $L(s)$-plane.

Therefore, consider a "typical" polar plot of $L(s = j\omega)$ in such cases, as depicted in Figure 6.1. A unit circle centered at the origin has been drawn in order to identify the **gain crossover frequency** ω_g, which is the frequency at which the magnitude of the loop gain equals 1; that is, the condition

$$|L(j\omega_g)| = 1 \qquad (6.1.3)$$

defines ω_g and, in light of Eq. (4.3.4),

$$\phi_g \overset{\text{def}}{=} \angle L(j\omega_g) = \tan^{-1}\left\{ \frac{Im[L(j\omega_g)]}{Re[L(j\omega_g)]} \right\} \qquad (6.1.4)$$

defines the phase angle of the loop gain at the gain crossover frequency.

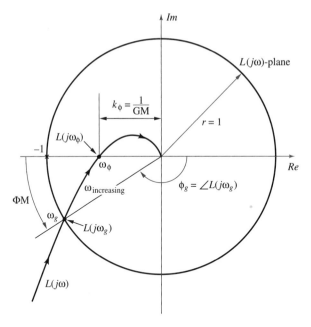

● **FIGURE 6.1**
A "Typical" Polar Plot of a Stable Closed-Loop System

The **phase crossover frequency** ω_ϕ also is identified in Figure 6.1 as the frequency at which $L(j\omega)$ crosses the negative real axis; that is, the condition

$$\angle L(j\omega_\phi) = -180° \qquad (6.1.5)$$

defines ω_ϕ, and

$$k_\phi \stackrel{\text{def}}{=} |L(j\omega_\phi)| \qquad (6.1.6)$$

defines the magnitude of the loop gain at the phase crossover frequency.

Now recall the definitions of the gain margin and the phase margin. If we were to increase the gain of $L(j\omega)$, without altering its phase, all points on the polar plot of $L(j\omega)$ would increase in magnitude along radial lines centered at the origin. Therefore, the closed-loop system would go unstable if the **phase crossover point,** $L(j\omega_\phi)$, were to move to the left of the critical -1 point. Looking at Figure 6.1, this would occur if the increase in the magnitude of the loop gain were to exceed the reciprocal of k_ϕ, which implies that the **gain margin**

$$\text{GM} = \frac{1}{k_\phi} = \frac{1}{|L(j\omega_\phi)|} \qquad (6.1.7)$$

For example, if $k_\phi = 0.5$, then the GM = 2, and the loop gain magnitude could be doubled before the closed-loop system would go unstable.

If we were to decrease the phase of $L(j\omega)$, without altering its gain, all points on a polar plot of $L(j\omega)$ would rotate clockwise about the origin an angular amount equal to the increase in phase lag. Therefore, the closed-loop system would go unstable if the **gain crossover point,** $L(j\omega_g)$, were to rotate beyond the critical -1 point. In light of Figure 6.1, this would occur if the increase in the phase lag of $L(j\omega)$ were to exceed $180° + \phi_g$, which implies that the **phase margin**

$$\Phi M = 180° + \phi_g \tag{6.1.8}$$

For example, if $\phi_g = \angle L(j\omega_g) = -135°$, then $45°$ (equaling the ΦM) of additional phase lag could be associated with $L(j\omega)$ before the closed-loop system would go unstable.

In general, the gain margin and the phase margin are mutually independent. For example, it is possible for $L(j\omega)$ to reflect an excellent gain margin but a poor phase margin. Such is the case depicted in Figure 6.2, where the GM = ∞ but the ΦM < 15°. In such a case, the phase crossover frequency ω_ϕ is undefined.   Conversely, $L(j\omega)$ can reflect a poor gain margin but an excellent phase margin, as is the case depicted in Figure 6.3, where the ΦM = ∞ but the GM is only slightly greater than 1. In such a case, the gain crossover frequency ω_g is undefined.

In light of these observations, gain and phase margins should be used in conjunction with one another when determining the robust stability

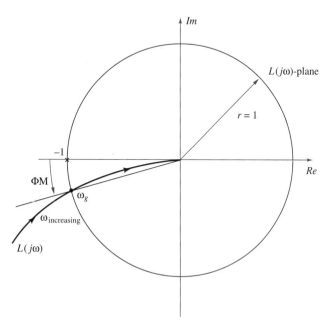

● **FIGURE 6.2**
Infinite GM, Poor ΦM

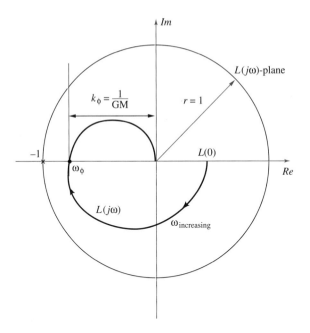

● **FIGURE 6.3**
Poor GM, Infinite ΦM

properties of a closed-loop system relative to the parameter variations that characterize its structured uncertainty. Together, they represent a measure of the "distance" of $L(j\omega)$ from the critical -1 point at the **mid-range frequencies,** which correspond to values of ω at and near the gain crossover frequency ω_g and the phase crossover frequency ω_ϕ. Clearly, the greater this distance, the more robust the stability of the closed-loop system. An acceptable nominal design usually is one that attains both a GM ≥ 2 and a ΦM $\geq +30°$.

Gain and Phase Margins from Bode Diagrams

In the minimum phase cases, gain and phase margins can be obtained directly from Bode diagrams of the loop gain. In particular, using Eq. (6.1.3), the gain crossover frequency ω_g is defined at the point where the magnitude plot crosses the 0 dB line, since

$$20 \log \underbrace{|L(j\omega_g)|}_{1} = 0 \text{ dB} \qquad (6.1.9)$$

The phase margin ΦM is then given on the corresponding phase plot as the angular distance of $\angle L(j\omega_g)$ above the $-180°$ line.

In light of Eq. (6.1.5), the phase crossover frequency ω_ϕ is defined on the phase plot of $L(j\omega)$ at the point where $\angle L(j\omega_\phi) = -180°$. The

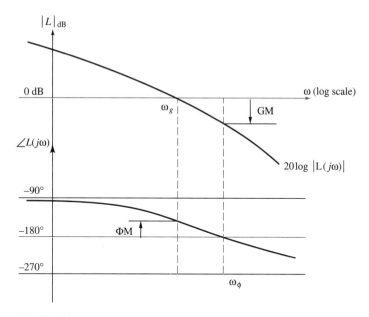

- **FIGURE 6.4**
GM and ΦM from the Bode Diagram of a "Typical" $L(j\omega)$

gain margin GM (in dB) is then given on the corresponding amplitude plot as the distance of $|L(j\omega_\phi)|_{dB}$ below the 0 dB line. The Bode diagram of a loop gain $L(j\omega)$, analogous to that depicted by the polar plot of Figure 6.1, is given in Figure 6.4 to illustrate these observations.

Sensitivity

Robust stability, with respect to plant parameter variations, can also be analyzed by the concept of sensitivity. Historically, the main reason for the development of the first negative feedback amplifiers at Bell Laboratories in the 1930s [9] was to render them relatively *insensitive* to the unpredictable changes in performance that characterized the vacuum tube devices of that time.

In a more general context, recall that a rational transfer function

$$G(s) = \frac{c(s)}{a(s)} \tag{6.1.10}$$

is often used to represent the dynamic performance of a nominal plant characterized by unpredictable internal parameter variations. In such cases, a more representative $G(s)$ is one that depends on one or more of these parameters α. When such a $G(s, \alpha)$ is compensated by the polynomial pair $h(s)$ and $k(s)$, as in Figures 5.2 and 5.3, it is of interest to quantify the effect of such compensation on the overall closed-loop performance

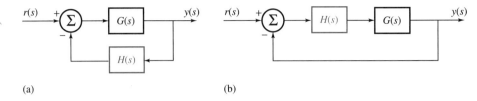

(a)   (b)

• **FIGURE 6.5**
Two Closed-Loop Configurations

of the system. As we will now show, the effect is the same whether the compensator

$$H(s) = \frac{h(s)}{k(s)}$$

is placed in the **return path**, as in Figure 6.5a, where the closed-loop transfer function is

$$T_r(s, \alpha) = \frac{y(s)}{r(s)} = \frac{G(s, \alpha)}{1 + G(s, \alpha)H(s)} \qquad (6.1.11)$$

or in the **feedforward path**, as in Figure 6.5b, where the unity feedback closed-loop transfer function is

$$T_f(s, \alpha) = \frac{y(s)}{r(s)} = \frac{G(s, \alpha)H(s)}{1 + G(s, \alpha)H(s)} \qquad (6.1.12)$$

In both cases, the effect of parameter variations can be characterized by the notion of "sensitivity," which was first defined by Bode [7]. In particular, the **sensitivity** of any transfer function $T(s, \alpha)$, with respect to a variable parameter α, is defined by the relationship:[1]

$$S_\alpha^T \stackrel{\text{def}}{=} \lim_{\Delta\alpha \to 0} \frac{\Delta T/T}{\Delta\alpha/\alpha} = \frac{dT/T}{d\alpha/\alpha} = \frac{\alpha}{T}\frac{dT}{d\alpha} \qquad (6.1.13)$$

which can be interpreted as the percentage change in $T(s, \alpha)$ caused by a differential percentage change in α. Therefore, if $T(s, \alpha) = G(s, \alpha)$, the open-loop plant transfer function, then

$$S_\alpha^G = \frac{\alpha}{G}\frac{dG}{d\alpha} \qquad (6.1.14)$$

Moreover, if $T(s)$ is a closed-loop transfer function, as defined by Eqs. (6.1.11) and (6.1.12), it follows that

$$\frac{dT_r}{d\alpha} = \frac{\partial T_r}{\partial G}\frac{dG}{d\alpha} = \frac{1}{(1+GH)^2}\frac{dG}{d\alpha} \qquad (6.1.15)$$

---

[1]The original definition given by Bode is actually the reciprocal of that given here, in that Bode defined sensitivity as the percentage change in α divided by the resulting percentage change in T.

and

$$\frac{dT_f}{d\alpha} = \frac{\partial T_f}{\partial G}\frac{dG}{d\alpha} = \frac{H}{(1+GH)^2}\frac{dG}{d\alpha} \tag{6.1.16}$$

respectively.

In light of Eq. (6.1.13), Eqs. (6.1.11), (6.1.15) and (6.1.14) then imply that

$$S_\alpha^{T_r} = \underbrace{\frac{\alpha(1+GH)}{G}}_{\dfrac{\alpha}{T_r}}\underbrace{\frac{1}{(1+GH)^2}\frac{dG}{d\alpha}}_{\dfrac{dT_r}{d\alpha}} = \frac{1}{1+GH}\underbrace{\frac{\alpha}{G}\frac{dG}{d\alpha}}_{S_\alpha^G} \tag{6.1.17}$$

while Eqs. (6.1.12), (6.1.16), and (6.1.14) imply that

$$S_\alpha^{T_f} = \underbrace{\frac{\alpha(1+GH)}{GH}}_{\dfrac{\alpha}{T_f}}\underbrace{\frac{H}{(1+GH)^2}\frac{dG}{d\alpha}}_{\dfrac{dT_f}{d\alpha}} = \frac{1}{1+GH}\underbrace{\frac{\alpha}{G}\frac{dG}{d\alpha}}_{S_\alpha^G} \tag{6.1.18}$$

The Sensitivity Function

Equations (6.1.17) and (6.1.18) show that the sensitivity of both of the closed-loop transfer functions, $S_\alpha^{T_r}$ and $S_\alpha^{T_f}$, relative to the sensitivity of the open-loop transfer function S_α^G, is defined by the same *sensitivity (transfer) function*, namely,

$$S(s) = \frac{1}{1+G(s)H(s)} = \frac{1}{1+L(s)} = \frac{a(s)k(s)}{a(s)k(s)+c(s)h(s)} = \frac{a(s)k(s)}{\delta(s)} \tag{6.1.19}$$

If $G(s) = G(s, \alpha)$, then $S(s) = S(s, \alpha)$ as well, although this dependence on α is not explicitly denoted in the *nominal representation* for $S(s)$ given by Eq. (6.1.19).

The sensitivity function $S(s)$ quantifies the effect of the loop compensator polynomials, $h(s)$ and $k(s)$, relative to unknown plant parameter variations, independent of the specific placement of $h(s)$ and $k(s)$ within the loop. At all frequencies where $|S(s = j\omega)| < 1$, the sensitivity of the closed-loop system to plant parameter variations is decreased by an amount equal to

$$|S(j\omega)| = \left|\frac{1}{1+G(j\omega)H(j\omega)}\right| = \left|\frac{a(j\omega)k(j\omega)}{\delta(j\omega)}\right| \tag{6.1.20}$$

Conversely, if $|S(j\omega)| \geq 1$, the parameter sensitivity of the closed-loop system is increased by an equivalent amount.

Note that the sensitivity function $S(s)$ equals the reciprocal of the return difference, $1+G(s)H(s) = 1+L(s)$. Therefore, a closed-loop design that is relatively insensitive to plant parameter variations would be insured by a large return difference amplitude at all frequencies. This usually is impossible to achieve in practice, however, because the transfer functions of most physical systems are low pass in nature, with $|G(j\omega)| \to 0$ as $\omega \to \infty$. If we attempted to counteract this effect by employing a compensator $H(s)$ whose gain increases with frequency, the resulting plant input $u(t)$ might saturate. Moreover, any sensory measurement noise would be amplified as well. Therefore, $H(s)$ generally is chosen to be a proper transfer function, so that

$$\lim_{\omega \to \infty} \{|L(j\omega)| = |G(j\omega)H(j\omega)|\} = 0 \qquad (6.1.21)$$

and, as a consequence,

$$\lim_{\omega \to \infty} |S(j\omega)| = 1 \qquad (6.1.22)$$

It often is impossible to ensure that $|S(j\omega)| \leq 1$ for all $\omega > 0$ (see Problem 6-12), and a $|S(j\omega)| > 1$ over some frequency range does imply acceptable performance in many cases. Indeed, we will show in Section 6.3 that frequency dependent trade-offs often must be made between the magnitudes of the sensitivity and the complementary sensitivity functions in order to ensure that *all* loop performance goals are met.

Finally, it is of interest to recall that the sensitivity transfer function $S(s)$ was first defined by Eq. (5.1.16) to describe the relationship between the measured output $y(t)$ and an output-additive disturbance input $d(t)$, that is,

$$S(s) = \frac{a(s)k(s)}{\delta(s)} = \frac{1}{1 + G(s)H(s)} = \frac{y(s)}{d(s)} \qquad (6.1.23)$$

We now have shown why this particular transfer function was termed the "sensitivity" transfer function.

Robust Stability Margins from the Sensitivity Function

As we will now show, the maximum amplitude of $S(s = j\omega)$ for all $\omega \geq 0$; that is, $\max_\omega |S(j\omega)|$ can be used to obtain simple bounds on both the gain and the phase margins of stable, closed-loop systems. Indeed, a single bound on $\max_\omega |S(j\omega)|$ can be used as a measure of robust stability in virtually all closed-loop stable cases, including the nonminimum phase and open-loop unstable cases in which both the GM and the ΦM are ill-defined.

We begin with the definition of the so-called "infinity-norm" of any stable, proper, rational transfer function $T(s)$. As shown by Francis [21], the (**Hardy space**) $\mathbf{H}_\infty$**-norm**, or simply the ∞**-norm** of a rational function $T(s)$, which is both analytic and bounded in the half-plane $Re(s) > 0$, is defined by the relationship:

$$\|T\|_\infty \overset{\text{def}}{=} \max_\omega |T(j\omega)| \qquad (6.1.24)$$

and simply represents *the maximum amplitude attained by its frequency response.*

We observe next that the frequency response magnitude of the return difference, $|1 + L(j\omega)| = |1 + G(j\omega)H(j\omega)|$, equals the length of a vector drawn from -1 to $L(j\omega)$ in the complex $L(j\omega)$-plane; that is, $|1 + L(j\omega)| = |S(j\omega)|^{-1}$ represents the distance from the loop gain $L(j\omega)$ to the critical -1 point in the complex $L(j\omega)$-plane for all $\omega \geq 0$. Therefore, the inverse of the ∞-norm of the sensitivity function, namely, $\|S\|_\infty^{-1}$ *represents the minimum distance from $L(j\omega)$ to the -1 point.*

Since Eq. (6.1.22) implies that

$$\|S\|_\infty = \max_\omega |S(j\omega)| \overset{\text{def}}{=} \bar{S} \geq 1 \qquad (6.1.25)$$

it follows that

$$\|S\|_\infty^{-1} = \bar{S}^{-1} = \min_\omega |1 + L(j\omega)| \leq 1 \qquad (6.1.26)$$

Therefore, a polar plot of $L(j\omega)$ will just contact, but fail to penetrate, a circle of radius $\bar{S}^{-1}$, centered at -1 in the complex $L(j\omega)$-plane, as depicted in Figure 6.6. Note that $\bar{S}^{-1}$ (hence $\bar{S}$) can be determined from a polar plot of $L(j\omega)$ by continuously increasing the radius of a circle centered at -1 until it just contacts the $L(j\omega)$ contour. The radius of the resulting circle is then equal to $\bar{S}^{-1}$.

Minimum bounds on both the gain margin and the phase margin of a system characterized by a minimum phase $L(s)$ can be expressed directly

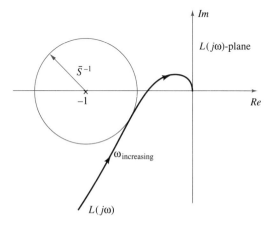

● **FIGURE 6.6**
The Loop Gain and the $\bar{S}^{-1}$ Circle

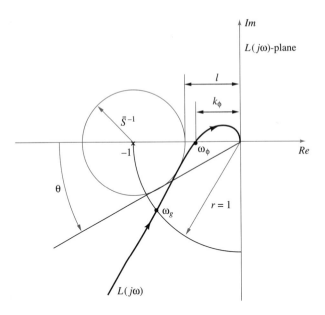

• **FIGURE 6.7**
GM and ΦM from the $\bar{S}^{-1}$ Circle

as functions of $\bar{S}$, in light of Figure 6.7, where Figure 6.6 has been embellished with an arc of radius 1, centered at the origin, to define the angle θ. Note that any loop gain point $L(j\omega)$ in Figure 6.7 will lie outside the circle of radius $\bar{S}^{-1}$, so that $|L(j\omega_\phi)| = k_\phi \leq l$. Therefore, the distance $l = 1 - \bar{S}^{-1}$ depicted in Figure 6.7 represents a lower bound on the gain margin of the system, in the sense that

$$\text{GM} \geq \frac{1}{l} = \frac{\bar{S}}{\bar{S} - 1} \qquad (6.1.27)$$

Furthermore, the angle θ depicted in Figure 6.7 represents a lower bound on the phase margin of the system, in the sense that

$$\Phi\text{M} \geq \theta = 2 \, \sin^{-1}\left(\frac{1}{2\bar{S}}\right) \qquad (6.1.28)$$

(See Problem 6-3.)

To ensure an acceptable nominal design that attains a GM ≥ 2 and a ΦM $\geq 30°$, as noted earlier, we may now require that

$$\bar{S} \leq 2 \approx 6 \text{ dB or that } \bar{S}^{-1} = \min_\omega |1 + L(j\omega)| \geq 0.5 \qquad (6.1.29)$$

so that the GM ≥ 2 and the ΦM $\geq 29° \approx 30°$, in view of Eqs. (6.1.27) and (6.1.28), respectively. However, such a requirement may be conservative, since larger values of $\bar{S}$ can imply both a GM > 2 and a ΦM $> 30°$ (see Problems 6-9 and 6-19).

Note that the single sensitivity function requirement defined by Eq. (6.1.29) can be used to replace both the GM and the ΦM requirements not only in the minimum phase cases but also in the unstable and nonminimum phase cases as well. In particular, Eq. (6.1.29) will ensure that $L(j\omega)$ remains an acceptable, "marginal" distance away from the critical -1 point, irrespective of the number and the direction of encirclements required for closed-loop stability; that is, Eq. (6.1.29) will ensure robust stability with respect to plant parameter variations once nominal closed-loop stability has been obtained.

. . . .

EXAMPLE 6.1.30 To illustrate the preceding discussion, consider a system defined by the loop gain transfer function of Example 5.4.9,

$$L(s) = K\frac{2.4s^3 + 15.36s^2 + 19.2s + 12}{s^4 + 30.08s^3 + 2.44s^2 + 1.2s} = \frac{K(s + .7 \pm j\,.714)(s + 5)}{s(s + 0.04 \pm j\,.196)(s + 30)}$$

Recall that this system is closed-loop stable for positive values of K below 0.0098, where the traditional GM and ΦM definitions would be employed. However, suppose a larger loop gain magnitude is required. Reviewing Figures 5.15 and 5.16, a polar plot of $L(s = j\omega)$, when $K = 1$, implies two clockwise encirclements of the -1 point, hence two unstable closed-loop poles. As noted in Example 5.4.9, however, if K is increased beyond 1.156, the two unstable poles cross back into the stable half-plane and remain there, approaching the zeros at $s = -0.7 \pm j\,.714$ as $K \to \infty$.

Figure 6.8 displays a partial polar plot of $L(j\omega)$ when $K = 2.5$, which is a stable value of K. Note that the $L(j\omega)$ contour just intersects a superimposed circle of radius 0.5 centered at -1, so that $\bar{S}^{-1} \approx 0.5$ and

$$\bar{S} = \|S\|_\infty \approx 2$$

thus implying acceptable robust stability with respect to plant-parameter variations.

. . . .

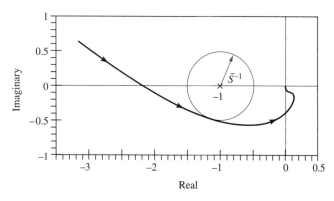

● **FIGURE 6.8**
$L(j\omega)$ and a Circle of Radius $0.5 \approx \bar{S}^{-1}$ in the $L(j\omega)$-plane

6.2   ROBUST STABILITY FOR UNMODELED DYNAMICS

Generally speaking, a nominal plant transfer function

$$G(s) = \frac{c(s)}{a(s)} \qquad (6.2.1)$$

represents a relatively good approximation of the actual plant transfer function, especially at the lower and mid-range frequencies, where Bode diagrams of $G(s = j\omega)$ often correspond to frequency response plots of the actual plant. This is not always the case at higher frequencies, where unmodeled dynamics can imply a substantial mismatch between $G(j\omega)$ and the actual frequency response of the plant.

For example, the elasticity of the structural members that comprise many mechanical systems, such as aircraft and space structures, are characterized by resonance or bending modes, which often are ignored in nominal system representations because their exact frequencies are difficult to determine. Electric motors have unmodeled poles that are located far into the stable half-plane $Re(s) < 0$. Such poles can cause unstable closed-loop performance if the loop gain (hence the bandwidth) is increased sufficiently (see Problem 6-4). Certain types of chemical and thermal processes also are characterized by uncertain high frequency behavior [50]. The selection of the loop compensator polynomials $h(s)$ and $k(s)$ should ensure *robust stability* with respect to such unmodeled high frequency dynamics as well. As we will show, a prudent design implies a sufficient attenuation of the loop gain at these higher values of ω.

A Model for Plant Uncertainty

Numerous control system investigations, which began in the early 1980s, have established several new and useful results relative to the control of systems characterized by high frequency dynamic uncertainty, many of which were motivated by the MIMO case and, therefore, extend beyond the scope of this text (see References [66] and [19]). However, we will now present one of the more relevant of these results in the SISO case.

We note first that a characterization of unmodeled, high frequency dynamic uncertainty often can be determined experimentally. Consider frequency response measurements that might be performed, as depicted in Figure 6.9, in which sinusoidal input signals $u(t) = \sin \omega t$ are applied simultaneously to both the actual plant and its nominal representation, as defined by the transfer function $G(s)$. The amplitude of the resulting output error signal $|e_a(\omega)|$ will generally increase in magnitude with increasing frequency, because unmodeled dynamic uncertainty almost always increases with frequency.

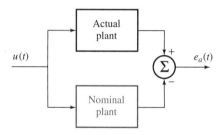

• **FIGURE 6.9**
An Experimental Determination of High Frequency Uncertainty

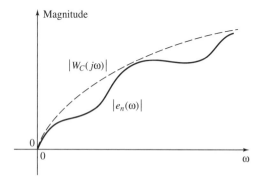

• **FIGURE 6.10**
Magnitude Plots of $|e_n(\omega)|$ and $|W_C(j\omega)|$

   If the corresponding phase information is difficult to determine, as is often the case, such unstructured uncertainty may be modeled by a **multiplicative perturbation** of the plant,[2] as defined by $\Delta(s)W_C(s)$. An **uncertainty weighting function** $W_C(s)$ can be experimentally selected to be a known, proper, stable transfer function whose frequency response amplitude represents an upper bound on that of the normalized error $e_n(\omega)$, in the sense that

$$|W_C(s = j\omega)| \geq \frac{|e_a(\omega)|}{|G(j\omega)|} \overset{\text{def}}{=} |e_n(\omega)| \quad \forall \quad \omega \geq 0 \qquad (6.2.2)$$

as depicted in Figure 6.10. Moreover, $\Delta(s)$ is a variable multiplier of $W_C(s)$, which satisfies the condition $\|\Delta\|_\infty \leq 1$, thus accounting for both phase uncertainty and amplitude variations.

   Typically,

$$|W_C(j\omega)| \begin{cases} \approx 0 \text{ at } \text{ low } \omega \\ \gg 1 \text{ at } \text{ high } \omega \end{cases} \qquad (6.2.3)$$

---

[2]It should be noted that other models for unstructured uncertainty are possible as well [19].

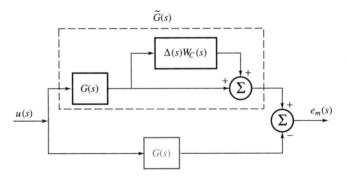

• FIGURE 6.11
The Replacement of the Actual Plant by $\tilde{G}(s)$

In such cases, the actual transfer function of the plant belongs to a group of transfer functions characterized by

$$\tilde{G}(s) = G(s)\,[1 + \Delta(s)W_C(s)] \tag{6.2.4}$$

The replacement of the actual plant by $\tilde{G}(s)$, as depicted in Figure 6.11, implies that

$$e_m(s) = \Delta(s)W_C(s)G(s) = e_a(s) \tag{6.2.5}$$

for an appropriate choice of $\Delta(s)$. However, such a $\Delta(s)$ need not be explicitly determined; that is, the mere existence of an appropriate $\Delta(s)$ is sufficient to establish the correspondence between $e_m(s)$ and $e_a(s)$, hence $e_m(t)$ and $e_a(t)$.

The High Frequency Loop Gain

The utilization of the high frequency loop gain transfer function

$$\tilde{L}(s) \overset{\text{def}}{=} \tilde{G}(s)H(s) = L(s)[1 + \Delta(s)W_C(s)] \tag{6.2.6}$$

to account for the unmodeled high frequency dynamics, then implies that a polar plot of

$$\tilde{L}(j\omega) = L(j\omega) + \Delta(j\omega)L(j\omega)W_C(j\omega) \tag{6.2.7}$$

lies within the **high frequency loop gain region** bounded by circular disks of radius

$$r(\omega) \overset{\text{def}}{=} |L(j\omega)W_C(j\omega)| \geq 0 \tag{6.2.8}$$

centered on $L(j\omega) = G(j\omega)H(j\omega)$, as depicted in Figure 6.12.

In view of the Nyquist stability criterion, any nominally stable closed-loop system characterized by such a $\tilde{L}(s = j\omega)$ would also be robustly

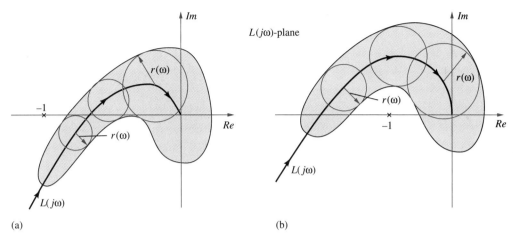

(a)   (b)

• **FIGURE 6.12**
The High Frequency Loop Gain Region

stable (relative to unmodeled high frequency dynamics) provided the bounded, high frequency loop gain region does not contact the critical -1 point. As depicted in Figure 6.12, this should be true both in case (a) when the nominal loop gain $L(j\omega)$ must not encircle the -1 point in order to ensure closed-loop stability and in case (b) when the nominal loop gain $L(j\omega)$ must encircle the -1 point in order to ensure closed-loop stability.

Since $|1+L(j\omega)|$ represents the distance from $L(j\omega)$ to the -1 point, a sufficient mathematical condition for robust closed-loop stability is that

$$r(\omega) = |L(j\omega)W_C(j\omega)| < |1+L(j\omega)| \quad \forall \quad \omega \geq 0 \qquad (6.2.9)$$

This latter relationship implies a loop gain condition for ensuring the robust closed-loop stability of a plant that is characterized by unmodeled high frequency dynamics. In particular, since $|1 + L(j\omega)| \approx 1$ at the higher values of ω, in light of Eq. (6.1.21), Eq. (6.2.9) implies that for $\omega \gg 1$,

$$|L(j\omega)W_C(j\omega)| = |L(j\omega)||W_C(j\omega)| < 1 \qquad (6.2.10)$$

It follows therefore that the high frequency loop gain condition

$$|L(j\omega)| < |W_C(j\omega)|^{-1} \ll 1 \qquad (6.2.11)$$

will ensure robust stability with respect to unmodeled high frequency dynamics. Note that Eq. (6.2.11) can be achieved by an appropriate reduction in the nominal loop gain $L(j\omega)$ at those higher frequencies where $|W_C(j\omega)| \gg 1$.

The Complementary Sensitivity Function

It is of interest to note that Eq. (6.2.9) also can be expressed in terms of the *complementary sensitivity function*

$$C(s) = \frac{L(s)}{1 + L(s)} = \frac{G(s)H(s)}{1 + G(s)H(s)} = \frac{c(s)h(s)}{\delta(s)} \qquad (6.2.12)$$

as defined by Eq. (5.1.29). Since

$$|C(j\omega)W_C(j\omega)| = \left| \frac{L(j\omega)W_C(j\omega)}{1 + L(j\omega)} \right| = \frac{|L(j\omega)W_C(j\omega)|}{|1 + L(j\omega)|} \qquad (6.2.13)$$

Eq. (6.2.9) will hold if and only if

$$|C(j\omega)W_C(j\omega)| < 1 \quad \forall \quad \omega \geq 0$$

or, equivalently, if and only if the ∞-norm of $C(s)W_C(s)$, namely,

$$\|CW_C\|_\infty < 1 \qquad (6.2.14)$$

Therefore, Eq. (6.2.14) also represents a sufficient condition for the robust closed-loop stability of a system characterized by unmodeled high frequency plant dynamics.

6.3 DISTURBANCE REJECTION AND NOISE ATTENUATION

Let us again recall the general control problem of Section 5.1, in which the two primary transfer functions associated with loop performance, namely, Eqs. (5.1.16) and (5.1.17) or Eqs. (5.1.28) and (5.1.29), served to define the *sensitivity function*

$$S(s) = \frac{1}{1 + L(s)} = \frac{a(s)k(s)}{\delta(s)} = \frac{y(s)}{d(s)} \qquad (6.3.1)$$

which represents the effect of an output-additive disturbance signal $d(t)$ on the output $y(t)$, and the *complementary sensitivity function*

$$C(s) = \frac{L(s)}{1 + L(s)} = \frac{c(s)h(s)}{\delta(s)} = \frac{y(s)}{\eta(s)} \qquad (6.3.2)$$

which represents the effect of the sensor noise $\eta(t)$ on the output $y(t)$.

Good loop performance requires a minimization of the effect of both of these undesirable, exogenous inputs on the controlled output $y(t)$. Using Eqs. (6.3.1) and (6.3.2), this implies minimizing $|S(j\omega)|$ over the band of frequencies that characterize $d(t)$, which will be termed **disturbance rejection**, while simultaneously minimizing $|C(j\omega)|$ over the band of frequencies that characterize $\eta(t)$, which will be termed **noise attenuation**. Recall, however, that $S(s)$ and $C(s)$ are mutually dependent because

$$S(s) + C(s) = \frac{1}{1 + L(s)} + \frac{L(s)}{1 + L(s)} = 1 \qquad (6.3.3)$$

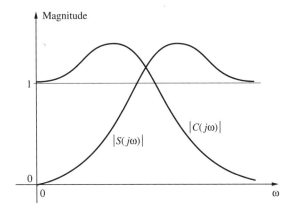

● **FIGURE 6.13**
Typical Amplitude Plots of $S(j\omega)$ and $C(j\omega)$

so that it is impossible to minimize the frequency response amplitudes of both of these functions over the same frequency bands.

Fortunately, external disturbances $d(t)$ are usually low frequency signals, such as wind forces on an antenna, waves on a ship, or terrain changes that affect the cruise control of an automobile. On the other hand, sensory noise signals $\eta(t)$ generally are high frequency in nature. Therefore, the mutually dependent loop goals of disturbance rejection and noise attenuation are not necessarily inconsistent with one another, since disturbance rejection can be attained if $|S(j\omega)| \approx 0$ at "low" frequencies, so that $|C(j\omega)| \approx 1$, and noise attenuation can be attained if $|C(j\omega)| \approx 0$ at "high" frequencies, so that $|S(j\omega)| \approx 1$. Figure 6.13 depicts typical amplitude plots of both the sensitivity function and the complementary sensitivity function that are consistent with the loop goals of disturbance rejection and noise attenuation.

Disturbance Rejection

Motivated by the uncertainty weighting function $W_C(s)$ and the ∞-norm condition for robust stability, as defined by Eq. (6.2.14), disturbance rejection can be defined by employing a frequency dependent **disturbance weighting function** $W_S(s)$, a rational function of s that characterizes the "frequency spectrum" of $d(t)$. In particular, $d(t)$ may be any signal produced at the output of a dynamic system, defined by the transfer function $W_S(s)$, whose input $d_i(t)$ depends on $d(t)$. If, for example, the disturbance is constant in nature, such as a sudden steady wind on the surface of a radar antenna, then

$$W_S(s) = \frac{K}{s}$$

with $d_i(t) = \delta(t)$, so that $d(t) = K r_s(t)$, a step function.

In general,

$$d(s) = W_S(s)d_i(s)$$

so that in light of Eq. (6.3.1),

$$\frac{y(s)}{d_i(s)} = S(s)W_S(s) \qquad (6.3.4)$$

Since disturbances are generally low frequency in nature, such a $W_S(s = j\omega)$ would be the converse of a $W_C(j\omega)$ defined by Eq. (6.2.3), namely,

$$|W_S(j\omega)| \begin{cases} \gg 1 \text{ at } \text{ low } \omega \\ \approx 0 \text{ at } \text{ high } \omega \end{cases} \qquad (6.3.5)$$

In view of Eqs. (6.2.14) and (6.3.4), the loop goal of **disturbance rejection** can then be defined by the condition

$$|S(j\omega)W_S(j\omega)| < 1 \quad \forall \quad \omega \geq 0$$

or, equivalently, the ∞-norm condition

$$\|SW_S\|_\infty < 1 \qquad (6.3.6)$$

Since

$$S(j\omega) = \frac{1}{1 + L(j\omega)} \qquad (6.3.7)$$

it follows that

$$|S(j\omega)W_S(j\omega)| = |S(j\omega)||W_S(j\omega)| = \frac{|W_S(j\omega)|}{|1 + L(j\omega)|} \qquad (6.3.8)$$

Therefore, Eq. (6.3.6) will hold if and only if

$$|W_S(j\omega)| < |S(j\omega)|^{-1} = |1 + L(j\omega)| \quad \forall \quad \omega \geq 0 \qquad (6.3.9)$$

A Graphical Interpretation

It is of interest to note that Eq. (6.3.9) has an illuminating graphical interpretation in the $L(j\omega)$-plane, which is analogous to that associated with Eq. (6.2.9). Recall that $|1 + L(j\omega)|$ represents the distance from $L(j\omega)$ to the critical -1 point. Therefore, Eq. (6.3.9) implies that

$$d(\omega) \overset{\text{def}}{=} |W_S(j\omega)| < |1 + L(j\omega)| \quad \forall \, \omega \geq 0 \qquad (6.3.10)$$

or that at every frequency ω, a disk of radius $d(\omega) = |W_S(j\omega)|$, centered on $L(j\omega)$, lies outside of the critical -1 point in the $L(j\omega)$-plane. Stated otherwise, Eq. (6.3.10) implies that the **low frequency loop gain region**, which encompasses the area bounded by circular disks of radius $d(\omega)$ $= |W_S(j\omega)|$ centered on $L(j\omega)$, does not contact the -1 point, as depicted in Figure 6.14.

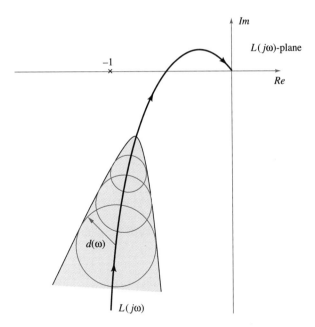

• **FIGURE 6.14**
The Low Frequency Loop Gain Region

To achieve disturbance rejection, Eq. (6.3.6) implies that the amplitude of the sensitivity function be decreased sufficiently at the lower frequencies. In light of Eq. (6.3.9) and Figure 6.14, this objective can be achieved by an appropriate amplification of the nominal loop gain $L(j\omega)$ at those lower frequencies where $|W_S(j\omega)| \gg 1$. In particular, if $|1 + L(j\omega)| \approx |L(j\omega)|$ at low values of ω, then Eq. (6.3.9) implies that the low frequency loop gain condition

$$|L(j\omega)| > |W_S(j\omega)| \gg 1 \qquad (6.3.11)$$

will ensure disturbance rejection.

Noise Attenuation

To achieve noise attenuation, Eq. (6.3.2) implies that $|C(j\omega)| \approx 0$ at those higher frequencies that characterize $\eta(t)$. As in the case of disturbance rejection, one way to ensure this loop objective is to define an appropriate **noise weighting function** $W_\eta(s)$, a stable, rational function of s that characterizes the "frequency spectrum" of $\eta(t)$. In particular, $\eta(t)$ may be any signal produced at the output of a dynamic system, defined by the transfer function $W_\eta(s)$, whose input $\eta_i(t)$ would be *white noise*; that is, a signal containing all frequencies with equal *spectral densities* [40]. In such cases,

$$\eta(s) = W_\eta(s)\eta_i(s)$$

and, in view of Eq. (6.3.2),

$$\frac{y(s)}{\eta_i(s)} = C(s)W_\eta(s) \tag{6.3.12}$$

Note that an appropriate $W_\eta(s = j\omega)$ would be analogous to the uncertainty weighting function $W_C(s)$ defined by Eqs. (6.2.2) and (6.2.3), in the sense that

$$|W_\eta(j\omega)| \begin{cases} \approx 0 & \text{at low} \quad \omega \\ \gg 1 & \text{at high} \quad \omega \end{cases} \tag{6.3.13}$$

Using Eqs. (6.2.14) and (6.3.12), the loop goal of **noise attenuation** can then be defined by the condition that $|C(j\omega)W_\eta(j\omega)| < 1 \; \forall \; \omega \ge 0$ or, equivalently, the ∞-norm condition:

$$\|CW_\eta\|_\infty < 1 \tag{6.3.14}$$

Note that Eq. (6.3.14) is analogous to Eq. (6.2.14), the condition for ensuring robust stability with respect to unmodeled high frequency dynamics. Therefore, in those cases in which unmodeled high frequency dynamics are not particularly relevant; for example, in certain "stiff" or inflexible dynamic systems, $W_C(s) = W_\eta(s)$ can be used to characterize the noise by means of Eq. (6.3.13). In light of Eq. (6.2.11), the high frequency loop gain condition:

$$|L(j\omega)| < |W_\eta(j\omega)|^{-1} \ll 1 \tag{6.3.15}$$

would then ensure noise attenuation.

Conversely, in those cases in which measurement noise is not particularly relevant, for example, in many process control situations [50], $W_C(s)$ can be employed, as in Section 6.2, to ensure robust stability with respect to the unmodeled high frequency dynamics.

In those cases in which both noise and unmodeled dynamics are relevant, the attainment of Eq. (6.2.14) may imply a $C(s = j\omega)$ that ensures adequate noise attenuation as well. In particular, the frequency dependent gain of $W_C(s)$ can be increased, as necessary, to ensure adequate noise attenuation while retaining the complementary goal of robust stabilization with respect to unmodeled high frequency dynamics.

6.4   ROBUST LOOP GOAL PERFORMANCE

The previous sections now serve to define **robust loop goal performance**, which implies the attainment of *all* of the loop goals that have been delineated. Figure 6.15 depicts a more representative version of the nominal control system depicted in Figure 5.2, which will serve as a focus for our discussions. Note that the nominal plant transfer function $G(s)$ of

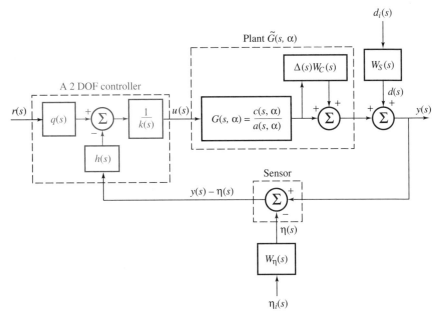

• **FIGURE 6.15**
A More Realistic Control System

Figure 5.2 has been replaced in Figure 6.15 by one that is characterized by plant parameter variations and multiplicative uncertainty, namely,

$$\tilde{G}(s, \alpha) = G(s, \alpha)\,[1 + \Delta(s)W_C(s)] \qquad (6.4.1)$$

for a known uncertainty weighting function $W_C(s)$. Moreover, both the disturbance input and the sensory noise are produced as outputs of dynamic systems defined by a known disturbance weighting function $W_S(s)$ and a known noise weighting function $W_\eta(s)$, respectively.

A primary design goal that must always be obtained is *nominal closed-loop stability*, which implies a choice of the loop compensator polynomials, $h(s)$ and $k(s)$, which ensures that

$$\delta(s) = a(s)k(s) + c(s)h(s) \qquad (6.4.2)$$

is stable. Robust loop goal performance can then be obtained if the nominal loop gain $L(s)$ also can be chosen to ensure appropriate *loop gain conditions* in each of the three frequency ranges noted earlier. Table 6.1 summarizes these loop gain conditions in each of the frequency ranges.

The three loop goal conditions defined by Eqs. (6.3.10), (6.1.29) and (6.2.9), which together imply robust loop goal performance, can be represented by the $L(j\omega)$-plane plots depicted in Figure 6.16, in both case (a) when the nominal loop gain $L(j\omega)$ must not encircle the -1 point to ensure closed-loop stability and in case (b) when the nominal loop gain $L(j\omega)$ must encircle the -1 point to ensure closed-loop stability.

TABLE 6.1 Loop Gain Conditions for Robust Loop Goal Performance

The Low Frequency Loop Gain Condition

The loop goal that is dominant at low values of ω is **disturbance rejection.** In light of Eq. (6.3.10), this objective can be achieved if

$$d(\omega) = |W_S(j\omega)| < |1 + L(j\omega)| \quad \forall\ \omega \geq 0 \tag{6.3.10}$$

so that at every frequency ω, the **low frequency loop gain region** lies outside of the critical -1 point in the complex $L(j\omega)$-plane, as depicted in Figure 6.14. In light of Eq. (6.3.11), this objective also can be obtained if, at low values of ω,

$$|L(j\omega)| > |W_S(j\omega)| \gg 1 \tag{6.3.11}$$

which represents the relevant, **low frequency loop gain condition.**

The Mid-Frequency Loop Gain Condition

The loop goal that is dominant at the mid-range values of ω is **robust stability for plant parameter variations**. In general, a bound on the magnitude of the sinusoidal sensitivity function $S(j\omega)$, such as that given by Eq. (6.1.29),

$$\bar{S} = \|S\|_\infty \leq 2 \approx 6\text{ dB} \quad \text{or a} \quad \bar{S}^{-1} = \min_\omega |1 + L(j\omega)| \geq 0.5 \tag{6.1.29}$$

will ensure that $L(j\omega)$ remains "far enough" away from the critical -1 point. The choice of a loop gain $L(j\omega)$ that satisfies Eq. (6.1.29) represents the relevant, **mid-frequency loop gain condition**.

The High Frequency Loop Gain Condition

The two loop goals that are dominant at high values of ω are **robust stability for unmodeled plant dynamics** and **noise attenuation**. In light of Eq. (6.2.9), these goals can be obtained if $W_\eta(s) = W_C(s)$, and

$$r(\omega) = |L(j\omega)W_C(j\omega)| < |1 + L(j\omega)| \quad \forall\ \omega \geq 0 \tag{6.2.9}$$

so that at every frequency ω, the **high frequency loop gain region** lies outside of the critical -1 point in the complex $L(j\omega)$-plane, as depicted in Figure 6.12. In light of Eq. (6.2.11), this objective also can be obtained if, at high values of ω,

$$|L(j\omega)| < |W_C(j\omega)|^{-1} \ll 1 \tag{6.2.11}$$

which represents the relevant, **high frequency loop gain condition**.

Using Eqs. (6.3.11) and (6.2.11), robust loop goal performance also implies that a Bode plot of the amplitude of the nominal loop gain $|L(j\omega)|$ must lie above $|W_S(j\omega)|$ at low frequencies and below $|W_C(j\omega)|^{-1}$ at high frequencies, as depicted in Figure 6.17. Stated otherwise, a Bode plot of $|L(j\omega)|$ must avoid the disturbance rejection region below $|W_S(j\omega)|$, as well as the unmodeled dynamics and noise attenuation region above

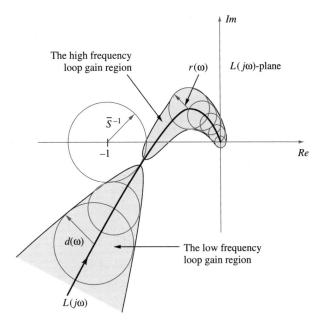

(a)

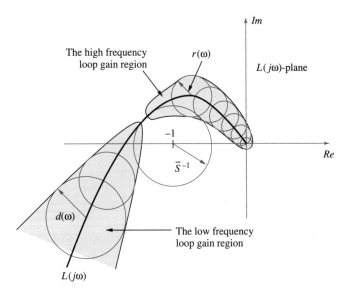

(b)

● **FIGURE 6.16**
Robust Loop Goal Performance in the $L(j\omega)$−Plane

$|W_C(j\omega)|^{-1}$, to ensure robust loop goal performance. Moreover, the phase of $L(j\omega)$ must be "consistent" with the stability margins implicit in Eq. (6.1.29). For example, $\angle L(j\omega_g)$ should be no less than $-180° + 30° = -150°$ in case (a) of Figure 6.16.

Design Trade-Offs

As we will show later, the selection of the loop compensator polynomials, $h(s)$ and $k(s)$, can be based on a frequency dependent modification of both the magnitude and the phase of the nominal loop gain $L(j\omega)$ at appropriate frequencies. Such designs are termed **loopshaping,** and the ultimate success of these procedures is dependent on both the magnitudes of and the separation between $|W_S(j\omega)|$ and $|W_C(j\omega)|^{-1}$ in Figure 6.17. Generally speaking, the greater this separation, the easier it is to "shape" $L(j\omega)$ so that all of the loop goals can be obtained. If the separation is too narrow, it may be impossible to achieve all of the loop goals, as stated, and design *trade-offs* may be required.

For example, it may be necessary to accept a smaller $|W_S(j\omega)|$ at the lower values of ω to ensure acceptable robust stability or noise attenuation. Conversely, if unmodeled high frequency dynamics are not particularly relevant, we may be willing to accept a more noisy output, by decreasing $|W_\eta(j\omega) = W_C(j\omega)|$ at the higher values of ω to obtain a desired amount of disturbance rejection.

In general, the choice of the weighting functions $W_S(s)$ and $W_C(s)$ $= W_\eta(s)$ is not at all straightforward in many situations. Indeed, most contemporary control texts, which emphasize SISO compensation, do not even define these functions, because acceptable control system designs can usually be obtained independently of their explicit utilization. In particular, SISO loopshaping designs very often imply a sinusoidal loop gain $L(j\omega)$ whose magnitude is high at low frequencies and low at high frequencies, essentially achieving the equivalent of Eqs. (6.3.11) and (6.2.11) for some implicit, undefined weighting functions $W_S(s)$ and $W_C(s)$. Nonetheless, mathematical weighting functions such as $W_C(s)$, $W_S(s)$, and $W_\eta(s)$ can serve to define the notion of robust loop goal performance in a precise mathematical manner, and it is primarily for this reason that they have been introduced here.

In the more complex MIMO cases, some significant and relatively recent advances in control system design have been based not only on the specification of explicit weighting functions but also on the subsequent optimization of appropriately "weighted" ∞-norms [66]. Although such procedures also are relevant in the SISO case [19], they will not be presented here because they require additional mathematical procedures that extend beyond the intended scope of this text. Moreover, there are more direct ways of obtaining quite acceptable closed-loop performance in the SISO case, as we will later show.

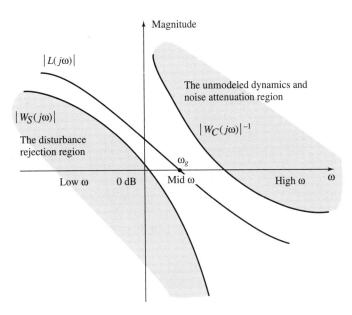

● **FIGURE 6.17**
A Bode Plot of $|L(j\omega)|$ for Robust Loop Goal Performance

Before any loop goal design techniques are detailed, however, we will first discuss the additional *response goals* that also are important to overall system performance. The attainment of robust loop goal performance does not necessarily imply acceptable response performance as well, as we will illustrate in the next chapter. Only after we have a thorough understanding of *all* of the characteristics that we would like a compensated system to possess will we be able to make informed design decisions and appropriate trade-offs.

6.5   SUMMARY

We have now outlined the various loop performance goals that should be attained by an appropriately designed, closed-loop control system—disturbance rejection, robust stability relative to both parameter variations and unmodeled dynamics, and noise attenuation.

As summarized in Table 6.1, these particular goals can be achieved if appropriate loop gain magnitude conditions are satisfied over various frequency ranges. Generally speaking, the magnitude of $L(s = j\omega)$ should be high at the lower range of frequencies and low at the higher range of frequencies. Furthermore, a polar plot of $L(j\omega)$ should avoid the critical -1 point at the mid-range frequencies, where the classical notions of gain margin (GM) and phase margin (ΦM) are defined; that is, in the minimum phase cases, a GM $\geq$ 2 and a ΦM $\geq +30°$ will generally ensure robust stability with respect to unknown plant parameter variations. Alternatively, a sensitivity function $S(s)$ whose ∞-norm $\leq$ 2, also will ensure robust stability with respect to plant-parameter variations in *all* cases.

The low and high frequency loop gain magnitude conditions for obtaining the other loop goals can be quantified more precisely using appropriate weighting functions, which are associated with some relatively new $\mathbf{H}_\infty$ results. As noted, however, the determination of these weighting functions is not at all straightforward in many situations, and most contemporary control texts do not explicitly define or employ them. Nonetheless, they can serve to define the notion of robust loop goal performance in a precise mathematical manner, and it is for this reason that they were introduced here.

PROBLEMS

6-1. Verify explicitly that for any two rational transfer functions, $G(s)$ and $H(s)$,

$$|G(j\omega)H(j\omega)| = |G(j\omega)||H(j\omega)|$$

6-2. * Consider a system defined by the minimum phase, loop gain transfer function:

$$L(s) = G(s)H(s) = \frac{20}{s^2 + 6s + 8}$$

(a) Determine the phase crossover frequency ω_ϕ and the GM.

(b) Determine the gain crossover frequency ω_g and the ΦM.

(c) Sketch a polar plot of this system, and determine its closed-loop stability properties using the Nyquist stability criterion, $Z = N + P$.

6-3. In light of Figure 6.7, show that

$$\theta = 2 \ \sin^{-1} \left(\frac{1}{2\bar{S}} \right) \leq \Phi M$$

where ΦM is defined in Figure 6.1.

6-4. Consider the systems defined by the loop gain transfer functions:

$$L_1(s) = \frac{K}{(\tau_1 s + 1)(\tau_2 s + 1)} \quad \text{and} \quad L_2(s) = \frac{K}{(\epsilon s + 1)(\tau_1 s + 1)(\tau_2 s + 1)}$$

with $0 < \epsilon \ll \tau_1 < \tau_2$. Sketch a root locus for both of these systems. Discuss the effect that the additional pole at $s = -\epsilon^{-1}$ has on the closed-loop stability properties of the systems as $K \to \infty$.

6-5. * Determine the locations of the closed-loop poles in Example 6.1.30 when the loop gain magnitude $K = 2.5$.

6-6.* Determine both the gain margin and the phase margin of a minimum phase system defined by the nominal transfer function

$$G(s) = L(s) = \frac{10}{(s+1)(0.1s+1)}$$

6-7.* Consider an open-loop system defined by the differential equation:

$$\ddot{y}(t) + \dot{y}(t) = Ku(t)$$

(a) Sketch a polar plot of $G(s) = \frac{y(s)}{u(s)}$.

(b) For what value of the gain K is the ΦM equal to 45°?

(c) Determine the gain crossover frequency ω_g for this value of K.

(d) Determine the GM when $K = 10$.

(e) Determine both ω_g and the ΦM when $K = 5$.

6-8. Consider a system whose nominal dynamic behavior is defined by the state-space equations:

$$\dot{x}_1(t) = x_2(t); \qquad \dot{x}_2(t) = Ku(t); \qquad y(t) = x_1(t) + x_2(t)$$

(a) Sketch a polar plot of $G(s) = \frac{y(s)}{u(s)}$.

(b) Determine the value of K for which the ΦM = 30°.

(c) Determine the gain crossover frequency ω_g when $K = 2$.

6-9.* Consider a system defined by the nominal loop gain transfer function

$$L(s) = \frac{K(0.05s+1)}{(s+1)(0.5s+1)(0.1s+1)}$$

(a) Determine graphically the approximate ω_g when $K = 10$ from an asymptotic amplitude Bode plot of the loop gain.

(b) Determine the corresponding approximate ΦM.

(c) Determine the actual phase crossover frequency, ω_ϕ, and the corresponding GM.

(d) Plot Bode diagrams of both $|S(j\omega)|$ and $|C(j\omega)|$ for $1 \le \omega \le 10$.

(e) Use $\bar{S} = \max_\omega |S(j\omega)|$ to obtain bounds on both the GM and the ΦM, and compare your results to the stability margins obtained in Parts (b) and (c).

6-10. Consider a system defined by a nominal, minimum phase, loop gain transfer function $L(s)$.

(a) Determine the real value of the loop gain magnitude at the phase crossover frequency, $|L(j\omega_\phi)|$, if the GM = +10 dB.

(b) Determine the phase angle of the loop gain at the gain crossover frequency, $\angle L(j\omega_g)$, if the ΦM = 50°.

6-11. Consider a system defined by a minimum phase $L(s)$. Show that when the ∞-norm of its sensitivity function equals 1, that is, when

$$\|S\|_\infty = \max_\omega |S(j\omega)| = \bar{S} = 1$$

then the closed loop system will be characterized by a GM $= \infty$ and a $\Phi M \geq 60°$.

6-12. Show that if a minimum phase system is defined by a nominal $L(s)$ of relative order ≥ 3, that is, if $L(s)$ has three or more poles than zeros, then the ∞-norm of its sensitivity function will be greater than 1.

6-13. Show that the condition $\|SW_S\|_\infty < 1$, as defined by Eq. (6.3.6), implies that at every frequency ω, the loop gain point $L(j\omega)$ lies outside of a circle of radius $d(\omega) = |W_S(j\omega)|$ centered at the -1 point in the $L(j\omega)$-plane.

6-14. If a minimum phase system must have a GM ≥ 3 and a $\Phi M \geq 45°$, determine a corresponding, upper bound on its $\bar{S} = \|S\|_\infty$.

6-15. Guided by your answer to the previous problem, determine a general expression for the maximum allowable $\bar{S}$ as a function of defined, simultaneous bounds on both the GM and the ΦM.

6-16.* Consider a system defined by the nominal open-loop transfer function

$$G(s) = \frac{1}{0.2s + 1} = \frac{5}{s + 5}$$

that is characterized by high frequency dynamic behavior, as defined by the actual transfer function

$$\tilde{G}(s) = G(s)[1 + \Delta(s)W_C(s)] = G(s)\frac{100(s - 10)^2}{s^2 + 20s + 10,000}$$

(a) Determine explicitly $W_C(s)$ by assuming a $\Delta(s) = 1$ (which reflects exact knowledge of $\tilde{G}(s)$).

(b) Verify that Eq. (6.2.3) holds in this case.

(c) Show that the nominal closed-loop system is stable, but that the actual closed-loop system is not.

(d) Use Bode plots to verify that the high frequency loop gain condition Eq. (6.2.11) is not satisfied in this case.

(e) Show that a choice of the loop compensator

$$H(s) = \frac{h(s)}{k(s)} = \frac{1}{s + 1}$$

reduces the actual loop gain $L(j\omega) = \tilde{G}(j\omega)H(j\omega)$ sufficiently at the higher frequencies to ensure closed-loop stability.

6-17. Suppose the -1 point in Figure 6.16 is enclosed by the low frequency loop gain region defined by circles of radius $d(\omega)$. Assuming a stable nominal $\delta(s)$, what effect would such an enclosure have on actual closed-loop stability?

6-18. Show that if $W_S(s)$ is a *type* $p > 0$ transfer function, that is, if

$$W_S(s) = \frac{1}{s^p} \hat{W}_S(s)$$

with $\hat{W}_S(0)$ both nonzero and finite, then the loop gain $L(s)$ also must be type p or higher to ensure disturbance rejection, as defined by Eq. (6.3.6).

6-19. Sketch a polar plot of a minimum phase $L(j\omega)$ that is characterized by a GM $= \infty$, a $\Phi M = 45°$, and a $\bar{S} = 4$. What measure of robust stability would be the appropriate one to use in this case?

6-20. Determine both $L(s)$ and $C(s)$ for a system characterized by a

$$S(s) = \frac{s^3 + 3s^2 + 2s + 2}{s^3 + 5s^2 + 3s + 1}$$

6-21. Determine $H(s)$ if

$$G(s) = \frac{2}{s^2 + 2s} \quad \text{and} \quad C(s) = \frac{4}{s^2 + 4s + 4}$$

6-22. Consider a plant defined by the nominal state-space representation: $\dot{\mathbf{x}}(t) = A\mathbf{x}(t) + Bu(t); \; y(t) = C\mathbf{x}(t)$, with

$$A = \begin{bmatrix} 0 & 1 & 0 \\ 0 & 0 & 1 \\ 0 & 3 & -2 \end{bmatrix}, \qquad B = \begin{bmatrix} 0 \\ 0 \\ 1 \end{bmatrix}, \qquad C = [2 \quad 1 \quad 0]$$

(a) Determine the nominal transfer function $G(s) = \frac{y(s)}{u(s)}$.

(b) Determine the transfer function of a loop compensator $H(s)$ that yields the desired sensitivity function:

$$S(s) = \frac{1}{1 + G(s)H(s)} = \frac{s^3 + 4s^2 + 3s}{s^3 + 4s^2 + 4s + 2}$$

(c) Discuss the controllability, the observability, and the stability properties of the resulting closed-loop system.

7

RESPONSE GOALS

7.1 OUTPUT REGULATION

Looking back at Figure 5.2, the **response performance** of a controlled system is determined by the manner in which its output $y(t)$ responds to the exogenous signals $r(t)$, $d(t)$, and $\eta(t)$. Generally speaking, good response performance is achieved when $y(t)$ follows the reference input $r(t)$ as closely as possible, despite the undesirable effects caused by both the disturbance signal $d(t)$ and the sensor noise $\eta(t)$.

We begin our discussion of the specific output response goals that we would like to achieve with the subject of *regulation*. The regulation properties of a stable system are characterized by the transition of its entire state to the zero state from any arbitrary set of nonzero initial conditions, in the absence of any external input signals. Therefore, using Eqs. (2.5.9) and (2.5.16),

$$\mathbf{x}(t) = e^{A(t-t_0)}\mathbf{x}(t_0) = V\left[e^{\Lambda(t-t_0)}\right]V^{-1}\mathbf{x}(t_0) \qquad (7.1.1)$$

so that the natural modes $e^{\lambda_i t}$ of a system, as defined by the eigenvalues λ_i of the state matrix A, determine state regulation. The same can be said of **output regulation** [6], which is defined as the manner in which a stable plant output

$$y(t) = C\mathbf{x}(t) = Ce^{A(t-t_0)}\mathbf{x}(t_0) = \sum \alpha_i e^{\lambda_i(t-t_0)}$$

returns to an equilibrium value of 0 from any $y(t_0) \neq 0$.

Now consider a stable closed-loop system, as depicted in Figure 5.2, so that

$$y(s) = \underbrace{\frac{c(s)q(s)}{\delta(s)}}_{T(s)} r(s) + \underbrace{\frac{a(s)k(s)}{\delta(s)}}_{S(s)} d(s) + \underbrace{\frac{c(s)h(s)}{\delta(s)}}_{C(s)} \eta(s) \qquad (7.1.2)$$

in view of Eqs. (5.1.12) through (5.1.17). Output regulation is determined by the closed-loop poles p_i of the system, that is, the roots of the polynomial

$$\delta(s) = a(s)k(s) + c(s)h(s) \qquad (7.1.3)$$

that correspond to the eigenvalues of A in Eq. (7.1.1). These poles define the natural modes of the closed-loop system, namely, $e^{p_i t}$. Clearly, the farther away from the $j\omega$-axis these poles are in the stable half-plane $Re(s = \sigma + j\omega) < 0$, the faster the response of $y(t) = \sum \alpha_i e^{p_i(t-t_0)}$ to 0 and the better the output regulation properties of the compensated system.

This observation may serve to motivate a compensator design based on a positioning of the roots of $\delta(s)$ "far enough" into the left-half s-plane to obtain excellent output regulation. However, such a design could increase the bandwidth of the system sufficiently to jeopardize the loop performance of the system or produce plant input $u(t)$ saturation. For example, a control valve can open only so far, fluid flow cannot be increased without limit, and an aircraft's control surfaces can be deflected only so much. Input commands that exceed actual physical limits often result in unpredictable nonlinear behavior. Therefore, design trade-offs may be necessary to achieve robust loop goal performance, in addition to acceptable output regulation.

An LQR Performance Index

An alternative way of obtaining desirable output regulation, without requiring an excessive plant input signal, is to employ a compensator that minimizes an appropriate "performance index." One of the most useful and commonly employed performance indexes is a so-called (output) **linear quadratic regulator** (LQR) **performance index**, such as that defined by the **H_2-norm** [21], or simply the 2-norm:

$$J_{\text{LQR}} = \int_0^\infty \left\{ \rho y^2(t) + u^2(t) \right\} dt \qquad (7.1.4)$$

The minimization of J_{LQR} implies a desire to minimize both excessive output $y(t)$ excursions and the control $u(t)$ effort required to prevent such excursions. The adjustable *weighting factor* $\rho > 0$ can be used to obtain appropriate trade-offs between these two conflicting goals. As we will later show, a controller that minimizes an LQR performance index implies an "optimal" positioning of the closed-loop system poles.

7.2   THE TRANSIENT TRACKING RESPONSE

Output regulation is an important response goal. In many control cases, however, the primary response goal is **output tracking**, which requires that the plant output $y(t)$ follow or "track" a nonzero reference input $r(t)$

as closely as possible. For example, a communications antenna may be required to change from one fixed pointing position to another to receive a transmission from a new source; the end effector of a robot often must move along a desired trajectory to perform a designated task, such as spray painting or seam welding; an automatic camera, recording the lift-off of a space vehicle, has to follow the acceleration trajectory of the vehicle; and certain computer-disk memory systems must be designed to follow a sinusoidal data track.

In light of Eq. (7.1.2), the output response of the nominal control system depicted in Figure 5.2, relative to a nonzero reference input $r(t)$, is defined by the relationship:

$$y(s) = T(s)r(s) = \frac{c(s)q(s)}{a(s)k(s) + c(s)h(s) = \delta(s)}r(s) \qquad (7.2.1)$$

irrespective of a disturbance $d(s)$ or sensor noise $\eta(s)$. In many physical situations, $r(t)$ is a known, nondiminishing signal, such as a step, ramp, or parabolic function, or a sinusoid of known frequency. In such cases, $r(t)$ can be generated as the impulse response of a system defined by the strictly proper transfer function

$$\frac{m_r(s)}{p_r(s)} = r(s) = \mathcal{L}[r(t)] \qquad (7.2.2)$$

whose poles, the roots of $p_r(s)$, lie on the $j\omega$-axis in the complex s-plane. These roots of $p_r(s)$ define the **modes** of the reference input $r(t)$.

Using Eq. (4.2.13), Eqs. (7.2.1) and (7.2.2) imply that

$$y(s) = T(s)r(s) = \frac{c(s)q(s)m_r(s)}{\delta(s)p_r(s)} = \underbrace{\frac{\hat{m}_{cq}(s)}{\delta(s)}}_{y_n(s)} + \underbrace{\frac{\hat{m}_r(s)}{p_r(s)}}_{y_f(s)}$$

so that the **output response** of the system to $r(t)$ is defined by

$$y(t) = \mathcal{L}^{-1}[T(s)r(s)] = \underbrace{\mathcal{L}^{-1}\left[\frac{\hat{m}_{cq}(s)}{\delta(s)}\right]}_{y_n(t)} + \underbrace{\mathcal{L}^{-1}\left[\frac{\hat{m}_r(s)}{p_r(s)}\right]}_{y_f(t)} \qquad (7.2.3)$$

the sum of the **natural response** $y_n(t)$ and the **forced response** $y_f(t)$. Since $\delta(s)$ is assumed to be stable, the *final value theorem* implies that

$$\lim_{t \to \infty} y_n(t) = \lim_{s \to 0}\left[\frac{s\hat{m}_{cq}(s)}{\delta(s)}\right] = 0 \qquad (7.2.4)$$

so that $y_n(t)$ defines the transient component, and $y_f(t)$ defines the nondiminishing component of the output tracking response.

As in the case of output regulation, the output transient response of a system also depends on the roots of $\delta(s)$, which define the natural modes of the closed-loop system. A desirable output transient response $y_n(t)$ is one that decays to zero as quickly as possible. For example, recall the typical unit step output response depicted in Figure 4.5. Some desirable transient step response goals to be achieved through a proper compensator design would be a fast rise time t_r, a small (possibly zero) maximum overshoot M_p, and a short settling time t_s. Such goals are consistent with good output regulation and, as a consequence, are subject to the same concerns noted earlier—the desire to achieve a fast transient response may compromise the loop performance of the system or cause plant input saturation.

A 2 DOF Design Procedure

In view of the preceding discussion, it would be useful to have a design procedure that maintains a desired response performance while varying the loop performance. Such a design is possible with a 2 DOF compensator, as we will now show. Consider a system defined by a rational transfer function

$$G(s) = \frac{y(s)}{u(s)} = \frac{c(s)}{a(s)} \tag{7.2.5}$$

which is both strictly proper and minimal (controllable and observable), so that the nominal plant polynomials $a(s)$ and $c(s)$ are coprime, with $\deg[a(s)] = n > \deg[c(s)] = m$.

The **pole placement algorithm** of Appendix B now can be used to determine a monic $(n-1)$ degree polynomial

$$k(s) = s^{n-1} + k_{n-2}s^{n-2} + \cdots + k_1 s + k_0 \tag{7.2.6}$$

and a corresponding $(n-1)$ degree polynomial

$$h(s) = h_{n-1}s^{n-1} + h_{n-2}s^{n-2} + \cdots + h_1 s + h_0 \tag{7.2.7}$$

which will ensure **arbitrary pole placement**; that is, the $n-1$ arbitrary coefficients of $k(s)$, namely, $k_0, k_1, \ldots, k_{n-2}$, plus the n arbitrary coefficients of $h(s)$, namely, $h_0, h_1, \ldots, h_{n-1}$, will define the $n-1+n = 2n-1$ coefficients of any desired monic $\delta(s) = a(s)k(s) + c(s)h(s)$ of degree $2n-1$.[1] Therefore, $h(s)$ and $k(s)$ can be chosen to solve the *Diophantine equation*

$$\delta(s) = a(s)k(s) + c(s)h(s) = \hat{\delta}(s)\hat{q}(s) \tag{7.2.8}$$

with

$$\deg[\hat{\delta}(s)] + \deg[\hat{q}(s)] = 2n-1 \tag{7.2.9}$$

and both $\hat{\delta}(s)$ and $\hat{q}(s)$ are arbitrary, stable, monic polynomials.

---

[1] Actually, $\delta(s)$ can be any arbitrary polynomial of degree $> 2n-1$, provided $k(s)$ and $h(s)$ are allowed to be of degree $> n-1$. For example, see Section 7.3 and Problem 7-16.

This observation now serves to motivate a **2 DOF design procedure** for a system defined by the open-loop $G(s)$ of Eq. (7.2.5), based on an initial choice of a desired, closed-loop transfer function

$$T(s) = \frac{y(s)}{r(s)} = \frac{\alpha c(s)}{\hat{\delta}(s)} \qquad (7.2.10)$$

with α an arbitrary scalar and $\deg[\hat{\delta}(s)] = n = \deg[a(s)]$. In particular, the determination of $h(s)$ and $k(s)$ using Eq. (7.2.8) for any arbitrary but stable $\hat{q}(s)$ of degree $n - 1 = \deg[k(s)]$, along with a choice of

$$q(s) = \alpha \hat{q}(s) \qquad (7.2.11)$$

implies the nominal closed-loop system depicted in Figure 7.1. Such a system is characterized by the desired $T(s)$; that is, using Eq. (7.2.1), Eqs. (7.2.8) and (7.2.11) imply that

$$T(s) = \frac{c(s)q(s)}{a(s)k(s) + c(s)h(s)} = \frac{\alpha c(s)\hat{q}(s)}{\hat{\delta}(s)\hat{q}(s)} = \frac{\alpha c(s)}{\hat{\delta}(s)} \qquad (7.2.12)$$

as desired. It may be noted that the pole-zero cancellation of the polynomial $\hat{q}(s)$ in $T(s)$ corresponds to a lack of closed-loop system controllability (see Problem 7-2).

Note that each choice for $\hat{q}(s)$, for any fixed α and $\hat{\delta}(s)$ (hence a fixed $T(s)$ as well) will produce a different but unique choice for $h(s)$ and $k(s)$ using Eq. (7.2.8). Therefore, by choosing $q(s) = \alpha \hat{q}(s)$ to "cancel" $n - 1$ of the nominal closed-loop poles of the system, it is possible to obtain the same output response transfer function for an infinite number of different loop compensators. A subsequent evaluation of the loop performance associated with each $\{h(s), k(s)\}$ pair can then be made to determine whether or not robust loop goal performance is achieved, along with the desired output response.

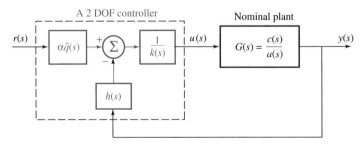

• **FIGURE 7.1**
A 2 DOF Design Procedure

. . . .

EXAMPLE 7.2.13 To illustrate the preceding discussion, consider an unstable, second-order system defined by the nominal state-space representation: $\dot{\mathbf{x}}(t) = A\mathbf{x}(t) + Bu(t)$; $y(t) = C\mathbf{x}(t)$, with

$$A = \begin{bmatrix} 0 & 1 \\ 0 & 1 \end{bmatrix} \qquad B = \begin{bmatrix} 0 \\ 1 \end{bmatrix} \qquad C = [3 \quad 0]$$

which implies the nominal plant transfer function

$$\frac{y(s)}{u(s)} = C(sI - A)^{-1}B = G(s) = \frac{3}{s^2 - s} = \frac{c(s)}{a(s)}$$

Suppose we wish to obtain the output response associated with the desired, second-order (since $n = 2$ in this case),

$$T(s) = \frac{\omega_n^2}{s^2 + 2\zeta\omega_n s + \omega_n^2} = \frac{2.44}{s^2 + 2.4s + 2.44} = \frac{\alpha c(s)}{\hat{\delta}(s)}$$

as defined by Eq. (7.2.10), which is characterized by closed-loop poles at $s = -1.2 \pm j1$, an undamped natural frequency $\omega_n = 1.562$, and a damping ratio $\zeta = 0.768$, with $\alpha = \omega_n^2/3 = 0.813$. Note that such a design implies a desirable transient step response, in view of Figure 4.7.

Three distinct choices for $\hat{q}(s) = s + \hat{q}_0$ will now be considered, depending on the value selected for $\hat{q}_0$—a low $\hat{q}_l(s) = s + 0.2$, a medium $\hat{q}_m(s) = s + 2$, and a high $\hat{q}_h(s) = s + 20$. In the low case, when $\hat{q}_l(s) = s + 0.2$, Eqs. (7.2.6), (7.2.7), and (7.2.8) imply that

$$\delta_l(s) = \underbrace{(s^2 - s)}_{a(s)}\underbrace{(s + k_{0l})}_{k_l(s)} + \underbrace{3}_{c(s)}\underbrace{(h_{1l}s + h_{0l})}_{h_l(s)} = \underbrace{(s^2 + 2.4s + 2.44)}_{\hat{\delta}(s)}\underbrace{(s + 0.2)}_{\hat{q}_l(s)}$$

or that

$$\delta_l(s) = s^3 + (k_{0l} - 1)s^2 + (3h_{1l} - k_{0l})s + 3h_{0l} = s^3 + 2.6s^2 + 2.92s + 0.488$$

so that $h_{0l} = 0.488/3 = 0.163$, $k_{0l} = 2.6 + 1 = 3.6$, and $h_{1l} = (2.92 + k_{0l} = 6.52)/3 = 2.17$. Therefore, the unique loop compensator polynomials are given by

$$h_l(s) = 2.17s + 0.163 \quad \text{and} \quad k_l(s) = s + 3.6$$

Analogous computations in the medium and high cases subsequently imply (see Problem 7-3) that for $\hat{q}_m(s) = s + 2$,

$$h_m(s) = 4.21s + 1.63 \quad \text{and} \quad k_m(s) = s + 5.4$$

while for $\hat{q}_h(s) = s + 20$,

$$h_h(s) = 24.61s + 16.26 \quad \text{and} \quad k_h(s) = s + 23.4$$

In all three cases, the unit step output response is given by

$$y(t) = \mathcal{L}^{-1}\left[T(s)r(s) = \frac{2.44}{s^3 + 2.4s^2 + 2.44s} \right] = 1 + 1.56e^{-1.2t}\sin[t - 2.45]$$

as depicted in Figure 7.2. Moreover, in view of Eq. (5.1.18), the corresponding plant input response is given by

$$u(t) = \mathcal{L}^{-1}\left[u(s) = \frac{a(s)q(s)}{\delta(s)}r(s) = \frac{\alpha a(s)\hat{q}(s)}{\hat{\delta}(s)\hat{q}(s)}r(s) = \frac{\alpha a(s)}{\hat{\delta}(s)}\frac{1}{s} \right] \qquad (7.2.14)$$

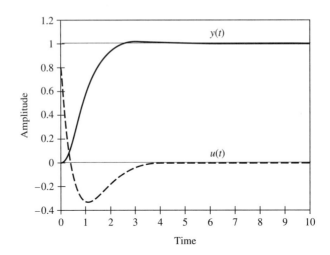

• FIGURE 7.2
A Plot of $y(t)$ and $u(t)$

in general, or

$$u(t) = \mathcal{L}^{-1}\left[\frac{0.813s - 0.813}{s^2 + 2.4s + 2.44}\right] = 1.97e^{-1.2t}\sin(t + 2.72)$$

in this example, which also is depicted in Figure 7.2.

Note that $u(t)$ attains its maximum value at the initial time $t = 0$, which can be determined using the initial value theorem. In particular, since

$$\lim_{t \to 0} u(t) = u(0) = \lim_{s \to \infty} su(s) \tag{7.2.15}$$

Eq. (7.2.14) implies that

$$\max_t u(t) = u(0) = \lim_{s \to \infty} s\frac{\alpha a(s)}{\hat{\delta}(s)}\frac{1}{s} = \lim_{s \to \infty} \frac{0.813(s^2 - s)}{s^2 + 2.4s + 2.44} = \alpha = 0.813$$

We will assume that this maximum value of $u(t)$ does not exceed any plant input saturation limits.

It is now of interest to compare amplitude Bode plots in the low, medium, and high cases of both the sensitivity function

$$|S(s = j\omega)| = \left|\frac{a(j\omega)k(j\omega)}{\delta(j\omega)}\right| = \left|\frac{a(j\omega)k(j\omega)}{\hat{\delta}(j\omega)\hat{q}(j\omega)}\right|$$

as depicted in Figure 7.3, and the loop gain

$$|L(s = j\omega)| = \left|\frac{c(j\omega)h(j\omega)}{a(j\omega)k(j\omega)}\right|$$

as depicted in Figure 7.4. In both sets of Bode plots, the low, medium, and high cases are represented by the solid, dashed, and dotted lines, respectively.

Note that the high case plots depict a system characterized by a wider loop gain $L(j\omega)$ bandwidth and a resulting lower overall sensitivity $S(j\omega) = [1 + L(j\omega)]^{-1}$ when compared with the low

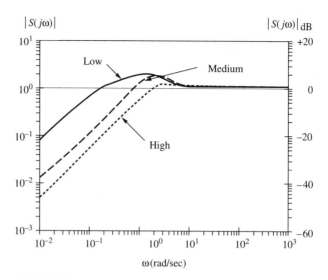

● **FIGURE 7.3**
Amplitude Bode Plots of $|S(j\omega)|$

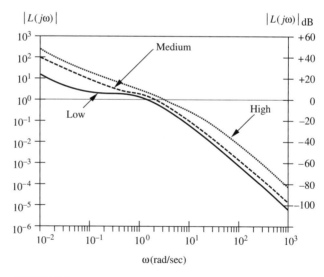

● **FIGURE 7.4**
Amplitude Bode Plots of $|L(j\omega)|$

and medium cases. This is to be expected, however, because larger loop gains are required to position the closed-loop poles farther into the left-half s-plane. Recall that in the high case, an $h_h(s) = 24.61s + 16.26$ is required to place the "free" closed-loop pole at $s = -20$, as compared with the $h_m(s) = 4.21s + 1.63$ and $h_l(s) = 2.17s + 0.163$ required for its placement at $s = -2$ and

$s = -0.2$ in the medium and low cases, respectively. Although larger loop gains generally imply a faster response, hence better output regulation and reduced sensitivity, they also imply an increase in the loop gain bandwidth, which could compromise the loop goal performance.

Also note that the ∞-norm of the sensitivity function in the low case, namely,

$$\|S_l\|_\infty = \max_\omega |S_l(j\omega)| = \bar{S}_l = 1.94$$

with $\bar{S}_m = 1.80$ and $\bar{S}_h = 1.16$ in the medium and high cases, respectively. Therefore, $\bar{S} \leq 2$ in all three cases, which implies robust stability with respect to plant parameter variations, in light of Eq. (6.1.29).

Satisfaction of the remaining loop goals depends on the disturbance weighting function $W_S(s)$ as well as the uncertainty weighting function $W_C(s)$, which are not explicitly defined in this example. It should be noted, however, that if $W_S(s)$ and $W_C(s)$ were known, a plot of $|W_S(j\omega)|$ and $|W_C(j\omega)|^{-1}$ superimposed on Figure 7.4, as in Figure 6.17, would enable the designer to determine which, if any, of the three designs achieves robust loop goal performance, in addition to the desired output response.

Whether or not $W_S(s)$ and $W_C(s)$ are known, note the obvious trade-offs between disturbance rejection and low sensitivity on the one hand and robust stability with respect to unmodeled dynamics and sensory noise attenuation on the other hand, as illustrated by the loop gain magnitude plots of Figure 7.4. Although the low case reflects the best unmodeled dynamic robust stability and noise attenuation, it also is characterized by the highest sensitivity as well as the poorest disturbance rejection and output regulation qualities. The converse is true in the high case, with the medium case representing a compromise between the other two.

The three cases considered are by no means the only ones possible—there are an infinite number of choices for $\hat{q}(s)$. For example, the designer may select a value for q_0 in $\hat{q}(s) = s + q_0$, hence corresponding values for $h(s)$ and $k(s)$, which result in a $|L(j\omega)|$ plot just below that of a known $|W_C(j\omega)|^{-1}$. Such a design not only would ensure robust stability with respect to unmodeled dynamics as well as acceptable noise attenuation but also would "optimize" the disturbance rejection and sensitivity properties of the closed-loop system relative to the desired output response. Once a desired $T(s)$ has been selected, it should be possible to determine a 2 DOF compensator that reflects acceptable loop performance as well by evaluating a reasonable number of alternative designs.

The 1 DOF Restriction

We have now introduced and illustrated a 2 DOF design procedure for obtaining a closed-loop system characterized by a desired, fixed, output response transfer function, but with variable loop performance. It is of interest to illustrate the degradation in the transient response performance that would result if the compensator were restricted to be a more traditional, 1 DOF error-driven type, as depicted by the unity feedback configuration of Figure 7.5.

If we were to employ any one of the three $\{h(s), k(s)\}$ loop compensator pairs derived in Example 7.2.13, but with $q(s) = h(s)$, rather than $\alpha\hat{q}(s)$, the complementary sensitivity function $C(s)$ then would define the

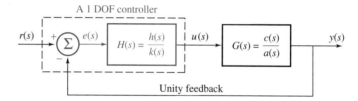

- **FIGURE 7.5**
A 1 DOF Unity Feedback Configuration

output response transfer function, as in Eq. (5.1.25); that is, Eq. (7.2.1) would imply that

$$y(s) = \underbrace{\frac{c(s)h(s)}{\delta(s)}}_{C(s)} r(s) = \frac{c(s)h(s)}{\hat{\delta}(s)\hat{q}(s)} r(s) \qquad (7.2.16)$$

and Eq. (7.2.14) would imply that

$$u(s) = \frac{a(s)h(s)}{\delta(s)} r(s) = \frac{a(s)h(s)}{\hat{\delta}(s)\hat{q}(s)} r(s) \qquad (7.2.17)$$

Time response plots of the resulting unit step output responses $y(t) = \mathcal{L}^{-1}[y(s)]$ and the unit step input responses $u(t) = \mathcal{L}^{-1}[u(s)]$ in the three cases considered in Example 7.2.13 are depicted in Figure 7.6. As in the $|S(j\omega)|$ and $|L(j\omega)|$ Bode plots of Figures 7.3 and 7.4, the low, medium, and high cases are represented by the solid, dashed, and dotted lines, respectively. The 2 DOF $y(t)$ and $u(t)$ plots of Figure 7.2 also are depicted in Figure 7.6 by the dash-dotted lines for comparison purposes. These plots display a considerable $y(t)$ overshoot in all three cases, which is due to the numerator zero term $h(s)$ in Eq. (7.2.16), as discussed in Section 4.2. The low case also is characterized by an excessive settling time, which is due to the presence of the uncanceled $\hat{q}_l(s)$ output mode $e^{-0.2t}$.

The maximum initial 1 DOF values for $u(t)$ also exceed the maximum 2 DOF $u(t)$ value of 0.813 by substantial amounts. In particular, in light of Eqs. (7.2.15) and (7.2.17),

$$u(0) = \lim_{s \to \infty} \left[\frac{a(s)h(s)}{\delta(s)} = \frac{(s^2 - s)(h_1 s + h_0)}{(s^2 + 2.4s + 2.44)(s + q_0)} \right] = h_1$$

so that in the 1 DOF cases,

$$u_l(0) = 2.17, \quad u_m(0) = 4.21, \quad \text{and} \quad u_h(0) = 24.61$$

as shown by the $u(t)$ plots of Figure 7.6. Therefore, the 1 DOF designs could imply plant input saturation.

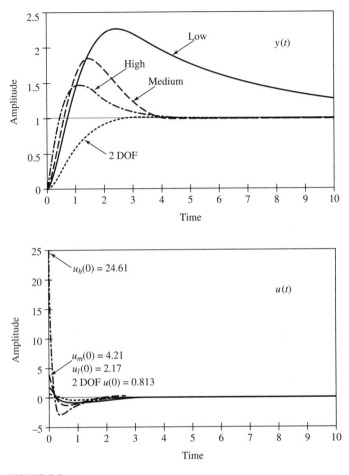

● **FIGURE 7.6**
1 DOF Plots of $y(t)$ and $u(t)$

Plant Parameter Variations

The 2 DOF design procedure for obtaining a fixed response performance, while varying the loop performance, depends on an exact cancellation of the pole-zero pair defined by $\hat{q}(s)$ in Eq. (7.2.12). This can rarely be achieved in practice. Variations in the nominal plant polynomials, $a(s)$ and $c(s)$, will generally imply a change in all $(2n-1)$ roots of $\delta(s) = a(s)k(s) + c(s)h(s)$, so that the fixed compensator polynomial $q(s) = \alpha\hat{q}(s)$ will not divide $\delta(s)$ exactly. As a consequence, natural closed-loop modes "close to" those defined by the $(n-1)$ roots of $\hat{q}(s)$ will appear in the output response $y(t)$. However, their effect may be negligible if, for example, their defining poles are almost cancelled by the zeros of $q(s)$ or they are relatively fast. Therefore, the choice of $\hat{q}(s)$ in Eq. (7.2.8) should not result in its roots being positioned too close to the $j\omega$-axis, especially when there are significant plant parameter variations, because poor output regulation and excessive settling times could result (see Problem 7-8).

7.3   THE STEADY-STATE TRACKING RESPONSE

Desirable output tracking properties generally imply not only a well-behaved transient response, hence good output regulation, but also a minimal error, as defined by Eq. (5.1.12),

$$e(s) = \frac{a(s)k(s) + c(s)[h(s) - q(s)]}{a(s)k(s) + c(s)h(s) = \delta(s)} r(s) - \underbrace{\frac{a(s)k(s)}{\delta(s)}}_{S(s)} d(s) - \underbrace{\frac{c(s)h(s)}{\delta(s)}}_{C(s)} \eta(s)$$

(7.3.1)

Clearly, a very desirable response goal is to obtain a **robust, zero steady-state error**, which is defined by the condition that

$$e_{ss}(t) = \lim_{t \to \infty} [e(t) = r(t) - y(t)] = 0 \tag{7.3.2}$$

despite any potential plant parameter variations. This objective will be the focus of this section.

The error response of the nominal control system of Figure 7.1, relative to a nondiminishing reference input $r(t)$, is defined by the **error response** transfer function relationship

$$e(s) = \frac{\overbrace{\{a(s)k(s) + c(s)[h(s) - q(s)]\}}^{\stackrel{\text{def}}{=} m_e(s)} m_r(s)}{\delta(s)p_r(s)} = \underbrace{\frac{\hat{m}_e(s)}{\delta(s)}}_{e_n(s)} + \underbrace{\frac{\hat{m}_r(s)}{p_r(s)}}_{e_f(s)} \tag{7.3.3}$$

using Eqs. (7.3.1) and (7.2.2). If $\delta(s)$ is stable, $\mathcal{L}^{-1}[e_n(s)] = e_n(t) \to 0$ as $t \to \infty$, so that the *forced error response*

$$e_f(t) = \mathcal{L}^{-1}\left[e_f(s) = \frac{\hat{m}_r(s)}{p_r(s)}\right] = e_{ss}(t) \tag{7.3.4}$$

Therefore, in order to ensure a zero steady-state error, as defined by Eq. (7.3.2), it is necessary that $\hat{m}_r(s) = 0$.

Blocking Zeros

To ensure a $\hat{m}_r(s) = 0$ in Eq. (7.3.3), $m_e(s)m_r(s) = \hat{m}_e(s)p_r(s)$. Since $m_r(s)$ and $p_r(s)$ are coprime, $p_r(s)$ must therefore divide $m_e(s)$, the numerator of the error response transfer function. Note that if this is the case,

$$m_e(s) = a(s)k(s) + c(s)[h(s) - q(s)] = \tilde{m}_e(s)p_r(s) \qquad (7.3.5)$$

for some polynomial $\tilde{m}_e(s)$. The substitution of Eq. (7.3.5) for $m_e(s)$ in Eq. (7.3.3) will then imply that

$$e(s) = \underbrace{\frac{\tilde{m}_e(s)m_r(s)p_r(s)}{\delta(s)p_r(s)} = \frac{\hat{m}_e(s) = \tilde{m}_e(s)m_r(s)}{\delta(s)}}_{e_n(s)} + \underbrace{\frac{\hat{m}_r(s) = 0}{p_r(s)}}_{e_f(s)} \qquad (7.3.6)$$

or that

$$e_f(t) = \mathcal{L}^{-1}[e_f(s)] = e_{ss}(t) = 0 \qquad (7.3.7)$$

When Eqs. (7.3.5) and (7.3.6) hold, the modes of the reference input $r(t)$ will be zeroed or completely "blocked" by the so-called **blocking zeros** of the error response transfer function (recall Problem 4-5), that correspond to the $p_r(s)$ zeros of $m_e(s)$.

In the 1 DOF case of Figure 7.5, $q(s) = h(s)$, so that

$$e(s) = S(s)r(s) = \frac{\overbrace{a(s)k(s)}^{m_e(s)} m_r(s)}{\delta(s)p_r(s)} \qquad (7.3.8)$$

in light of Eqs. (7.3.1) and (7.3.3). Therefore, if

$$a(s)k(s) = m_e(s) = \bar{m}_e(s)p_r(s) \qquad (7.3.9)$$

despite any possible changes in the plant polynomial $a(s)$, then Eq. (7.3.5) will hold, with $\bar{m}_e(s) = \tilde{m}_e(s)$, and a robust $e_{ss}(t) = 0$ would be assured by Eq. (7.3.7).

The Internal Model Principle (IMP)

This latter observation now can be stated as the **internal model principle** (IMP) [22]: A robust $e_{ss}(t) = 0$ is assured by a stable 1 DOF unity feedback control system if a *model* of $p_r(s)$ is present in the denominator $a(s)k(s)$ of the *internal* loop gain transfer function

$$L(s) = G(s)H(s) = \frac{c(s)h(s)}{a(s)k(s)} = \frac{c(s)h(s)}{\bar{m}_e(s)p_r(s)} \qquad (7.3.10)$$

because the $p_r(s)$ poles of $L(s)$ become error response blocking zeros when the loop is closed. As we will show now, an IMP design implies the unobservability of the reference input modes defined by the blocking zeros of $p_r(s)$.

Figure 7.7 illustrates an IMP design in the unity feedback case. Note that if the loop is "broken" just after $e(s)$, as indicated, modes that correspond to the zeros of $p_r(s)$ would be generated by both $r(t) = \mathcal{L}^{-1}[r(s)]$

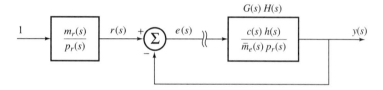

● **FIGURE 7.7**
A 1 DOF (IMP) Design for a Robust $e_{ss}(t) = 0$

and the natural $y(t) = \mathcal{L}^{-1}[y(s)]$ response. Since it would be impossible to distinguish whether $r(t)$ or $y(t)$ generated these modes at the "output" $e(t)$ (recall Problem 3-12), it follows that *the IMP modes, which correspond to the blocking zeros of $p_r(s)$, are unobservable at the error.*

In view of Eq. (7.3.9), the blocking zeros of $p_r(s)$ may correspond to the roots of either $a(s)$ or $k(s)$. To ensure a robust $e_{ss}(t) = 0$ in the former case, plant parameter changes cannot alter any zeros of $a(s)$ that correspond to those of $p_r(s)$. In the latter case, a choice for

$$k(s) = \hat{k}(s)p_r(s) \tag{7.3.11}$$

with $\hat{k}(s)$ arbitrary, will always ensure a robust $e_{ss}(t) = 0$, despite any plant parameter variations, since $k(s)$ is fixed by the designer.

System Type

The input $r(t)$ that must be tracked in many practical situations is one of the *three common reference inputs*, namely, a step, a ramp, or a parabolic function, respectively. These were defined in Section 4.2 by

$$r(t) = \begin{cases} 0 & \text{for } t < 0 \\ Kt^{\bar{n}} & \text{for } t > 0 \end{cases} \tag{7.3.12}$$

with $\bar{n} = 0, 1,$ and 2, respectively, which implies a resulting

$$r(s) = \mathcal{L}[r(t)] = \frac{m_r(s)}{p_r(s)} = \frac{K\bar{n}!}{s^{\bar{n}+1}} \tag{7.3.13}$$

We now recall (see Section 5.4, and Eq. (**??**)) that a system defined by the loop gain

$$L(s) = G(s)H(s) = \frac{c(s)h(s)}{a(s)k(s)} \tag{7.3.14}$$

is said to be *type p* if $L(s)$ has exactly p poles at $s = 0$ or, using Eq. (7.3.9), if

$$a(s)k(s) = \bar{m}_e(s)s^p, \quad \text{with} \quad \bar{m}_e(0) \neq 0 \tag{7.3.15}$$

To ensure a robust $e_{ss}(t) = 0$ for a stable closed-loop system that is subjected to an external step, ramp, or parabolic input $r(t)$ (when $p_r(s) =$

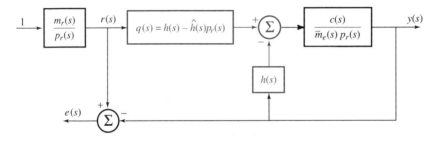

FIGURE 7.8
A 2 DOF (IMP) Design for a Robust $e_{ss}(t) = 0$

s^1, s^2, or s^3, respectively), a classical control system design principle requires that the loop gain of the system be at least type 1, 2, or 3, respectively. It can now be noted that this classic principle represents a special case of the modern internal model principle, which implies the unobservability at $e(t)$ of the reference input modes that are defined by the blocking zeros of $p_r(s)$.

The 2 DOF Case

In the closed-loop stable 2 DOF case, a robust $e_{ss}(t) = 0$ is ensured by Eq. (7.3.5), which will hold provided *both* the IMP is satisfied, that is, if

$$a(s)k(s) = \bar{m}_e(s)p_r(s) \qquad (7.3.16)$$

as in Eq. (7.3.9), *and* if

$$h(s) - q(s) = \hat{h}(s)p_r(s) \qquad (7.3.17)$$

for some polynomial $\hat{h}(s)$. In particular, Eqs. (7.3.16) and (7.3.17) together imply that

$$a(s)k(s) + c(s)[h(s) - q(s)] = \overbrace{[\bar{m}_e(s) + c(s)\hat{h}(s)]}^{\tilde{m}_e(s)} p_r(s)$$

Note that $k(s)$ can always be chosen to ensure Eq. (7.3.16) and $q(s)$ can always be chosen to ensure Eq. (7.3.17),[2] despite any possible plant parameter variations. Therefore, a robust $e_{ss}(t) = 0$ IMP design, as depicted in Figure 7.8, also can be implemented in the 2 DOF case.

The required choice for $q(s)$ defined by Eq. (7.3.17) is not entirely arbitrary and, as a consequence, the response performance may be constrained somewhat, depending on the complexity of $p_r(s)$. For example,

---

[2]For example, a 1 DOF choice of $q(s) = h(s)$ will ensure Eq. (7.3.17) for $\hat{h}(s) = 0$. However, other "more appropriate" choices do exist, as we will show.

in the case of step changes in the reference input, when $p_r(s) = s$, a robust $e_{ss}(t) = 0$ will be assured for step inputs if the IMP condition defined by Eq. (7.3.16) is satisfied, and if

$$q(0) = q_0 = h(0) = h_0 \qquad (7.3.18)$$

since $s = p_r(s)$ will then divide $h(s) - q(s)$. In the case of ramp inputs, $p_r(s) = s^2$ will divide $h(s) - q(s)$ if both

$$q_0 = h_0 \quad \text{and} \quad q_1 = h_1 \qquad (7.3.19)$$

Therefore, a robust $e_{ss}(t) = 0$ will be assured for ramp inputs if both Eq. (7.3.16) and Eq. (7.3.19) hold. Neither of these cases imply an overly restrictive $q(s)$.

Complete Disturbance Rejection

Note that a robust $e_{ss}(t) = 0$ IMP design also implies the complete rejection of any output disturbance modes $d(s)$ that correspond to those of the reference input. In particular, in light of Eq. (7.1.2),

$$y(s) = S(s)d(s) = \frac{a(s)k(s)}{\delta(s)}d(s) = \frac{\bar{m}_e(s)p_r(s)}{\delta(s)}d(s) \qquad (7.3.20)$$

when Eq. (7.3.16) holds. Therefore, if

$$d(s) = \frac{m_d(s)}{p_d(s)} \qquad (7.3.21)$$

is characterized by nondiminishing disturbance modes, as defined by the roots of $p_d(s)$, the inclusion of the corresponding blocking zeros in $a(s)k(s)$ will ensure their complete rejection at $y(t)$ (see Problem 7-12).

Sensor (Feedback) Dynamics

Consider a system defined by the rational, strictly proper, minimal transfer function of Eq. (7.2.5):

$$G(s) = \frac{y(s)}{u(s)} = \frac{c(s)}{a(s)}, \quad \text{with} \quad \deg[a(s)] = n \qquad (7.3.22)$$

Suppose the system is characterized by dynamics in the feedback path (often attributed to the sensor), as defined by the proper, minimal transfer function

$$M(s) = \frac{m(s)}{p(s)}, \quad \text{with} \quad \deg[p(s)] = n_p \qquad (7.3.23)$$

In such cases, a 2 DOF controller also can be employed to obtain a fixed, desired output response for an infinite number of different loop compensators, as in the $M(s) = 1$ case.

In particular, the pole placement algorithm of Appendix B can be used to determine a monic $k(s)$ and a corresponding $h(s)$, both of degree $(n + n_p - 1)$, such that

$$\delta(s) = a(s)p(s)k(s) + c(s)m(s)h(s) = \bar{\delta}(s)\bar{q}(s) \qquad (7.3.24)$$

where $\bar{\delta}(s)$ and $\bar{q}(s)$ are arbitrary stable polynomials of degree $n + n_p$ and $n + n_p - 1$, respectively.

If $k(s)$ and $h(s)$ are then employed as the loop polynomials in the 2 DOF configuration of Figure 7.9, where the loop gain transfer function

$$L(s) = \frac{c(s)m(s)h(s)}{a(s)p(s)k(s)} \qquad (7.3.25)$$

the resulting output response transfer function will be given by

$$T(s) = \frac{y(s)}{r(s)} = \frac{c(s)p(s)q(s)}{\bar{\delta}(s)\bar{q}(s)} \qquad (7.3.26)$$

Therefore, a choice of $q(s) = \alpha\bar{q}(s)$ will cancel $n + n_p - 1$ closed-loop poles, thus implying a nominal

$$T(s) = \frac{\alpha c(s)p(s)}{\bar{\delta}(s)} \qquad (7.3.27)$$

Each choice for $\bar{q}(s)$, for any fixed α and $\bar{\delta}(s)$ (hence a fixed $T(s)$ as well) will produce a different, but unique choice for $h(s)$ and $k(s)$ using Eq. (7.3.24). Therefore, as in the nondynamic $M(s) = 1$ case, it is possible to obtain the same output response transfer function for an infinite number of different loop compensators.

Note that when $p(s)$ is stable, n_p roots of $\bar{\delta}(s)$ can be chosen equal to the n_p roots of $p(s)$, so that

$$\bar{\delta}(s) = \hat{\delta}(s)p(s) \qquad (7.3.28)$$

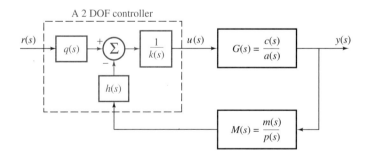

• **FIGURE 7.9**
2 DOF Dynamic Feedback Compensation

with $\hat{\delta}(s)$ an arbitrary stable polynomial of degree n. Such a choice will imply an $h(s) = \bar{h}(s)p(s)$, in light of Eq. (7.3.24), and a resulting nominal

$$T(s) = \frac{\alpha c(s)p(s)}{\hat{\delta}(s)p(s)} = \frac{\alpha c(s)}{\hat{\delta}(s)} \tag{7.3.29}$$

which is completely analogous to the $T(s)$ defined by Eq. (7.2.12).

We note finally that when $M(s) \neq 1$, the error response transfer function

$$\frac{e(s)}{r(s)} = 1 - \frac{y(s)}{r(s)} = \frac{a(s)p(s)k(s) + c(s)[m(s)h(s) - p(s)q(s)]}{a(s)p(s)k(s) + c(s)m(s)h(s) = \delta(s)} \tag{7.3.30}$$

Therefore, if $r(s)$ is a nondiminishing input, as defined by Eq. (7.2.2), Eq. (7.3.5) implies that

$$a(s)p(s)k(s) + c(s)[m(s)h(s) - p(s)q(s)] = \tilde{m}_e(s)p_r(s) \tag{7.3.31}$$

to ensure a robust $e_{ss}(t) = 0$. Note, however, that if the parameters that define the feedback transfer function $M(s)$ vary arbitrarily, that is, if the coefficients of $m(s)$ and $p(s)$ vary in an unknown manner, then Eq. (7.3.31) cannot be satisfied for any fixed $p_r(s)$. As a consequence, it is impossible to ensure a robust $e_{ss}(t) = 0$ for any known, nondiminishing reference input $r(t)$ when there are unknown variations in $M(s)$.

7.4 PERFECT NOMINAL CONTROL

In certain situations, it is possible to achieve what might be termed **perfect nominal control** [50], namely, a nominal zero steady-state error

$$e_{ss}(t) = \lim_{t \to \infty} [r(t) - y(t)] = 0 \tag{7.4.1}$$

for *any* known, sufficiently differentiable reference input $r(t)$. Consider a system defined by a rational transfer function

$$G(s) = \frac{y(s)}{u(s)} = \frac{c(s)}{a(s)}$$

which is both strictly proper and minimal (controllable and observable), so that the nominal plant polynomials $a(s)$ and $c(s)$ are coprime, with $\deg[a(s)] = n > \deg[c(s) = c_m s^m + \cdots + c_1 s + c_0] = m$.

If a 2 DOF compensator is employed, Eq. (7.3.1) implies that

$$e(s) = \frac{a(s)k(s) + c(s)[h(s) - q(s)]}{\delta(s)} r(s) \tag{7.4.2}$$

Therefore, if the compensator polynomials $h(s)$, $k(s)$, and $q(s)$ can be chosen such that

$$a(s)k(s) + c(s)[h(s) - q(s)] = 0 \tag{7.4.3}$$

then $e(s) = 0$, independent of $r(s)$. Note that Eq. (7.4.3) is equivalent to the relationship

$$c(s)q(s) = a(s)k(s) + c(s)h(s) = \delta(s) \qquad (7.4.4)$$

Since $\delta(s)$ must be stable to ensure a zero steady-state error, Eq. (7.4.4) implies that both $c(s)$ and $q(s)$ must be stable for Eq. (7.4.1) to hold. Since $a(s)$ and $c(s)$ are coprime, Eq. (7.4.4) also implies that

$$k(s) = c(s)\tilde{k}(s) \qquad (7.4.5)$$

for some polynomial $\tilde{k}(s)$, with a resulting

$$q(s) = \frac{a(s)c(s)\tilde{k}(s) + c(s)h(s)}{c(s)} = a(s)\tilde{k}(s) + h(s) \qquad (7.4.6)$$

Therefore, to achieve perfect nominal control, we first choose

$$\hat{q}(s) = s^{2n-m-1} + \hat{q}_{2n-m-2}s^{2n-m-2} + \cdots + \hat{q}_1 s + \hat{q}_0 \qquad (7.4.7)$$

to be any arbitrary, stable, monic polynomial of degree $2n - m - 1$. A monic $k(s) = c(s)\tilde{k}(s)$ of degree $n - 1$, and a corresponding $h(s)$, also of degree $n - 1$, can then be chosen so that

$$a(s)\tilde{k}(s) + h(s) = q(s) \stackrel{\text{def}}{=} c_m^{-1}\hat{q}(s) \qquad (7.4.8)$$

which will imply Eq. (7.4.4).

A direct 2 DOF implementation of the resulting $h(s)$, $k(s)$, and $q(s)$ could present some problems, however, since

$$\deg[q(s)] - \deg[k(s)] = 2n - m - 1 - n + 1 = n - m > 0 \qquad (7.4.9)$$

As a consequence, $q(s)/k(s)$ will be an improper transfer function, contrary to the Section 5.1 requirement that controllers be proper in order to avoid excessive noise amplification and possible plant input saturation. However, if $r(t)$ is a "sufficiently differentiable" function of time, this requirement can be relaxed, as we now will show.

We first express $q(s)/k(s)$ (uniquely) as the sum of its strictly proper part and quotient, as in Eq. (3.2.6), that is,

$$\frac{q(s)}{k(s)} = \frac{\tilde{q}(s)}{k(s)} + g_q(s) \qquad (7.4.10)$$

so that

$$q(s) = \tilde{q}(s) + g_q(s)k(s) \qquad (7.4.11)$$

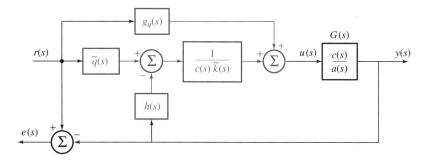

• **FIGURE 7.10**
An Implementation of an Alternative 2 DOF Compensator

A 2 DOF perfect control compensator can then be defined by the relationship

$$k(s)u(s) = -h(s)y(s) + q(s)r(s) = -h(s)y(s) + [\tilde{q}(s) + g_q(s)k(s)]r(s)$$

$$(7.4.12)$$

or

$$u(s) = \frac{1}{k(s) = c(s)\tilde{k}(s)} \left[-h(s)y(s) + \tilde{q}(s)r(s) \right] + g_q(s)r(s) \quad (7.4.13)$$

Therefore, the $g_q(s)k(s)$ "improper part" of $q(s)$ can be implemented by a direct $g_q(s)$ feedforward path from the reference input $r(s)$ to the plant input $u(s)$.

To summarize, if all of the zeros of a given plant transfer function $G(s)$ are stable, and if $h(s)$ and $\tilde{k}(s)$ are chosen to satisfy Eq. (7.4.8), for any stable $\hat{q}(s)$ defined by Eq. (7.4.7), then Eq. (7.4.13) will imply a 2 DOF perfect control compensator, as depicted in Figure 7.10, which will ensure the nominal $e_{ss}(t) = 0$ defined by Eq. (7.4.1).

Note that this 2 DOF design requires an ability to generate the first $n - m$ derivatives of the reference input $r(t)$, as defined by $g_q(D)r(t)$, as a bounded "feedforward part" of the plant input $u(t)$. This is not always possible. In certain applications, however, such as computer generated robotic motion [63], $r(t)$ can be a well-defined analytic function, for example, a sinusoid or a polynomial, such as

$$r(t) = r_0 + r_1 t + r_2 t^2 + r_3 t^3 + \cdots \qquad (7.4.14)$$

which represents a desired link angle trajectory. In such cases, perfect control is feasible, since a trajectory planning program in the control computer could generate and store both $r(t)$ and its first $n - m$ derivatives for subsequent playback during actual robotic motion.

. . . .

EXAMPLE 7.4.15 To illustrate the preceding discussion, consider the system defined in Example 7.2.13, which is characterized by the nominal plant transfer function

$$G(s) = \frac{3}{s(s-1)} = \frac{c(s)}{a(s)}$$

with $c(s) = 3 = c_m$, a stable "polynomial" of degree $m = 0$. To design a 2 DOF compensator so that $e_{ss}(t) = 0$, irrespective of $r(t)$, $\hat{q}(s)$ first is chosen to be any arbitrary polynomial of degree $2n - m - 1 = 2n - 1 = 3$. For example, if

$$\hat{q}(s) = (s+2)(s^2 + 2.4s + 2.44)$$

Eq. (7.4.8) then implies that

$$\underbrace{(s^2 - s)}_{a(s)} \underbrace{(0.333s + 1.8)}_{\tilde{k}(s)} + \underbrace{(4.21s + 1.63)}_{h(s)} = \underbrace{0.333s^3 + 1.467s^2 + 2.413s + 1.63}_{q(s) = c_m^{-1}\hat{q}(s)}$$

or that

$$h(s) = 4.21s + 1.63 \quad \text{and} \quad k(s) = c(s)\tilde{k}(s) = s + 5.4$$

as in the *medium case* of Example 7.2.13.

The division of $q(s)$ by $k(s) = s + 5.4$, as in Eq. (7.4.10), implies next that

$$\tilde{q}(s) = -21.12 \quad \text{and} \quad g_q(s) = 0.333s^2 - 0.333s + 4.213$$

The subsequent requirement to generate

$$g_q(D)r(t) = 0.333\ddot{r}(t) - 0.333\dot{r}(t) + 4.213r(t)$$

as the feedforward part of the plant input $u(t)$, would require that $r(t)$, $\dot{r}(t)$, and $\ddot{r}(t)$ all be well defined and bounded in this case. For example, if $r(t) = 2 - t^2$, so that $Dr(t) = \dot{r}(t) = -2t$ and $D^2r(t) = \ddot{r}(t) = -2$, then

$$g_q(D)r(t) = -0.666 + 0.666t + 8.426 - 4.213t^2 = 7.76 + 0.666t - 4.213t^2$$

would represent the feedforward signal from the reference input to $u(t)$. As a result of this input (see Problem 7-11),

$$y(t) = 2 - t^2 - 0.732e^{-2t} - 3.24e^{-1.2t}\sin(t + 22.8°)$$

so that $\lim_{t\to\infty} y(t) = 2 - t^2 = r(t)$, thus implying perfect nominal control.

It may be noted that the sensitivity function $S(s)$ and the loop gain $L(s)$, associated with this nominal $e_{ss}(t) = 0$ design, would remain the same as in the medium case of Example 7.2.13, since $h(s)$ and $k(s)$ are identical in both examples.

7.5 SUMMARY

We have now outlined the various response goals that should be attained by an appropriately designed, closed-loop control system, namely, desirable output regulation, as characterized by a fast transient response with minimal overshoot and constrained plant input excursions, and a robust, zero steady-state error between the reference input and the plant output. As noted, the farther the closed-loop poles of a system are moved into the stable half-plane, the faster the system response, although a compensator design that positions the roots of $\delta(s)$ "too far" into the left-half s-plane could increase the bandwidth of the system enough to jeopardize the loop performance of the system or produce plant input $u(t)$ saturation.

This latter observation served to motivate a 2 DOF controller design that maintains a desired response performance, while varying the loop performance. The procedure is based on a desired, fixed set of n closed-loop poles, which characterize the output response transfer function, and an adjustable set of $n - 1$ closed-loop poles, which are cancelled by corresponding zeros external to the loop.

The ability to obtain a robust, zero steady-state tracking error for a variety of nondiminishing reference inputs was then addressed. The presence of an internal loop model of the input modes, as defined by the roots of $p_r(s)$, was shown to insure a zero tracking error in the stable, 1 DOF unity feedback cases. In particular, if $p_r(s)$ defines certain of the open-loop system poles, these poles subsequently become blocking zeros of the error response transfer function when the loop is closed. Alternatively, such an IMP design was shown to imply the unobservability of the reference input modes defined by the blocking zeros of $p_r(s)$.

These results were then extended to include 2 DOF closed-loop designs as well. It was shown that for both step and ramp inputs, a robust $e_{ss}(t) = 0$ can be assured in the 2 DOF cases without compromising the benefits of such a design. If feedback dynamics are present, a 2 DOF design can still be used to ensure a fixed, desired response performance, while varying the loop performance. However, in both the 1 DOF and the 2 DOF cases, a robust $e_{ss}(t) = 0$ cannot be achieved if there are unknown variations in the feedback dynamics.

Finally, it was shown that in certain situations, it is possible to achieve perfect nominal control, namely, a nominal $e_{ss}(t) = 0$ for any known but sufficiently differentiable reference input $r(t)$. In particular, if all of the transfer function zeros of a given system are stable, and if the reference input is a known, analytic function of time, then a 2 DOF compensator can be designed that acts essentially as an inverse of the plant (see Problem 7-15), thus implying a nominal $y(s) = r(s)$.

PROBLEMS

7-1.* Sketch an extended polar plot of the nominal loop gain $L(s) = G(s)H(s)$ in the medium case of Example 7.2.13, and show why the closed-loop system is robustly stable in light of the Nyquist stability criterion.

7-2. Express the dynamic behavior of the 2 DOF controller system depicted in Figure 7.1 in differential operator form. Show that the roots of $\hat{q}(s)$ in Eq. (7.2.12) imply closed-loop system modes that are uncontrollable by the reference input $r(t)$.

7-3. Verify that in Example 7.2.13, $h(s) = 4.21s + 1.63$ and $k(s) = s + 5.4$ in the medium case defined by $\hat{q}(s) = s + 2$, and that $h(s) = 24.61s + 16.26$ and $k(s) = s + 23.4$ in the high case defined by $\hat{q}(s) = s + 20$.

7-4. Consider a system defined by the nominal state-space representation of Example 7.2.13. Assume that the magnitude of the external reference input $r(t)$ does not exceed 1 and that the pole at $s = 0$ remains invariant. To prevent plant input saturation, assume also that the magnitude of $u(t)$ cannot exceed 1.

    (a) Define a zeroless, second-order $T(s)$, whose poles are located as far away from the $j\omega$-axis as possible (consistent with the requirement that $|u(t)| \leq 1$), which will imply a robust $e_{ss}(t) = 0$ for step inputs using the 2 DOF design procedure of Section 7.2.

    (b) Determine the unique loop compensator polynomials $h(s)$ and $k(s)$, if $\hat{q}(s) = s + \hat{q}_0 = s + 5$.

    (c) Determine the maximum value that $u(t)$ attains if a 1 DOF unity feedback design is employed, with $q(s) = h(s)$, instead of the 2 DOF design.

7-5. Consider the 2 DOF design procedure of Section 7.2. In light of Eq. (7.2.10), note that a choice of $\hat{\delta}(s) = c(s)\bar{\delta}(s)$, with $\bar{\delta}(s)$ arbitrary, will imply a

$$T(s) = \frac{\alpha c(s)}{\hat{\delta}(s)} = \frac{\alpha}{\bar{\delta}(s)}$$

Determine the effect that such a choice for $\hat{\delta}(s)$ would have on the controllability and observability properties of the closed-loop system. When would such a choice imply closed-loop stability?

7-6. Show that the three 2 DOF designs presented in Example 7.2.13 imply a robust $e_{ss}(t) = 0$ for step inputs, provided the plant pole at $s = 0$ remains invariant.

7-7. Consider a system defined by the nominal transfer function

$$G(s) = \frac{3(s + \beta)}{(s^2 - s)(s + \gamma)}$$

Let $h_a(s)$, $k_a(s)$, and $q_a(s)$ define the 2 DOF compensator polynomials of any one of the three cases considered in Example 7.2.13.

(a) Show that a choice here of the 2 DOF compensator polynomials $h(s) = (s+\gamma)h_a(s)$, $k(s) = (s+\beta)k_a(s)$, and $q(s) = (s+\gamma)q_a(s)$ will imply the exact same nominal plots as those depicted in Figures 7.2, 7.3, and 7.4.

(b) Determine the effect that such a choice for the compensator polynomials would have on the controllability and observability properties of the closed-loop system.

(c) When would such a design imply closed-loop stability?

(d) Under what circumstances would it be possible to replace $s + \beta$ and $s + \gamma$ by polynomials of arbitrary degree?

7-8.* Suppose that the nominal value of $a(s)$ changes from $s(s - 1)$ to $s(s - 1.2)$ in Example 7.2.13. Determine the resulting changes in the closed-loop poles in the three cases considered. Plot the unit step response of the closed-loop system in all three cases, and explain the differences in the settling times of these "nonnominal" plots in light of the changes in the closed-loop poles.

7-9. Consider a stable 1 DOF unity feedback control system, such as that depicted in Figure 7.5. Determine whether $e_{ss}(t)$ is

(a) a robust 0,

(b) a nominal, finite $K \neq 0$, or

(c) ∞,

in response to a step, a ramp, and a parabolic input, that is, when $r(s) = \dfrac{1}{p_r(s)}$, with $p_r(s) = s^1$, s^2, and s^3, respectively, if

$$L(s) = G(s)H(s) = \frac{c(s)h(s)}{a(s)k(s) = m_e(s)}$$

$$= \frac{(c_m s^m + \cdots + c_1 s + c_0)(h_{n-1}s^{n-1} + \cdots + h_1 s + h_0)}{s^{2n-1} + \cdots + m_{e1}s + m_{e0}}$$

is type 0, 1, and 2, as defined by Eq. (7.3.15). Use your results to determine the entries of Table 7.1, and express any nominal, finite $e_{ss}(0) = K \neq 0$

TABLE 7.1 A 1 DOF Steady-State Error Table for Problem 7-9

	$r(t) = $ Step	$r(t) = $ Ramp	$r(t) = $ Parabolic
Type $p = 0$			
Type $p = 1$			
Type $p = 2$			

as functions of the polynomial coefficients that define $L(s)$ (assuming unit magnitude inputs).

7-10. Prove that if the loop gain transfer function of a system is type p, as defined by Eq. (7.3.15), then its closed-loop sensitivity function $S(s)$ satisfies the relationship

$$\lim_{s \to 0} \frac{S(s)}{s^k} = 0 \quad \text{for} \quad 0 \le k < p$$

7-11. If $r(t) = 2 - t^2$ in Example 7.4.15, verify explicitly that

$$y(t) = 2 - t^2 - 0.732e^{-2t} - 3.24e^{-1.2t} \sin(t + 22.8°)$$

7-12. Assume that the output of the system defined in Example 7.2.13 is subjected to a sinusoidal disturbance signal of frequency $\omega = 2$, so that $d(s) = W_S(s)d_i(s)$, for a disturbance weighting function

$$W_S(s) = \frac{K}{s^2 + 4} = \frac{m_d(s)}{p_d(s)}$$

To reject this disturbance, in light of Eq. (7.3.20), the compensator polynomial $k(s)$ is chosen to include the $p_d(s)$ factor $s^2 + 4$.

(a) Determine the unique choice of $k(s) = (s^2 + 4)(s + k_0) = s^3 + k_0 s^2 + 4s + k_0$ and $h(s) = h_3 s^3 + h_2 s^2 + h_1 s + h_0$ that will position the nominal closed-loop poles of the system at $s = -1.2 \pm j$ and -2, as in the medium case of Example 7.2.13, as well as at $s = -2 \pm j$, so that

$$\delta(s) = a(s)k(s) + c(s)h(s)$$
$$= (s^2 + 2.4s + 2.44)(s + 2)(s^2 + 4s + 5)$$
$$= s^5 + 8.4s^4 + 29.84s^3 + 55.84s^2 + 55.72s + 24.4$$

(b) Show that a corresponding choice for

$$q(s) = \underbrace{0.813}_{\alpha} \underbrace{(s + 2)(s^2 + 4s + 5)}_{\hat{q}(s)} = 0.813(s^3 + 6s^2 + 13s + 10)$$

will imply the same nominal closed-loop transfer function as that obtained in Example 7.2.13, namely, a

$$T(s) = \frac{c(s)q(s)}{\delta(s)} = \frac{\omega_n^2}{s^2 + 2\zeta\omega_n s + \omega_n^2} = \frac{2.44}{s^2 + 2.4s + 2.44}$$

(c) Verify that $q(0) = \alpha\hat{q}(0) = h(0)$ in this case, so that a robust $e_{ss}(t) = 0$ is assured for step inputs as well, provided the plant pole at $s = 0$ remains invariant.

7-13. Consider the 2 DOF, perfect control compensator of Section 7.4, with $c(s)q(s) = \delta(s)$, as in Eq. (7.4.4). Prove that the roots of $c(s)$ define closed-loop modes that are unobservable at $y(t)$, and that the roots of $q(s)$ define closed-loop modes that are uncontrollable by $r(t)$.

7-14. Verify that the division of $q(s) = c_m^{-1}\hat{q}(s) = 0.333s^3 + 1.467s^2 +2.413s +1.627$ by $k(s) = s + 5.4$, as in Eq. (7.4.10), will imply the $\tilde{q}(s) = -21.12$ and $g_q(s) = 0.333s^2 - 0.333s + 4.213$ obtained in Example 7.4.15.

7-15. Show that when Eq. (7.4.4) holds, the closed-loop transfer function

$$T(s) = \frac{y(s)}{r(s)} = \frac{c(s)q(s)}{\delta(s) = c(s)q(s)} = 1$$

so that a 2 DOF perfect nominal controller acts as an **inverse plant**, that is, an open-loop feedforward control system defined by $G^{-1}(s)$, thus implying a $y(s) = r(s)$. Under what conditions, if any, will $G^{-1}(s)$ itself represent a perfect nominal control system?

7-16. Consider the unstable, second-order system of Example 7.2.13, as defined by the nominal transfer function

$$\frac{y(s)}{u(s)} = G(s) = \frac{3}{s^2 - s} = \frac{c(s)}{a(s)}$$

Assume that a nonunity sensor defined by the nominal transfer function

$$M(s) = \frac{2s + 1}{s + 4} = \frac{m(s)}{p(s)}$$

is employed to measure the output. Show that it is possible to obtain the same output response transfer function here as that obtained in Example 7.2.13, namely,

$$T(s) = \frac{y(s)}{r(s)} = \frac{\omega_n^2}{s^2 + 2\zeta\omega_n s + \omega_n^2} = \frac{2.44}{s^2 + 2.4s + 2.44} = \frac{\alpha c(s)}{\hat{\delta}(s)}$$

In particular, in light of the 2 DOF controller of Figure 7.9 and Eq. (7.3.24), let $\bar{q}(s) = (s+1)(s+2) = s^2 + 3s + 2$, and determine the unique, second-degree $k(s)$, and the corresponding, unique $h(s) = \bar{h}(s)p(s)$, also of degree 2, such that

$$a(s)k(s) + c(s)m(s)\bar{h}(s) = \hat{\delta}(s)\bar{q}(s)$$

Verify that the resulting loop polynomials $h(s)$ and $k(s)$, together with a $q(s) = \alpha\bar{q}(s)$, with $\alpha = 2.44/3 = 0.813$, do imply the desired nominal $T(s)$.

PART III

COMPENSATION

Courtesy of NASA

CHAPTER 8

8

CLASSICAL CONTROL TECHNIQUES

8.1 PID COMPENSATION

In Part I of the text, we focused on the various ways the behavior of a dynamic system can be modeled. In Part II we outlined a variety of techniques that can be used to define the loop and response goals that we would expect to achieve as the result of an appropriate control system design. The focus of Part III will be on the actual design of compensators that attain the desired performance goals.

Generally speaking, it is difficult to define precisely and hence to obtain the "best" possible compensator for a given application. If we were to assign the same set of performance goals to different control system engineers, the resulting designs would most likely vary because of the many trade-offs associated with not only the explicit design goals but also the implicit goals of minimizing cost and complexity, while maximizing reliability. One engineer may decide to implement a relatively simple design, although it may only marginally satisfy the stated performance goals, because simplicity generally implies both low cost and reliability. Another engineer may take an entirely different approach, arriving at a more complex, but "optimal" design that might minimize an LQR performance index, such as in Eq. (7.1.4).

1 DOF Designs

The approach to design taken in this section will stress simplicity first, introducing additional complexity as needed to achieve more stringent objectives. As we will illustrate, one degree-of-freedom (1 DOF) designs often produce acceptable closed-loop performance, especially in the easier to control situations when the plant is stable and minimum phase. Such designs involve the selection of a single loop compensator defined by a proper, rational transfer function

$$H(s) = \frac{h(s)}{k(s)}$$

which completely determines both the loop and the response performance characteristics of the system. $H(s)$ is often placed in series with the plant, as defined by some nominal transfer function

$$G(s) = \frac{c(s)}{a(s)}$$

with the resulting configuration then closed under unity feedback, as was depicted in Figure 7.5.

In the 1 DOF unity feedback case the output response transfer function

$$T(s) = \frac{y(s)}{r(s)} = \frac{G(s)H(s)}{1 + G(s)H(s)} = \frac{c(s)h(s)}{a(s)k(s) + c(s)h(s)} \qquad (8.1.1)$$

that, in light of Eq. (5.1.25), corresponds to the complementary sensitivity function $C(s)$, while the error response transfer function

$$\frac{e(s)}{r(s)} = 1 - \frac{y(s)}{r(s)} = 1 - C(s) = \frac{a(s)k(s)}{\delta(s)} \qquad (8.1.2)$$

that, looking back at Eq. (5.1.24), corresponds to the sensitivity function $S(s)$.

It may be noted that 1 DOF unity feedback designs form the basis of what commonly is referred to as **classical control** or the **frequency-domain** approach. These methods, which are based on frequency response measurements and transfer function models, were among the first systematic techniques developed for the analysis and design of control systems, as noted in Chapter 1. They proliferated during World War II, in large part due to the need for more accurate, error-driven systems or **servomechanisms**, which are defined by Bower and Schultheiss [8] as follows:

> The device controls some physical quantity by comparing its actual value C with its desired value R and uses the difference (or error) $R-C$ to drive C into correspondence with R.

The 1 DOF servomechanism-based designs were popularized before the advent of the so-called *modern* or *state-space* methods, which introduced such notions as state feedback and state observers, as well as the concepts of complete state controllability and observability. As we will show in the next chapter, modern control system designs are often based on the minimization of a performance index, such as the $\mathbf{H}_2$-norm defined by Eq. (7.1.4).

Although modern methods often produce superior closed-loop performance, especially in the more difficult to control situations, classical techniques are likely to retain a certain degree of popularity, because they imply relatively simple designs that frequently result in acceptable closed-loop performance. Moreover, they can often be implemented without explicit knowledge of a transfer function $G(s)$, state-space representation, or a differential operator representation of the plant.

Proportional (P) Compensation

Undoubtedly, the simplest 1 DOF control system that can be employed is a **proportional (P) compensator**, which is an adjustable gain

$$K_P \stackrel{\text{def}}{=} H_P(s) \tag{8.1.3}$$

generally placed in the forward path in series with $G(s)$ and subsequently closed under unity feedback. The closed-loop performance of such a system is defined by the output response transfer function

$$\frac{y(s)}{r(s)} = T(s) = \frac{H_P(s)G(s)}{1 + H_P(s)G(s)} = \frac{K_P c(s)}{a(s) + K_P c(s)} \tag{8.1.4}$$

In certain situations, an appropriate choice for K_P can result in acceptable closed-loop performance. For example, if the primary design goal is to track constant set-point changes in the reference input $r(t)$ with "sufficient" accuracy, and

$$G(s) = \frac{c(s)}{a(s)} = \frac{c_0}{s^2 + a_1 s + a_0} \tag{8.1.5}$$

then using Eq. (8.1.4),

$$\frac{y(s)}{r(s)} = \frac{K_P c_0}{s^2 + a_1 s + a_0 + K_P c_0} = \tilde{G} \frac{\omega_n^2}{s^2 + 2\zeta\omega_n s + \omega_n^2} \tag{8.1.6}$$

as in Eq. (4.2.25). Therefore, the resulting proportional design will imply a closed-loop system characterized by an undamped natural frequency

$$\omega_n = \sqrt{a_0 + K_P c_0} \tag{8.1.7}$$

and a damping ratio

$$\zeta = \frac{a_1}{2\sqrt{a_0 + K_P c_0}} \tag{8.1.8}$$

Moreover, for a unit step input $r(s) = \frac{1}{s}$, the error

$$e(s) = r(s) - y(s) = \left[1 - \frac{K_P c_0}{s^2 + a_1 s + a_0 + K_P c_0} \right] \frac{1}{s} \tag{8.1.9}$$

so that the closed-loop steady-state error

$$e_{ss}(t) = \lim_{t \to \infty} e(t) = \lim_{s \to 0} se(s) = 1 - \frac{K_P c_0}{a_0 + K_P c_0} = \frac{a_0}{a_0 + K_P c_0} \quad (8.1.10)$$

compared to an open-loop

$$e_{ss}(t) = \lim_{s \to 0} s[e(s) = u(s) - y(s)] = 1 - \frac{c_0}{a_0} = \frac{a_0 - c_0}{a_0} \quad (8.1.11)$$

Clearly, the larger that K_P is in Eq. (8.1.10), the less sensitive $e_{ss}(t)$ will be to changes in the plant parameters a_0 and c_0. Therefore, if K_P can be chosen large enough to achieve a small enough steady-state error, despite the anticipated plant parameter variations, as well as a "reasonable" damping ratio ζ, proportional control alone can produce acceptable response performance. The loop performance of the system can then be evaluated independently, relative to any explicitly stated loop goals. Note that the root locus represents an excellent means of determining the variation of the closed-loop poles of a system as a function of a single adjustable gain (proportional controller), as in Example 5.3.30.

. . . .

EXAMPLE 8.1.12 To illustrate the preceding discussions, consider the design of a controller for maintaining a desired level H_{2d} in the second tank of the two-tank liquid level system, which was outlined in Section 2.5 and is depicted in Figure 8.1, by automatically adjusting the input flow rate to tank one. Referring to Eqs.(2.2.6) and (2.2.7), the dynamic behavior of the open-loop system can be defined by the state-space representation

$$\begin{bmatrix} \dot{H}_1(t) \\ \dot{H}_2(t) \end{bmatrix} = \underbrace{\begin{bmatrix} -\dfrac{1}{A_1 R_1} & \dfrac{1}{A_1 R_1} \\ \dfrac{1}{A_2 R_1} & -\dfrac{R_1 + R_2}{A_2 R_1 R_2} \end{bmatrix}}_{A} \begin{bmatrix} H_1(t) \\ H_2(t) \end{bmatrix} + \underbrace{\begin{bmatrix} \dfrac{1}{A_1} \\ 0 \end{bmatrix}}_{B} Q_i(t)$$

$$y(t) = \underbrace{[0 \ 1]}_{C} \begin{bmatrix} H_1(t) \\ H_2(t) \end{bmatrix} = H_2(t)$$

Therefore, if $A_1 = 0.3$, $A_2 = 0.4$, $R_1 = 1.667$, and $R_2 = 1$, so that

$$A = \begin{bmatrix} -2 & 2 \\ 1.5 & -4 \end{bmatrix} \quad \text{and} \quad B = \begin{bmatrix} 3.333 \\ 0 \end{bmatrix}$$

the nominal performance of the open-loop system can be defined by its transfer function

$$\frac{y(s)}{u(s)} = G(s) = \underbrace{[0 \ 1]}_{C} \underbrace{\frac{\begin{bmatrix} s+4 & 2 \\ 1.5 & s+2 \end{bmatrix}}{(s+1)(s+5)}}_{(sI-A)^{-1}} \underbrace{\begin{bmatrix} 3.333 \\ 0 \end{bmatrix}}_{B} = \frac{5}{s^2 + 6s + 5} = \frac{c(s)}{a(s)}$$

as in Eq. (8.1.5), with open-loop poles at $s = -1$ and -5.

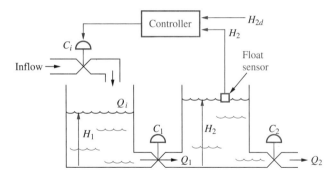

A Two-Tank Liquid Level Control System

In light of Figure 8.1, an appropriate controller may compare the actual (measured output) level $H_2(t) = y(t)$ with the desired (reference input) level $H_{2d} = r(t)$, and use the difference, or error, $e(t) = H_{2d} - H_2(t)$ to automatically alter the input flow rate $Q_i(t) = u(t)$ to the system by means of control valve C_i. A block diagram of such a 1 DOF, unity feedback control system is depicted in Figure 8.2. We will assume that the control system must maintain an $e_{ss}(t) < 10\%$ for step changes in the input (desired tank two level H_{2d}) despite flow resistance factors (plant parameters), R_1 and R_2, which can vary by more than 10%, and that $G(s)$ is also subjected to low frequency output disturbances due to such factors as evaporation and possible leaks. If this were not the case, there would be no need for feedback compensation, since Eq. (8.1.11) would imply a nominal, open-loop

$$e_{ss}(t) = \frac{a_0 - c_0}{a_0} = \frac{5 - 5}{5} = 0$$

Therefore, to ensure a relatively insensitive $e_{ss}(t) < 0.1$ for step changes in the input, a proportional compensator defined by an

$$H_P(s) = K_P = 10$$

will be chosen so that

$$e_{ss}(t) = \frac{a_0}{a_0 + K_P c_0} = \frac{5}{5 + 50} = 0.091$$

using Eq. (8.1.10). The resulting output response transfer function is then given by

$$T(s) = \frac{y(s)}{r(s)} = \frac{50}{s^2 + 6s + 55}$$

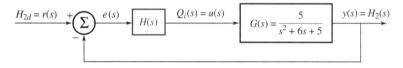

A Block Diagram of a 1 DOF Liquid Level Control System

with a damping ratio

$$\zeta = \frac{a_1}{2\sqrt{a_0 + K_P c_0}} = \frac{6}{2\sqrt{5 + 50}} = 0.4$$

using Eq. (8.1.8).

Figure 8.3 depicts a unit step response plot of $y(t) = H_2(t)$ in this case, which also would characterize the response of $H_2(t)$ to arbitrary step changes in the desired level H_{2d}. Note that the selection of $K_P = 10$ represents a *design trade-off*, since a larger value of $K_P > 10$ would further reduce the steady-state error at the expense of a more oscillatory response, hence the potential for tank overflow, while a smaller value of $K_P < 10$ would have the reverse effect.

A Bode plot of the loop gain magnitude

$$|L(j\omega)| = |K_P G(s = j\omega)| = \left| \frac{10}{(j\omega + 1)(0.2j\omega + 1)} \right|$$

is depicted by the solid line in Figure 8.4, along with the sensitivity function magnitude, which is the dashed line. Note that at the lower frequencies,

$$|S(j\omega)| = \frac{1}{|1 + L(j\omega)|} = 0.1$$

that, we will assume, ensures adequate rejection of any external disturbances.

The gain margin GM = ∞ in this case, and at the gain crossover frequency $\omega_g = 6.2$, $\angle L(j\omega_g) = \angle K_P G(j\omega_g) = -132°$, so that the resulting $\Phi M = 48°$ will imply adequate robust stability with respect to the plant parameter variations. A maximum value of $|S(j\omega)| = \bar{S} = 1.567$ at $\omega = 8.28$ confirms these stability margins in light of Eq. (6.1.29). We observe finally that the relatively narrow bandwidth associated with this system will be sufficient to ensure adequate attenuation of the float sensor noise caused by the sloshing of the liquid in tank two.

....

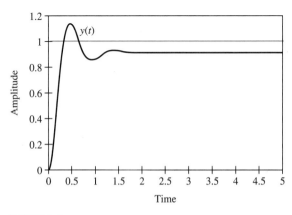

• **FIGURE 8.3**
The Unit Step Response of Proportional Compensation

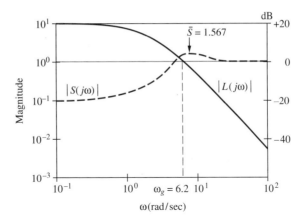

FIGURE 8.4
Bode Plots of $|L(j\omega)|$ and $|S(j\omega)|$ for Proportional Compensation

Proportional-Integral (PI) Compensation

Although proportional compensation is simple to analyze and design, it has obvious limitations. One additional level of control complexity combines a proportional gain K_P with an **integral (I) compensator**, which is defined by the transfer function

$$\frac{K_I}{s} \overset{\text{def}}{=} H_I(s) \tag{8.1.13}$$

to produce a **proportional-integral (PI) compensator**, namely,

$$K_P + \frac{K_I}{s} \overset{\text{def}}{=} H_{PI}(s) = \frac{K_P s + K_I}{s} = \frac{K_I \left(\dfrac{K_P}{K_I} s + 1 \right)}{s} \tag{8.1.14}$$

whose Bode diagram is depicted in Figure 8.5.

At low values of ω, the asymptotic magnitude graph is characterized by a slope of -20 dB/decade that crosses the 0 dB line at $\omega = K_I$. At the corner frequency $\omega = \dfrac{K_I}{K_P}$, the slope becomes and remains zero (horizontal) at a magnitude of $20 \log K_P$. The corresponding phase angle $\angle H_{PI}(j\omega)$ changes continuously from a value of $-90°$ at $\omega = 0$ to $0°$ as $\omega \to \infty$, crossing through $-45°$ at the corner frequency.

PI compensation increases the loop gain at the lower values of ω and decreases the loop gain at the higher values of ω, without compromising the stability margins; that is, although PI compensation increases the phase lag of the loop gain $L(j\omega) = G(j\omega) H_{PI}(j\omega)$ at values of ω near and below $\dfrac{K_I}{K_P}$, these frequencies generally are chosen considerably below the mid-range frequencies where the stability margins are defined. Hence, PI compensation improves many of the loop performance goals of a closed-

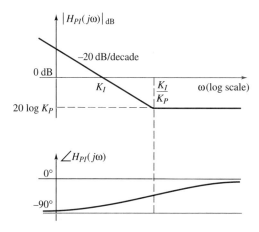

● **FIGURE 8.5**
A Bode Diagram of a PI Compensator $H_{PI}(j\omega)$

loop system, although it generally produces a more sluggish response, because of the reduction in the loop gain bandwidth.

Another reason for employing integral compensation in combination with proportional compensation is to improve the steady-state tracking accuracy of a closed-loop system. For example, an $H_{PI}(s)$ placed in series with $G(s)$ increases the system type by one, which ensures a robust $e_{ss}(t) = 0$ for step changes in the reference input $r(t)$, provided the closed-loop system remains stable.

. . . .

EXAMPLE 8.1.15 To illustrate the preceding, consider the liquid level control system of Example 8.1.12, whose nominal performance is defined by the open-loop transfer function

$$G(s) = \frac{y(s)}{u(s)} = \frac{5}{(s+1)(s+5)} = \frac{1}{(s+1)(0.2s+1)}$$

Suppose a PI compensator is employed, with $K_P = K_I = 1$, and the resulting

$$H_{PI}(s) = \frac{s+1}{s}$$

which cancels the pole at $s = -1$, is cascaded with $G(s)$ under unity feedback, as in Figure 8.2. A Bode plot of the resulting loop gain magnitude

$$|L(j\omega)| = |G(j\omega)H_{PI}(j\omega)| = \left| \frac{1}{(s+1)(0.2s+1)} \frac{s+1}{s} \right|_{s=j\omega} = \left| \frac{1}{j\omega(0.2j\omega+1)} \right|$$

is depicted by the dashed line in Figure 8.6. A plot of the loop gain magnitude associated with the P compensation of Example 8.1.12 is displayed by the solid line for comparison purposes.

Note that the PI compensated $L(j\omega)$ has a larger magnitude at the lower values of $\omega < 0.1$ than that of the P compensated system, which implies improved disturbance rejection. Moreover, the PI

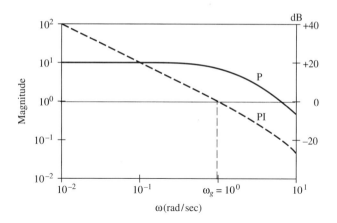

• **FIGURE 8.6**
Bode Plots of $|L(j\omega)|$ for P and PI Compensation

compensated loop gain magnitude is considerably smaller at the higher values of ω, which implies better noise attenuation and improved robust stability with respect to any unmodeled high frequency dynamics. Also, the GM remains $= \infty$, and at the new gain crossover frequency $\omega_g = 1$,

$$\angle L(j\omega_g) = \angle G(j\omega_g)H_{PI}(j\omega_g) = \angle \frac{1}{j(1+0.2j)} = -90° - 11.3° = -101.3°$$

so that the new $\Phi M = 78.7°$, which represents a significant increase in the $48°$ of ΦM obtained using only proportional compensation. Therefore, improvement is obtained relative to virtually all of the loop performance goals.

Figure 8.7 depicts a unit step response plot of the PI compensated system (the dashed trajectory), which is defined by the closed-loop transfer function

$$\frac{y(s)}{r(s)} = \frac{G(s)H_{PI}(s)}{1 + G(s)H_{PI}(s)} = \frac{5}{s^2 + 5s + 5}$$

The step response of the P compensated system of Example 8.1.12 (the solid trajectory) is superimposed for comparison purposes. Note that PI compensation reflects a damping ratio $\zeta = 1.12 > 1$ in this case, hence no overshoot, as well as a robust $e_{ss}(t) = 0$. However, because of the reduction in the loop gain bandwidth, its response is more sluggish than that associated with P compensation alone, which is characterized by an $\omega_n = \sqrt{55} = 7.42$, compared with an $\omega_n = \sqrt{5} = 2.24$ in the case of PI compensation.

Ideal Proportional-Derivative (PD) Compensation

In those cases where neither a relatively large proportional K_P nor a PI compensator is required to ensure adequate tracking accuracy, we may elect to combine a relatively small proportional gain K_P with an ideal

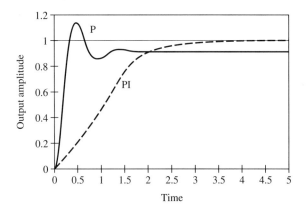

• **FIGURE 8.7**
Unit Step Responses of P and PI Compensation

derivative (D) compensator, which is defined by the transfer function

$$K_D s \overset{\text{def}}{=} H_D(s) \qquad (8.1.16)$$

to produce an ideal **proportional-derivative (PD) compensator**, namely,

$$K_P + K_D s \overset{\text{def}}{=} H_{PD}(s) = K_P \left(\frac{K_D}{K_P} s + 1 \right) \qquad (8.1.17)$$

whose Bode diagram is depicted with the solid lines in Figure 8.8.

At low values of ω, the asymptotic magnitude graph of a PD compensator remains at a magnitude of $20 \log K_P$. At frequencies greater

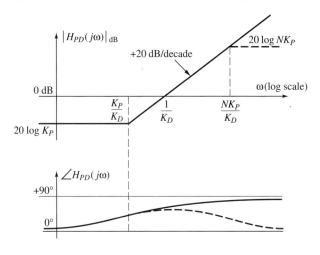

• **FIGURE 8.8**
Bode Diagrams of Ideal and Practical PD Compensation

than the corner frequency $\omega = \dfrac{K_P}{K_D}$, the magnitude graph is characterized by a slope of +20 dB/decade that crosses the 0 dB line at $\omega = \dfrac{1}{K_D}$. The corresponding phase angle $\angle H_{PD}(j\omega)$ changes continuously from a value of $0°$ at $\omega = 0$ to $+90°$ as $\omega \to \infty$, crossing through $+45°$ at the corner frequency. Hence PD compensation increases the phase lead of the loop gain $L(j\omega) = G(j\omega)H_{PD}(j\omega)$ at the medium and high values of ω, thus improving the stability margins relative to plant parameter variations. PD compensation also increases the loop gain bandwidth of the system, thereby producing a faster response. Unfortunately, any sensor noise is amplified as well.

Practical PD Compensation

In general, an ideal derivative compensator is difficult to construct.[1] Moreover, since its magnitude increases without bound as $\omega \to \infty$, an ideal differentiator produces an undesirable amplification of any high frequency noise that may be present within the loop. In the case of our liquid level control system, for example, the sloshing of the liquid in tank two would make it quite difficult to obtain an accurate measurement of the velocity signal $\dot{H}_2(t)$. Furthermore, the increase in the loop bandwidth associated with ideal PD compensation could cause stability problems caused by unmodeled high frequency dynamics.

Therefore, actual derivative compensation often is implemented by introducing a pole at a frequency generally between three and ten times higher than the corner frequency $\dfrac{K_P}{K_D}$, that is, at $\omega = \dfrac{NK_P}{K_D}$, where $3 \leq N \leq 10$ [5]. As a consequence, a physical PD compensator is characterized by the proper transfer function

$$H_{PD}(s) = \dfrac{K_P \left(\dfrac{K_D}{K_P}s + 1 \right)}{\dfrac{K_D}{NK_P}s + 1} \qquad (8.1.18)$$

whose Bode diagram is depicted with the dashed lines in Figure 8.8 along with that of the ideal PD compensator defined by Eq. (8.1.17). Note that at frequencies below the so-called *roll-off corner frequency* $\omega = \dfrac{NK_P}{K_D}$, the asymptotic amplitude graphs are identical and the phase plots are nearly identical.

---

[1] A notable exception is a **tachometer**, which is a special physical device that directly measures the velocity, hence the time derivative of the position, of a motor. **Optical encoders** also can be used to produce velocity feedback signals in certain low-noise cases (see Problem 8-5).

Ideal Proportional-Integral-Derivative (PID) Compensation

To obtain the benefits of both PI and PD compensation, a **proportional-integral-derivative (PID) compensator** may be employed. A PID compensator is defined "ideally" by the transfer function

$$H_{PID}(s) \overset{\text{def}}{=} K_P + \frac{K_I}{s} + K_D s = \frac{K_D s^2 + K_P s + K_I}{s} \qquad (8.1.19)$$

or as

$$H_{PID}(s) = K_P \left(1 + \frac{1}{T_I s} + T_D s \right) \qquad (8.1.20)$$

with

$$T_I \overset{\text{def}}{=} \frac{K_P}{K_I} \quad \text{and} \quad T_D \overset{\text{def}}{=} \frac{K_D}{K_P}$$

However, as in the case of practical PD compensation, a practical PID compensator usually includes an additional "roll-off" pole, as we will later show.

A PID compensator implies added design flexibility, when compared to a P, PI, or PD compensator, since it involves three arbitrarily adjustable terms. This form of compensation is widely used in industrial process control applications, such as petroleum refining, paper making, and metal forming, where it often is called a **three-term process controller**. In many such instances, the nominal plant transfer function $G(s)$ is unknown, and PID design is based on a step response analysis of the process, with the *proportional gain K_P*, the *integral (or reset) time T_I*, and the *derivative time T_D*, manually adjusted or tuned on-line to obtain the "best" performance. Well-established techniques, such as the Ziegler-Nichols tuning rules [67], can be used to determine nominal values for the three PID terms in such cases.

In industrial applications, integral control often is termed **reset control**, because without it, an operator must manually alter or "reset" the set-point value to achieve a desired output value. Integral control automatically ensures an output value equal to the set-point value, that is, a robust $e_{ss}(t) = 0$ for set-point changes in the reference input, without the requirement for manual reset.

. . . .

EXAMPLE 8.1.21 To illustrate ideal PID control, consider the liquid level control system of Examples 8.1.12 and 8.1.15, whose performance is defined by the nominal, open-loop transfer function

$$G(s) = \frac{y(s)}{u(s)} = \frac{1}{(s+1)(0.2s+1)} = \frac{5}{s^2 + 6s + 5}$$

In this case, the three adjustable terms of a series PID controller can be used to completely and arbitrarily position all three poles of the closed-loop system, since

$$\frac{y(s)}{r(s)} = \frac{G(s)H_{PID}(s)}{1 + G(s)H_{PID}(s)} = \frac{5(K_D s^2 + K_P s + K_I)}{s^3 + (6 + 5K_D)s^2 + (5 + 5K_P)s + 5K_I}$$

Alternatively, both of the stable, nominal poles of

$$G(s) = \frac{5}{s^2 + 6s + 5} = \frac{5}{(s + 1)(s + 5)}$$

at $s = -1$ and -5 can be "cancelled" by the zeros of $H_{PID}(s)$. For example, if K_D, K_P, and K_I are chosen such that

$$H_{PID}(s) = \frac{K_D s^2 + K_P s + K_I}{s} = \frac{\alpha\,(s + 1)(s + 5)}{5}\frac{1}{s} = \frac{0.2\alpha s^2 + 1.2\alpha s + \alpha}{s}$$

for any arbitrary scalar α, it follows that the corresponding, nominal, loop gain transfer function

$$L(s) = G(s)H_{PID}(s) = \frac{\alpha}{s}$$

which implies a closed-loop

$$\frac{y(s)}{r(s)} = \frac{L(s)}{1 + L(s)} = \frac{\alpha}{s + \alpha}$$

In particular, consider the utilization of an ideal PID compensator, with $\alpha = 2$, so that

$$H_{PID}(s) = \frac{2}{5}\frac{(s + 1)(s + 5)}{s} = \frac{0.4s^2 + 2.4s + 2}{s} = \frac{K_D s^2 + K_P s + K_I}{s}$$

The resulting loop gain transfer function

$$L(s) = G(s)H_{PID}(s) = \frac{2}{s}$$

which implies a GM $= \infty$ and a ΦM $= 90°$, defined at an $\omega_g = 2$. We will assume that such a choice for $H_{PID}(s)$ reflects an acceptable loop gain bandwidth for the system.

A Bode plot of the PID loop gain magnitude

$$|L(j\omega)| = \frac{2}{\omega}$$

is depicted by the dotted line in Figure 8.9. Corresponding plots of the PI loop gain magnitude of Example 8.1.15 (the dashed line) and the P loop gain magnitude of Example 8.1.12 (the solid line) are superimposed for comparison purposes.

The unit step response of the ideal PID compensated system, which is defined by the closed-loop transfer function

$$\frac{y(s)}{r(s)} = \frac{2}{s + 2} = \frac{1}{0.5s + 1}$$

is depicted by the dotted trajectory in Figure 8.10. The unit step responses of the P compensated system of Example 8.1.12 and the PI compensated system of Example 8.1.15 are also displayed for comparison purposes by the solid and dashed lines, respectively. Note that PID compensation produces a faster response than PI compensation, without the overshoot associated with P compensation alone, while retaining a robust $e_{ss}(t) = 0$.

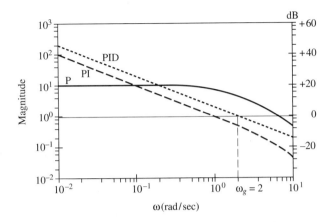

● **FIGURE 8.9**
Bode Plots of $|L(j\omega)|$ for P, PI, and PID Compensation

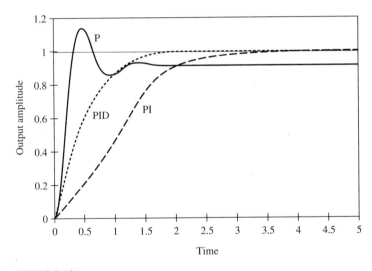

● **FIGURE 8.10**
The Unit Step Responses of P, PI, and PID Compensation

Practical PID Compensation

A PID compensator need not be implemented with the derivative term in the forward path, especially when there are step changes in the reference input that could produce plant input saturation; for example, an actual implementation of the ideal $H_{PID}(s)$ of Example 8.1.21, using the unity feedback configuration of Figure 8.2, would imply that

$$\frac{u(s)}{r(s)} = \frac{1}{G(s)} \frac{y(s)}{r(s)} = \frac{s^2 + 6s + 5}{5} \frac{2}{s+2} = \frac{0.4s^2 + 2.4s + 2}{s+2}$$

which is an improper transfer function. As a consequence, step changes in $r(s)$, the desired level of tank two, would demand impulsive input flow rates of infinite magnitude to tank one, a physical impossibility.

In such cases, a more practical implementation of PID compensation is that depicted in Figure 8.11, namely, a derivative term with a roll-off pole,

$$H_D(s) = \frac{K_D s}{\dfrac{K_D}{N} s + 1} \tag{8.1.22}$$

implemented by means of a *minor loop* around the plant, and

$$H_{PI}(s) = K_P + \frac{K_I}{s} \tag{8.1.23}$$

in the forward path, so that

$$y(s) = G(s)H_{PI}(s)r(s) - G(s)[H_{PI}(s) + H_D(s)]y(s) \tag{8.1.24}$$

Alternatively, ideal derivative (velocity) feedback, with $H_D(s) = K_D s$, could be used in those situations where a tachometer or an optical encoder is employed. Note that all configurations will imply the same PID loop gain transfer function

$$L(s) = G(s)[H_{PI}(s) + H_D(s)] \tag{8.1.25}$$

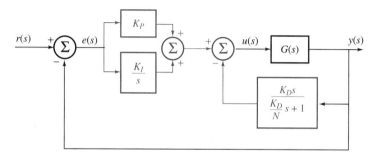

• **FIGURE 8.11**
A Practical Minor Loop Implementation of a PID Controller

. . . .

EXAMPLE 8.1.26 To illustrate the preceding discussion, a practical, minor loop implementation of the ideal PID controller of Example 8.1.21 would employ the Figure 8.11 configuration, with $K_P = 2.4$, $K_I = 2$ and $K_D = 0.4$, as in Example 8.1.21. If $N = 10$, Eq. (8.1.25) then implies a loop gain

$$L(s) = \frac{c(s)}{a(s)} \left[K_P + \frac{K_I}{s} + \frac{K_D s}{\frac{K_D}{N} s + 1} \right] = \frac{5}{s^2 + 6s + 5} \left[2.4 + \frac{2}{s} + \frac{0.4s}{0.04s + 1} \right]$$

$$= \frac{62s^2 + 310s + 250}{(s^2 + 6s + 5)(s^2 + 25s)} = \frac{250(0.99s + 1)(0.25s + 1)}{125s(s + 1)(0.2s + 1)(0.04s + 1)} \approx \frac{2}{s(0.04s + 1)}$$

which is characterized by a GM $= \infty$ and a ΦM $\approx 85°$, defined at an $\omega_g \approx 2$, analogous to the ideal PID compensator design of Example 8.1.21.

In view of Eq. (8.1.24), the closed-loop transfer function of the system is given by

$$\frac{y(s)}{r(s)} = T(s) = \frac{G(s)H_{PI}(s)}{1 + G(s)[H_{PI}(s) + H_D(s)]} = \frac{\dfrac{12s + 10}{s(s^2 + 6s + 5)}}{1 + \dfrac{62s^2 + 310s + 250}{s(s^2 + 6s + 5)(s + 25)}}$$

$$= \frac{12s^2 + 310s + 250}{s^4 + 31s^3 + 217s^2 + 435s + 250} = \frac{12(s + 25)(s + 0.833)}{(s + 22.02)(s + 6.16)(s + 1.803)(s + 1.023)}$$

$$= \frac{250(0.04s + 1)(1.2s + 1)}{250(0.05s + 1)(0.16s + 1)(0.55s + 1)(0.98s + 1)} \approx \frac{(1.2s + 1)}{(0.55s + 1)(0.98s + 1)}$$

The step response of this system approximates that of the ideal PID compensator of Example 8.1.21, as verified in Figure 8.12 by the solid and dashed trajectories, respectively. The lack of an exact cancellation of the pole-zero pair defined by $(0.98s + 1)$ and $(1.2s + 1)$ in $T(s)$ results in the explicit presence of the slowest modal term $+0.51e^{-1.02t}$ in the practical PID case (see Problem 8-7). This accounts for the slight overshoot and the slower settling time of $y(t)$ in the system controlled by practical PID compensation.

. . . .

2 DOF PID Compensation

The practical, minor loop PID compensator of Figure 8.11 now serves to motivate a 2 DOF PID compensator that retains the loop performance of a PID controller but allows adjustments to the response performance. In particular, Figure 8.11 implies that

$$u(s) = \left(\frac{K_P s + K_I}{s} \right) [r(s) - y(s)] - \frac{K_D s}{\frac{K_D}{N} s + 1} y(s)$$

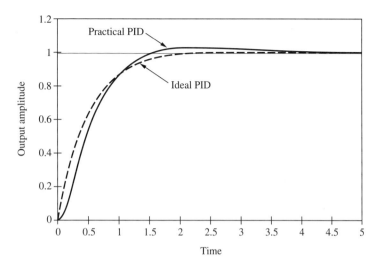

• **FIGURE 8.12**
Step Responses of the Ideal and Practical PID Compensators

or that

$$s\left(s + \frac{N}{K_D}\right)u(s) = (K_P s + K_I)\left(s + \frac{N}{K_D}\right)[r(s) - y(s)] - Ns^2 y(s)$$

Therefore,

$$k(s)u(s) = q(s)r(s) - h(s)y(s)$$

as in the 2 DOF configuration depicted in Figure 7.1, with

$$k(s) = s\left(s + \frac{N}{K_D}\right) \tag{8.1.27}$$

$$h(s) = (N + K_P)s^2 + \left(K_I + \frac{NK_P}{K_D}\right)s + \frac{NK_I}{K_D}$$

$$\approx Ns^2 + \frac{NK_P}{K_D}s + \frac{NK_I}{K_D} \tag{8.1.28}$$

for larger values of N, and a "constrained"

$$q(s) = K_P s^2 + \left(K_I + \frac{NK_P}{K_D}\right)s + \frac{NK_I}{K_D} = h(s) - Ns^2 \tag{8.1.29}$$

However, since it does not affect the loop performance of the system, $q(s)$ need not be defined by Eq. (8.1.29), and it therefore can be chosen to improve the response performance of the system, as we will now illustrate.

In the case of 2 DOF PID compensation, $k(s)$ and $h(s)$ will be defined by Eqs. (8.1.27) and (8.1.28), respectively, and $q(s)$ can be any arbitrary

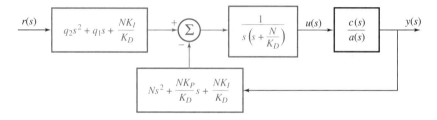

● **FIGURE 8.13**
A Practical 2 DOF PID Compensator

polynomial of degree 2 or less. Therefore, a choice of

$$q(s) = q_2 s^2 + q_1 s + \frac{N K_I}{K_D}$$ (8.1.30)

so that

$$q(0) = q_0 = h(0) = \frac{N K_I}{K_D}$$ (8.1.31)

will ensure a robust $e_{ss}(t) = 0$ for step changes in the reference input, in
light of Eq. (7.3.18).

Figure 8.13 depicts this **practical 2 DOF PID compensator**, which
is characterized by an output response transfer function

$$\frac{y(s)}{r(s)} = \frac{c(s)q(s)}{\delta(s)}$$ (8.1.32)

referring to Eq. (5.1.15), and an input response transfer function

$$\frac{u(s)}{r(s)} = \frac{a(s)}{c(s)} \frac{y(s)}{r(s)} = \frac{a(s)q(s)}{\delta(s)}$$ (8.1.33)

with $q(s)$ given by Eq. (8.1.30), and

$$\delta(s) = a(s)s \left(s + \frac{N}{K_D} \right) + c(s) \left(Ns^2 + \frac{N K_P}{K_D} s + \frac{N K_I}{K_D} \right)$$ (8.1.34)

using Eqs. (5.1.13), (8.1.27) and (8.1.28).

....

EXAMPLE 8.1.35 To illustrate the preceding discussion, consider the liquid level control system
defined in our earlier examples by a nominal

$$G(s) = \frac{c(s)}{a(s)} = \frac{5}{s^2 + 6s + 5}$$

If $K_P = 2.4$, $K_I = 2$, $K_D = 0.4$, and $N = 10$, as in Examples 8.1.21 and 8.1.26, Eq. (8.1.27)
implies that

$$k(s) = s \left(s + \frac{N}{K_D} \right) = s^2 + 25s$$

and Eq. (8.1.28) implies that

$$h(s) = Ns^2 + \frac{NK_P}{K_D}s + \frac{NK_I}{K_D} = 10s^2 + 60s + 50$$

In this case, the loop gain

$$L(s) = \frac{c(s)h(s)}{a(s)k(s)} = \frac{50(s^2 + 6s + 5)}{(s^2 + 6s + 5)(s^2 + 25s)} = \frac{2}{s(0.04s + 1)}$$

which is analogous to that obtained using the minor loop PID compensator of Example 8.1.26.

However, looking at Eqs. (8.1.32) and (8.1.34), the output response transfer function of the closed-loop system is defined by

$$\frac{y(s)}{r(s)} = \frac{c(s)q(s)}{\delta(s)} = \frac{5q(s)}{\underbrace{s^4 + 31s^3 + 205s^2 + 425s + 250}_{(s+1)(s+2.19)(s+5)(s+22.81)}}$$

and, referring to Eq. (8.1.33), the input response transfer function is given by

$$\frac{u(s)}{r(s)} = \frac{a(s)q(s)}{\delta(s)} = \frac{(s+1)(s+5)q(s)}{(s+1)(s+2.19)(s+5)(s+22.81)} = \frac{q(s)}{(s+2.19)(s+22.81)}$$

Three distinct choices for $q(s)$ will now be considered, each with

$$q(0) = q_0 = h(0) = \frac{NK_I}{K_D} = 50$$

which will ensure a robust $e_{ss}(t) = 0$ for step changes in $r(t)$. In particular,

$$q_a(s) = \frac{50}{22.81}(s+1)(s+22.81) = 2.19s^2 + 52.14s + 50$$

which cancels the two closed-loop poles at $s = -1$ and -22.81,

$$q_b(s) = 50(s+1) = 50s + 50$$

which cancels the slowest closed-loop pole at $s = -1$, and

$$q_c(s) = 30s + 50$$

The step responses of the three resulting 2 DOF implementations, when compared to that of the practical, minor loop PID implementation of Example 8.1.26, reflect interesting differences in both the unit step output $y(t) = H_2(t)$ response plots of Figure 8.14 and the unit step plant input $u(t) = Q_i(t)$ response plots of Figure 8.15. In both figures, the solid trajectories correspond to the minor loop PID implementation of Example 8.1.26, while the dashed, dotted, and dash-dotted trajectories reflect the practical 2 DOF PID implementations for $q_a(s) = 2.19s^2 + 52.14s + 50$, $q_b(s) = 50s + 50$, and $q_c(s) = 30s + 50$, respectively.

Note that the $y(t)$ responses of Figure 8.14 reflect only a slight difference between the $q_a(s)$ and $q_b(s)$ cases. This is due to the relatively small s^2 coefficient, namely, 2.19, of $q_a(s)$, which implies a $q_a(s) \approx q_b(s)$. Moreover, both of these cases display a superior step response, in the sense of no overshoot and a faster settling time, when compared with the more traditional minor loop PID implementation.

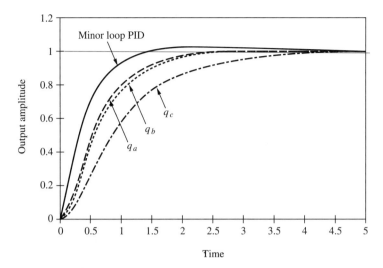

• **FIGURE 8.14**
The Unit Step Output $y(t)$ Responses

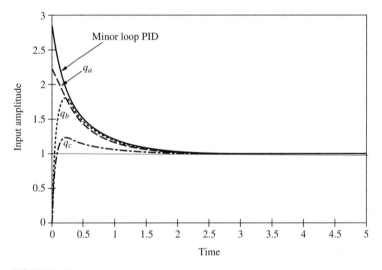

• **FIGURE 8.15**
The Unit Step Plant Input $u(t)$ Responses

Figure 8.15 reveals that the control effort $u(t)$, that is, the tank one input flow rate $Q_i(t)$ required in all three 2 DOF PID cases, is less than that required in the minor loop PID case. However, it can be shown that even moderate increases in the s^2 coefficient of $q(s)$ will produce potentially saturating $u(t)$ values. For example, although a relatively fast unit step $y(t)$ response would be produced by a

$$q(s) = \frac{50}{2.19}(s+1)(s+2.19)$$

which cancels the two slowest closed-loop poles, a tank one input flow rate $Q_i(t) = u(t)$ characterized by a relatively large maximum initial value of $50/2.19 = 22.83$ would be required (see Problem 8-8).

The $q_b(s) = 50s + 50$ and $q_c(s) = 30s + 50$ cases display a *design trade-off* between the speed of response of the liquid level in tank two $H_2(t) = y(t)$ and the required magnitude of the input flow rate $Q_i(t) = u(t)$ to tank one. For example, a control system designer could vary the coefficient q_1, in an otherwise fixed $q(s) = q_1 s + 50$, without affecting the loop properties of the system until an appropriate response is obtained. Such a procedure would imply a general 2 DOF **four-term process controller**, as depicted in Figure 8.13, with $q_2 = 0$, q_1 the variable fourth term, and $q_0 = q(0)$ fixed at $h(0)$.

....

It is finally of interest to note that Example 8.1.35 represents the "converse" of Example 7.2.13. In particular, in Example 7.2.13 we retained a desired response performance while varying the loop performance. Here, we have retained a desired loop performance while varying the response performance. The ability to achieve the specific trade-offs associated with both examples is due to the design flexibility afforded by 2 DOF compensation.

8.2 LAG AND LEAD COMPENSATION

Other popular types of classical control are lag, lead, or a combination of the two, which is termed lag-lead compensation. As we will show, lag compensation is analogous to proportional-integral (PI) compensation with a low frequency loop gain roll-off; lead compensation is analogous to proportional-derivative (PD) compensation with a high frequency loop gain roll-off; and lag-lead compensation is analogous to PID compensation with a loop gain roll-off at both high and low values of ω.

All three of these controllers are characterized by a proper, rational transfer function $H(s)$, generally placed in the forward path in cascade with the plant $G(s)$, which is then closed under unity feedback in a standard 1 DOF configuration to alter or "reshape" the loop gain frequency response

$$L(s = j\omega) = G(j\omega)H(j\omega) \tag{8.2.1}$$

For this reason, this form of compensator design often is referred to as **loopshaping**. Loopshaping techniques were first developed by Bode [7] for the design of feedback amplifiers, and they were later modified for control system design. In this section, we will first discuss lag, then lead, concluding with the design of combined lag-lead compensators.

In all cases, appropriate compensators will be illustrated using the DC servomotor that was depicted in Figure 2.6, whose dynamic behavior can be defined by the "generic" transfer function

$$G(s) = \frac{K_a}{s(s+1)(0.1s+1)} = \frac{10K_a}{s^3 + 11s^2 + 10s} \tag{8.2.2}$$

in light of Problem 3-19, with K_a an adjustable amplifier gain.

DC servomotors, which are direct current electrical motors employed in servomechanisms, are used extensively in a variety of industrial applications. Low power DC servomotors are used in fine-positioning instruments, such as X-Y plotters and rotor winders, as well as computer-related equipment, such as tape and disk drives, mechanical printers, and word processors. Medium and high power DC servomotors are employed in robot arm drive systems, numerically controlled milling machines, and antenna tracking systems.

In most applications, positional errors between a reference input and the plant output should be small (often zero), especially in the steady-state as $t \to \infty$. In other applications, errors between a constant velocity (ramp) input and the time-varying plant output must be minimized. The design procedures presented in this section will illustrate some of the more traditional methods that have been developed for achieving desired tracking accuracies for the defined DC servomotor. However, the techniques presented are generally applicable to many other controlled systems as well.

Lag Compensation

A **lag compensator** is characterized by an adjustable pole-zero pair, generally placed close to the origin on the negative real axis, together with a variable gain, as defined by the rational transfer function

$$K\frac{Ts+1}{\alpha Ts+1} = \frac{K}{\alpha}\frac{\left(s+\dfrac{1}{T}\right)}{\left(s+\dfrac{1}{\alpha T}\right)} \stackrel{\text{def}}{=} H_{la}(s), \text{ with } \alpha > 1 \tag{8.2.3}$$

so that the pole at $s = -\dfrac{1}{\alpha T}$ lies closer to the origin than the zero at $s = -\dfrac{1}{T}$. The frequency response of $H_{la}(j\omega)$, with corner frequencies at $\omega = \dfrac{1}{\alpha T}$ and $\dfrac{1}{T}$, is depicted by the asymptotic Bode diagram of Figure 8.16.

Lag compensation can be used to alter the magnitude of the loop gain, often when a closed-loop plant $G(s)$ either would be unstable without compensation or when its frequency response is characterized by an unacceptable gain or phase margin. In such cases, lag compensation reduces

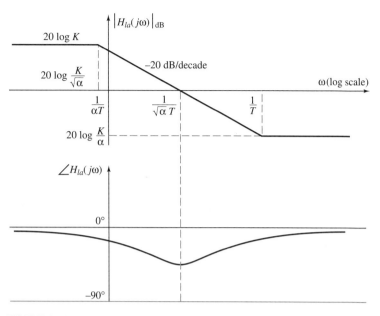

• **FIGURE 8.16**
The Bode Diagram of a Lag Compensator

the amplitude of the loop gain $G(j\omega)H_{la}(j\omega)$ at the mid-range and higher frequencies to produce acceptable stability margins. Such an amplitude reduction also improves both the noise attenuation characteristics of the system and its robust stability properties with respect to unmodeled high frequency dynamics. However, the resulting transient response may be sluggish because of the reduction in the loop bandwidth.

In such applications K often is set equal to 1, so that $|H_{la}(j\omega)| = 0$ dB at low values of ω, thereby preserving the low frequency loop gain properties of the uncompensated system. The negative (potentially destabilizing) phase angles associated with lag compensation are of little consequence, since they generally peak at frequencies significantly below the mid-range frequencies where the stability margins are determined.

. . . .

EXAMPLE 8.2.4 To illustrate lag compensation, let us assume that the DC servomotor defined by Eq. (8.2.2) is required to track step inputs with zero steady-state errors, and ramp inputs with steady-state errors no greater than 1%. Since

$$G(s) = \frac{K_a}{s(s+1)(0.1s+1)} = \frac{10K_a}{s^3 + 11s^2 + 10s}$$

is type 1, a 1 DOF unity feedback design will ensure a robust $e_{ss}(t) = 0$ for step inputs, and a "high enough" gain K_a will ensure a steady-state ramp error $\leq 1\%$, provided the closed-loop system remains stable.

In particular, if the loop compensator $H(s) = 1$,

$$\frac{e(s)}{r(s)} = \frac{1}{1 + G(s)} = \frac{s(s^2 + 11s + 10)}{s^3 + 11s^2 + 10s + 10K_a}$$

Therefore, for a unit ramp input, when $r(s) = \frac{1}{s^2}$,

$$\lim_{t \to \infty} e(t) = e_{ss}(t) = \lim_{s \to 0} \frac{s^2(s^2 + 11s + 10)}{s^3 + 11s^2 + 10s + 10K_a} \frac{1}{s^2} = \frac{10}{10K_a} = \frac{1}{K_a}$$

which implies that $K_a \geq 100$ to ensure a steady-state ramp tracking error $\leq 1\%$, as desired.

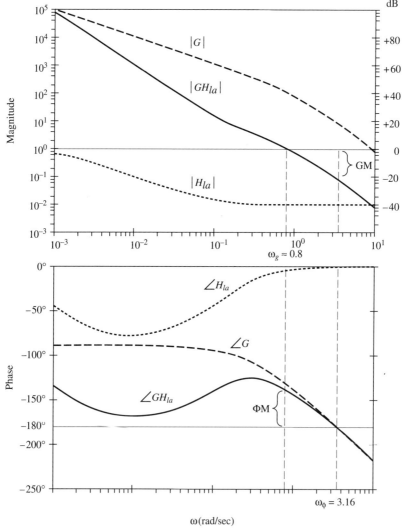

• **FIGURE 8.17**
A Bode Diagram of Lag Compensation

A Bode diagram of $G(s)$, with $K_a = 100$, is depicted by dashed lines in Figure 8.17. Note that the system would be unstable if closed under unity feedback with $H(s) = 1$, because at the phase crossover frequency $\omega_\phi = \sqrt{10} = 3.16$, $|G(j3.16)| = 1000/110 = 9.09 > 1$ (see Problem 8-9). However, a 40 dB reduction in the loop gain would reduce $|G(j3.16)|$ to $0.0909 < 1$, which would imply a GM = 11, or nearly 21 dB. The resulting gain crossover frequency $\omega_g \approx 0.8$, would be characterized by a phase angle $\approx -135°$, hence a $\Phi M \approx 45°$.

To obtain these stability margins, a lag compensator with $K = 1$, $T = 10$, and $\alpha = 100$ can be employed, resulting in a

$$H_{la}(s) = K\frac{Ts+1}{\alpha Ts+1} = \frac{10s+1}{1000s+1} = \frac{0.01(s+0.1)}{s+0.001}$$

with corner frequencies at $\omega = 0.001$ and 0.1. The Bode diagram of this $H_{la}(j\omega)$ is depicted by the dotted lines in Figure 8.17. An $\alpha = 100$ is employed to produce the 40 dB gain attenuation, and a value of $T = 10$ is chosen so that the second corner frequency occurs at a low enough value of $\omega = 0.1$ to minimize the phase lag introduced by $H_{la}(j\omega)$ at the gain crossover frequency $\omega_g \approx 0.8$.

A Bode diagram of the loop gain of the compensated system, $G(j\omega)H_{la}(j\omega)$ is depicted by the solid lines in Figure 8.17. A MATLAB computation [46] was used to determine that the actual GM = 9.77 (or 19.8 dB) at a phase crossover frequency $\omega_\phi = 2.99$, with a corresponding $\Phi M = 39.97°$ at a gain crossover frequency $\omega_g = 0.793$.

Figure 8.18 depicts a unit step response of the resulting closed-loop system, which is defined by the transfer function

$$\frac{y(s)}{r(s)} = \frac{G(s)H_{la}(s)}{1+G(s)H_{la}(s)} = \frac{10s+1}{s^4+11.001s^3+10.011s^2+10.01s+1}$$

$$= \frac{10(s+0.1)}{(s+0.111)(s+0.391\pm j.861)(s+10.108)}$$

The response is characterized by a maximum overshoot M_p of nearly 35%, and a long settling time t_s, or "tail" [59]. In general, this phenomenon is caused by the mode defined by the pole closest to

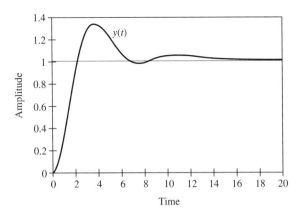

● **FIGURE 8.18**
A Unit Step Response of Cascade Lag Compensation

the origin requiring an excessive amount of time to decay to 0. In this case, the mode $e^{-0.111t}$, due to the pole at $s = -0.111$, implies a 2% settling time t_s greater than 14 seconds, with $y(t) = 1.013$ at $t = 20$ seconds, which may be unacceptable in certain applications. Such an occurrence can be remedied by a 2 DOF implementation in which a zero of $q(s)$ cancels the tail-producing pole (see Problem 8-11).

Lag and PI Compensation

As we noted at the beginning of this section, *lag compensation is analogous to PI compensation.* In particular, consider the lag compensator defined by Eq. (8.2.3). If we "fix" the ratio

$$\frac{K}{\alpha} \overset{\text{def}}{=} K_P \quad \Longrightarrow \quad K = \alpha K_P \tag{8.2.5}$$

and further define

$$K_I = \frac{K}{\alpha T} \quad \Longrightarrow \quad T = \frac{K_P}{K_I} \tag{8.2.6}$$

it follows that in the limiting case as $\frac{1}{\alpha T} \to 0$, or as $\alpha \to \infty$ and $K \to \infty$ in Figure 8.16,

$$\lim_{\alpha \to \infty} H_{la}(s) = \lim_{\alpha \to \infty} K_P \frac{s + \frac{1}{T}}{s + \frac{1}{\alpha T}} = \frac{K_I\left(\frac{K_P}{K_I}s + 1\right)}{s} = H_{PI}(s) \tag{8.2.7}$$

as defined by Eq. (8.1.14) and depicted in Figure 8.5. This observation serves to illustrate an alternative use for lag compensation, namely *to increase the low frequency loop gain in order to improve the steady-state accuracy of the system and to enhance the disturbance rejection properties of the closed-loop system.*

EXAMPLE 8.2.8 To illustrate this alternative use for lag compensation, consider the DC servomotor with $K_a = 1$, as defined by the open-loop transfer function

$$G(s) = \frac{10}{s(s + 1)(s + 10)}$$

which is $0.01 \times G(s)$ of Example 8.2.4. If closed under unity feedback with $H(s) = 1$, the stability margins would be analogous to those given by the $G(s)H_{la}(s)$ of Example 8.2.4. This type 1 system would also imply a robust $e_{ss}(t) = 0$ for step changes in the reference input. However, the steady-state unit ramp error would be an unacceptable $\frac{1}{K_a} = 1$, or 100%.

The introduction of a series lag compensator defined by an

$$H_{la}(s) = K\frac{Ts+1}{\alpha Ts+1} = 100\frac{10s+1}{1000s+1} = \frac{s+0.1}{s+0.001} = \underbrace{\frac{K}{\alpha}}_{K_P=1}\frac{\left(s+\frac{1}{T}\right)}{\left(s+\frac{1}{\alpha T}\right)}$$

which is $100 \times H_{la}(s)$ of Example 8.2.4, would yield the exact same loop gain transfer function $G(s)H_{la}(s)$ as in that example. The resulting increase in the low frequency loop gain would enhance the disturbance rejection properties of the compensated system, as well as ensure the desired 1% steady-state accuracy for ramp inputs.

Moreover, in the limiting case defined by Eq. (8.2.7),

$$\lim_{\alpha\to\infty} H_{la}(s) = \lim_{\alpha\to\infty} \underbrace{K_P}_{1}\frac{\left(s+\frac{1}{T}\right)}{\left(s+\frac{1}{\alpha T}\right)} = \frac{s+0.1}{s}$$

would achieve a robust $e_{ss}(t) = 0$ for ramp inputs as well, since such an $H_{la}(s)$ would become a PI compensator, and the loop gain transfer function $G(s)H_{la}(s) = G(s)H_{PI}(s)$ would be type 2. The stability margins and the transient response would be affected only slightly by this increase in the low frequency gain.

· ·

In Example 8.2.8, the asymptotic $|H_{la}(j\omega)| = 1$ at frequencies above $\omega = 0.1$. As a consequence, the loop gain of the compensated system is unaffected at the medium and high frequencies, which is the exact opposite of the loop gain changes obtained in Example 8.2.4. Indeed, differences in the magnitude of the plant transfer function gain K_a could require any one of an infinite number of corresponding amplitude choices for $H_{la}(s)$ to maintain the same loop gain $L(s) = G(s)H_{la}(s)$.

For example, if $K_a = 10$, so that

$$G(s) = \frac{100}{s(s+1)(s+10)}$$

then the choice of an

$$H_{la}(s) = K\frac{Ts+1}{\alpha Ts+1} = 10\frac{10s+1}{1000s+1} = \frac{0.1(s+0.1)}{s+0.001}$$

would ensure the same desired loop gain transfer function $G(s)H_{la}(s)$ as that obtained in Examples 8.2.4 and 8.2.8. Such an $H_{la}(s)$ would be characterized by an asymptotic $|H_{la}(j\omega)| = 10$, or +20 dB, at low values of ω, and a $|H_{la}(j\omega)| = 0.1$, or −20 dB, at the higher values of ω, so that the lag compensator would amplify the lower frequencies and attenuate the higher frequencies.

Lead Compensation

As in the case of lag compensation, a **lead compensator** also is character-ized by an adjustable pole-zero pair, generally placed some distance away from the origin on the negative real axis, together with a variable gain, as defined by the rational transfer function

$$K\frac{Ts+1}{\alpha Ts+1} = \frac{K}{\alpha}\frac{\left(s+\dfrac{1}{T}\right)}{\left(s+\dfrac{1}{\alpha T}\right)} \stackrel{\text{def}}{=} H_{le}(s), \text{ with } \alpha < 1 \qquad (8.2.9)$$

so that the zero at $s = -\dfrac{1}{T}$ lies closer to the origin than the pole at $s = -\dfrac{1}{\alpha T}$. The frequency response of $H_{le}(j\omega)$, with corner frequencies at $\omega = \dfrac{1}{T}$ and $\dfrac{1}{\alpha T}$, is depicted by the asymptotic Bode diagram of Figure 8.19.

Lead compensation can be used to alter the phase angle of the loop gain, often when a closed-loop plant $G(s)$ either would be unstable with-out compensation or when its frequency response is characterized by an unacceptable gain or phase margin. Looking at Figure 8.19, a key to the choice of the parameters that define a lead compensator is the observation that the maximum additional phase lead contributed by $H_{le}(j\omega)$ occurs at

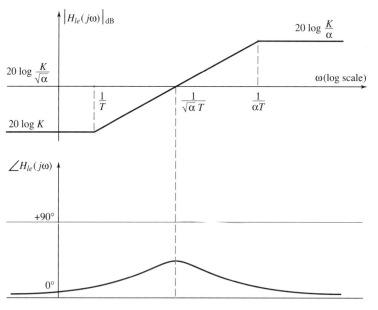

● **FIGURE 8.19**
The Bode Diagram of a Lead Compensator

a frequency midway between the two corner frequencies at[2]

$$\omega^* \overset{\text{def}}{=} \frac{1}{\sqrt{\alpha}T} \tag{8.2.10}$$

where

$$|H_{le}(j\omega^*)| = \frac{K}{\sqrt{\alpha}} \tag{8.2.11}$$

One way to design a lead compensator, for a fixed value of α, is to choose T and K so that ω^* corresponds to the gain crossover frequency of the compensated system, because a maximum amount of additional phase lead then would be added to the phase margin.

   To illustrate such a procedure, T could first be chosen in Eq. (8.2.10) so that ω^* corresponds to the phase crossover frequency ω_ϕ^G of $G(j\omega)$. In particular, a choice of

$$T = \frac{1}{\sqrt{\alpha}\omega_\phi^G}, \quad \text{where} \quad \angle G(j\omega_\phi^G) = -180° \tag{8.2.12}$$

will imply that

$$\omega^* = \frac{1}{\sqrt{\alpha}T} = \omega_\phi^G \tag{8.2.13}$$

   In view of Eq. (8.2.11), a corresponding choice of

$$K = \frac{\sqrt{\alpha}}{|G(j\omega_\phi^G)|} \tag{8.2.14}$$

then will imply that

$$|H_{le}(j\omega^*)| = |H_{le}(j\omega_\phi^G)| = \frac{K}{\sqrt{\alpha}} = \frac{1}{|G(j\omega_\phi^G)|} \tag{8.2.15}$$

As a consequence,

$$|G(j\omega^*)H_{le}(j\omega^*)| = 1 \tag{8.2.16}$$

and $\omega^* = \omega_\phi^G$, the *phase crossover frequency* of the uncompensated system would become the *gain crossover frequency* ω_g of the compensated system, with a resultant loop gain phase angle at $\omega_g = \omega^*$ given by

$$\phi_g \overset{\text{def}}{=} \angle G(j\omega^*)H_{le}(j\omega^*) = \underbrace{\angle G(j\omega_\phi^G)}_{-180°} + \angle H_{le}(j\omega^*) \tag{8.2.17}$$

In light of Eq. (6.1.8), the phase margin of the compensated system, namely

$$\Phi M = 180° + \phi_g = \angle H_{le}(j\omega^*) = \max_\omega \angle H_{le}(j\omega) \tag{8.2.18}$$

would be nearly 55° if $\alpha = 0.1$.

---

[2]If two points lie at d and βd on a log scale, then the midway point between them is at $\sqrt{\beta}d$.

EXAMPLE 8.2.19 To illustrate the preceding, consider the DC servomotor defined in Example 8.2.4 by the $K_a = 100$ open-loop transfer function

$$G(s) = \frac{100}{s(s+1)(0.1s+1)} = \frac{1000}{s(s+1)(s+10)}$$

Recall that this $G(s)$ would be unstable if closed under unity feedback with $H(s) = 1$, because at the phase crossover frequency $\omega_\phi = \sqrt{10} = 3.16$, $|G(j3.16)| = 9.09 > 1$. However, if we employ

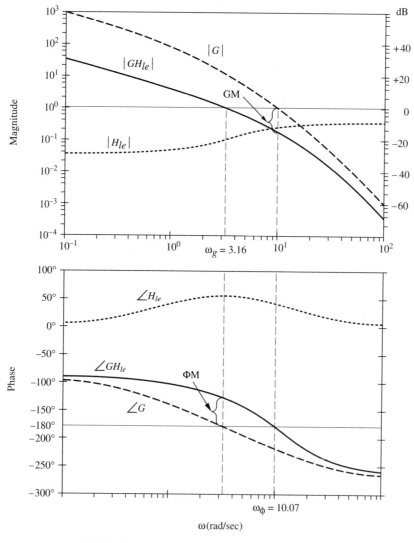

● **FIGURE 8.20**
A Bode Diagram of Lead Compensation

a lead compensator with $\alpha = 0.1$, Eqs. (8.2.12) and (8.2.14) will imply that

$$T = \frac{1}{\sqrt{\alpha}\omega_\phi^G} = \frac{1}{\sqrt{0.1}\sqrt{10}} = 1 = \frac{1}{T} \quad \text{and} \quad K = \frac{\sqrt{\alpha}}{|G(j\omega_\phi^G)|} = \frac{0.316}{9.09} = 0.0347$$

Therefore, referring to Eq. (8.2.9), the lead compensator is defined by

$$H_{le}(s) = K\frac{Ts+1}{\alpha Ts+1} = 0.0347\frac{s+1}{0.1s+1} = \frac{0.347(s+1)}{s+10}$$

with corner frequencies at $\omega = 1$ and 10. It may be noted that in this case the zero of $H_{le}(s)$ at $s = -1$ cancels a corresponding pole of $G(s)$.

Bode diagrams of $G(j\omega)$, $H_{le}(j\omega)$, and the resulting loop gain $G(j\omega)H_{le}(j\omega)$ of the compensated system are depicted in Figure 8.20 by the dashed, dotted, and solid lines, respectively. Note that the gain crossover frequency of the compensated system, namely $\omega_g = 3.16$, corresponds to the phase crossover frequency ω_ϕ^G of the uncompensated system. Moreover, $\omega = 3.16 = \omega_g$ represents the frequency where the stabilizing phase lead contributed by $H_{le}(s)$ peaks at nearly 55°. A MATLAB computation was used to determine that the actual GM $= 5.67$ (or 15.07 dB) at a phase crossover frequency $\omega_\phi = 10.07$.

Figure 8.21 depicts a unit step response of the closed-loop system, as defined by the closed-loop transfer function

$$\frac{y(s)}{r(s)} = \frac{G(s)H_{le}(s)}{1+G(s)H_{le}(s)} = \frac{347}{s^3+20s^2+100s+347} = \frac{347}{(s+2.58\pm j4.09)(s+14.84)}$$

The response is characterized by a maximum overshoot $M_p \approx 13\%$ and a 2% settling time $t_s = 1.28$ seconds, which is much faster than that associated with the lag compensator unit step response of Figure 8.18.

Since $0.0347 = K < \alpha = 0.1$ in this case, the lead compensator attenuates the loop gain at all frequencies, but especially at the lower frequencies where the asymptotic $|H_{le}(j\omega)| = 0.0347$, or -29.19 dB. This reduction in the low frequency loop gain reduces the disturbance rejection properties of the system significantly. It also implies an unacceptable, steady-state ramp tracking accuracy of

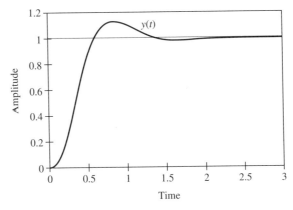

● **FIGURE 8.21**
A Unit Step Response of Cascade Lead Compensation

1/3.47, or 28.8%. These "deficiencies" can be corrected by the addition of a lag compensator, as we will show in Example 8.2.30.

Lead and PD Compensation

As we noted at the beginning of this section, *lead compensation is analogous to PD compensation*. In particular, consider the lead compensator defined by Eq. (8.2.9). If we define

$$K_P = K, \ K_D = KT, \text{ and } N = \frac{1}{\alpha} \implies T = \frac{K_D}{K_P} \text{ and } \alpha = \frac{1}{N} \quad (8.2.20)$$

it follows that

$$H_{le}(s) = K\frac{Ts+1}{\alpha Ts+1} = \frac{K_P\left(\frac{K_D}{K_P}s+1\right)}{\frac{K_D}{NK_P}s+1} = H_{PD}(s) \quad (8.2.21)$$

as defined by Eq. (8.1.18) and depicted in Figure 8.8. In a practical PD compensator, NK_P usually is considerably greater than 1. Therefore, if

$$NK_P = \frac{K}{\alpha} = \lim_{\omega\to\infty} |H_{le}(j\omega)| \gg 1 \quad (8.2.22)$$

then a lead compensator would be equivalent to a practical PD compensator.

This observation serves to illustrate an alternative use for lead compensation, namely, *to increase the loop bandwidth in order to produce a faster response*. Such would be the case if, for example,

$$G(s) = \frac{10}{s(s+1)(s+10)}$$

as in Example 8.2.8, and a lead compensator were chosen to achieve the same loop gain product $G(s)H_{le}(s)$ as in Example 8.2.19, hence the same stability margins. The required

$$H_{le}(s) = 3.47\frac{s+1}{0.1s+1} = \frac{34.7(s+1)}{s+10}$$

which is $100 \times H_{le}(s)$ of Example 8.2.19, would be characterized by a $\frac{K}{\alpha} = 34.7 \gg 1$. As in Example 8.2.19, however, the steady-state ramp tracking accuracy would be an unacceptable 28.8%.

Combined Lag-Lead Compensation

If a lag compensator is used in conjunction with a lead compensator, then the benefits of both can be obtained simultaneously. In particular, a **lag-lead compensator** is characterized by two adjustable pole-zero pairs, together with a variable gain, as defined by the rational transfer function

$$H_{ll}(s) = K \underbrace{\left(\frac{T_1 s + 1}{\alpha_1 T_1 s + 1} \right)}_{\text{lag}} \underbrace{\left(\frac{T_2 s + 1}{\alpha_2 T_2 s + 1} \right)}_{\text{lead}} = \frac{K}{\alpha_1 \alpha_2} \frac{\left(s + \dfrac{1}{T_1} \right) \left(s + \dfrac{1}{T_2} \right)}{\left(s + \dfrac{1}{\alpha_1 T_1} \right) \left(s + \dfrac{1}{\alpha_2 T_2} \right)},$$

(8.2.23)

with $\alpha_1 > 1$, $\alpha_2 < 1$, and $T_1 > T_2$. The frequency response of such a $H_{ll}(j\omega)$, with increasing corner frequencies at

$$\omega = \frac{1}{\alpha_1 T_1}, \quad \frac{1}{T_1}, \quad \frac{1}{T_2}, \quad \text{and} \quad \frac{1}{\alpha_2 T_2} \qquad (8.2.24)$$

is depicted by the Bode diagram of Figure 8.22. It may be noted that the lag-lead compensator defined by Eq. (8.2.23) is a "restrictive" second-order compensator because both of its poles and zeros are real.

When a plant is characterized by desirable low frequency and desirable high frequency loop gain properties, *lag-lead compensation can be used to attenuate the amplitude of the loop gain at the mid-range frequencies* to improve the stability margins of the compensated system. In such

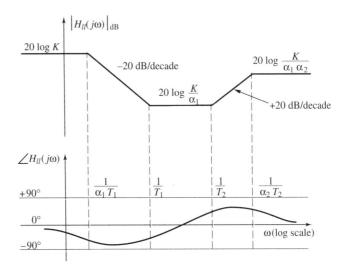

● **FIGURE 8.22**
The Bode Diagram of a Lag-Lead Compensator

cases, $K = \alpha_1\alpha_2 = 1$, so that

$$\lim_{\omega \to 0} |H_{ll}(j\omega)| = \lim_{\omega \to \infty} |H_{ll}(j\omega)| = 1, \text{ or 0 dB} \tag{8.2.25}$$

Figure 8.23 depicts the polar plot consequences of employing lag-lead compensation to reshape the mid-range frequency response of a potentially unstable closed-loop system to achieve acceptable stability margins.

Note that if $\alpha_1 = 10$, $\alpha_2 = 0.1$, and $T_1 = 10T_2$, then in light of Eq. (8.2.24), the corner frequencies will be exactly one decade apart at

$$\omega = \frac{1}{10T_1}, \quad \frac{1}{T_1}, \quad \frac{10}{T_1}, \quad \text{and} \quad \frac{100}{T_1}$$

with an asymptotic $|H_{ll}(j\omega)| = -20$ dB between the second and third corner frequencies. In such cases, the design of a lag-lead compensator simplifies to the selection of a single parameter $T_1 = 10T_2$, which corresponds to an appropriate placement of $H_{ll}(j\omega)$ on the ω-axis. Since the maximum additional phase lead associated with this lag-lead compensator occurs at

$$\omega^* = \frac{1}{\sqrt{\alpha_2}T_2} = \frac{1}{0.1\sqrt{0.1}T_1} = \frac{10\sqrt{10}}{T_1} = \frac{31.62}{T_1} \tag{8.2.26}$$

where $|H(j\omega^*)| = -10$ dB, ω^* can be chosen to correspond to the value of ω where $|G(j\omega)| = +10$ dB, so that $\omega^* = \omega_g$, the gain crossover frequency of the compensated system.

In particular, if the condition

$$|G(j\omega^*)| = +10 \text{ dB} \tag{8.2.27}$$

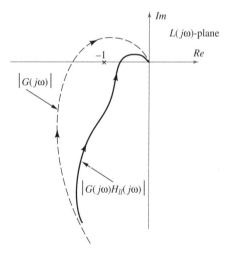

• **FIGURE 8.23**
The Polar Plot Consequences of Lag-Lead Compensation

defines ω^*, then the choice of

$$T_1 = \frac{31.62}{\omega^*} \quad \Longrightarrow \quad \frac{1}{T_1} = \frac{\omega^*}{31.62} \qquad (8.2.28)$$

completely defines a lag-lead compensator, namely,

$$H_{ll}(s) = \frac{(T_1 s + 1)(0.1 T_1 s + 1)}{(10 T_1 s + 1)(0.01 T_1 s + 1)} = \frac{\left(s + \dfrac{1}{T_1}\right)\left(s + \dfrac{10}{T_1}\right)}{\left(s + \dfrac{1}{10 T_1}\right)\left(s + \dfrac{100}{T_1}\right)}$$

$$= \frac{(s + 0.0316\omega^*)(s + 0.316\omega^*)}{(s + 0.00316\omega^*)(s + 3.16\omega^*)}$$

$$= \frac{s^2 + 0.348\omega^* s + 0.01\omega^{*2}}{s^2 + 3.163\omega^* s + 0.01\omega^{*2}} = \frac{h(s)}{k(s)} \qquad (8.2.29)$$

which yields a maximum amount of additional phase lead of $53.3°$ at the compensated gain crossover frequency $\omega_g = \omega^* = \dfrac{31.62}{T_1}$.

. . . .

EXAMPLE 8.2.30 To illustrate lag-lead compensation, consider the DC servomotor defined by Eq. (8.2.2), namely the transfer function

$$G(s) = \frac{K_a}{s(s+1)(0.1s+1)} = \frac{10 K_a}{s^3 + 11 s^2 + 10 s}$$

Assume that if $K_a = 100$ (as in Examples 8.2.4 and 8.2.19), the uncompensated loop gain $L(j\omega) = G(j\omega)$ will be characterized by both desirable low frequency and high frequency magnitudes. Recall, however, that if the $K_a = 100$ servomotor is closed under unity feedback, with $H(s) = 1$, the resulting system will be unstable.

This represents an ideal situation for employing a lag-lead compensator, analogous to the $H_{ll}(s)$ defined by Eq. (8.2.29), to reshape the mid-range loop gain $L(j\omega) = G(j\omega)H_{ll}(j\omega)$. Since $|G(j\omega)| = +10$ dB at $\omega = 5.4$, Eq. (8.2.27) implies that $\omega^* = 5.4$. Therefore, we will define

$$H_{ll}(s) = \frac{(s + 0.17)(s + 1.7)}{(s + 0.017)(s + 17)} = \frac{s^2 + 1.87 s + 0.289}{s^2 + 17.017 s + 0.289} = \frac{h(s)}{k(s)}$$

using Eq. (8.2.29). Bode diagrams of $G(j\omega)$, $H_{ll}(j\omega)$, and the resulting $L(j\omega) = G(j\omega)H_{ll}(j\omega)$ are depicted in Figure 8.24 by the dashed, dotted, and solid lines, respectively.

A MATLAB computation was used to determine that the actual GM $= 3.9$ (or 11.8 dB) at a phase crossover frequency $\omega_\phi = 12.2$, with a corresponding $\Phi M = 36.4°$ at a gain crossover frequency $\omega_g = 5.24$. Moreover, a 1 DOF unity feedback implementation of this $H_{ll}(s)$ implies a unit step response characterized by a robust $e_{ss}(t) = 0$, an overshoot of nearly 40%, and a 2% settling time t_s of 1.92 seconds, with $y(t = 5) = 1.007$, that is, a 0.7% "tail error" at $t = 5$ seconds, as depicted in Figure 8.25. The corresponding unit ramp response is characterized by an $e_{ss}(t) = (K_a)^{-1} = 0.01$, or 1%.

. . . .

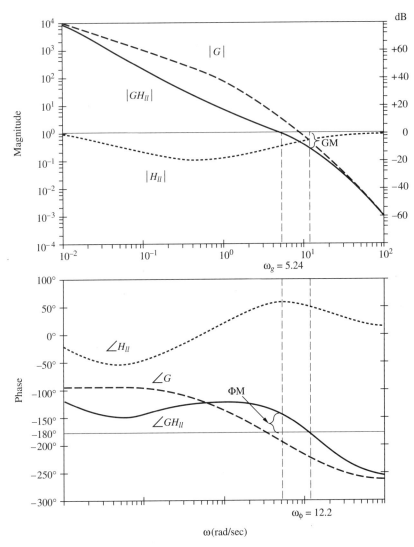

• **FIGURE 8.24**
A Bode Diagram of Lag-Lead Compensation

We note therefore that the individual benefits of both lag and lead compensation can be obtained using a combined lag-lead compensator, although further improvement is possible. We can ensure a robust $e_{ss}(t) = 0$ for ramp inputs by "converting" a $H_{ll}(s)$ to a PID compensator. If we then employ a 2 DOF implementation to cancel the pole that defines the slowest mode, the $y(t)$ tail can be reduced as well. These procedures will be illustrated in Example 8.2.37.

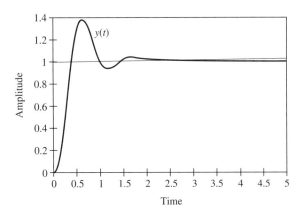

● **FIGURE 8.25**
The Unit Step Response of a Lag-Lead Controller

Lag-Lead and PID Compensation

As we noted at the beginning of this section, lag-lead compensation is analogous to PID compensation. Consider the lag-lead compensator defined by Eq. (8.2.23),

$$H_{ll}(s) = K \underbrace{\left(\frac{T_1 s + 1}{\alpha_1 T_1 s + 1} \right)}_{\text{lag}} \underbrace{\left(\frac{T_2 s + 1}{\alpha_2 T_2 s + 1} \right)}_{\text{lead}} = \frac{K}{\alpha_1 \alpha_2} \frac{\left(s + \frac{1}{T_1} \right) \left(s + \frac{1}{T_2} \right)}{\left(s + \frac{1}{\alpha_1 T_1} \right) \left(s + \frac{1}{\alpha_2 T_2} \right)}$$

If we "fix" the ratio

$$\frac{K}{\alpha_1} = F \qquad (8.2.31)$$

and subsequently define

$$K_P = \frac{F(T_1 + T_2)}{T_1}, \quad K_I = \frac{F}{T_1}, \quad K_D = F T_2, \text{ and } N = \frac{F}{\alpha_2} \qquad (8.2.32)$$

it follows that

$$\frac{1}{T_1 T_2} = \frac{K_I}{F} \frac{F}{K_D} = \frac{K_I}{K_D} \qquad (8.2.33)$$

$$\frac{1}{T_1} + \frac{1}{T_2} = \frac{T_1 + T_2}{T_1 T_2} = \frac{K_P}{F T_2} = \frac{K_P F}{F K_D} = \frac{K_P}{K_D} \qquad (8.2.34)$$

and

$$\frac{1}{\alpha_2 T_2} = \frac{N}{F} \frac{F}{K_D} = \frac{N}{K_D} \qquad (8.2.35)$$

In view of these relationships, Eq. (8.2.23) implies that in the limiting case as the corner frequency $\frac{1}{\alpha_1 T_1} \to 0$ and $K \to \infty$ in Figure 8.22,

$$\lim_{\alpha_1 \to \infty} H_{ll}(s) = \frac{F}{\alpha_2} \frac{\left(s + \frac{1}{T_1}\right)\left(s + \frac{1}{T_2}\right)}{\left(s + \frac{1}{\alpha_1 T_1}\right)\left(s + \frac{1}{\alpha_2 T_2}\right)} = \frac{N\left(s^2 + \frac{T_1 + T_2}{T_1 T_2}s + \frac{1}{T_1 T_2}\right)}{s\left(s + \frac{1}{\alpha_2 T_2}\right)}$$

$$= \frac{Ns^2 + \frac{NK_P}{K_D}s + \frac{NK_I}{K_D}}{s\left(s + \frac{N}{K_D}\right)} = \frac{h(s)}{k(s)} = H_{PID}(s) \qquad (8.2.36)$$

a practical PID compensator, as defined by Eqs. (8.1.27) and (8.1.28).

EXAMPLE 8.2.37 To illustrate the preceding discussion, recall the lag-lead compensator of Example 8.2.30,

$$H_{ll}(s) = \underbrace{\frac{K}{\alpha_1 \alpha_2}}_{1} \frac{\left(s + \frac{1}{T_1}\right)\left(s + \frac{1}{T_2}\right)}{\left(s + \frac{1}{\alpha_1 T_1}\right)\left(s + \frac{1}{\alpha_2 T_2}\right)} = \frac{(s + 0.17)(s + 1.7)}{(s + 0.017)(s + 17)}$$

Since $\frac{F}{\alpha_2} = N = 1$ in this case, Eq. (8.2.36) implies that

$$\lim_{\alpha_1 \to \infty} H_{ll}(s) = \frac{\left(s + \frac{1}{T_1}\right)\left(s + \frac{1}{T_2}\right)}{s\left(s + \frac{1}{\alpha_2 T_2}\right)} = \frac{(s + 0.17)(s + 1.7)}{s(s + 17)}$$

$$= \frac{s^2 + 1.87s + 0.289}{s(s + 17)} = \frac{Ns^2 + \frac{NK_P}{K_D}s + \frac{NK_I}{K_D}}{s\left(s + \frac{N}{K_D}\right)} = \frac{h(s)}{k(s)} = H_{PID}(s)$$

The utilization of this type 1 $H_{PID}(s) = H(s)$, in series with the $K_a = 100$ DC servomotor defined by

$$G(s) = \frac{1000}{s^3 + 11s^2 + 10s}$$

in a 1 DOF unity feedback configuration, would imply essentially the same stability margins and step response that were obtained using the lag-lead compensator $H_{ll}(s)$ of Example 8.2.30. In this case, however, *the ramp response is characterized by a robust* $e_{ss}(t) = 0$, because $G(s)H(s)$ is a type 2 transfer function.

Furthermore, if a 2 DOF configuration is employed, with

$$q(s) = q_2 s^2 + q_1 s + q_0 = q_2 s^2 + 1.87s + 0.289$$

so that

$$q_0 = h_0 = \frac{HK_I}{K_D} = 0.289 \quad \text{and} \quad q_1 = h_1 = \frac{NK_P}{K_D} = 1.87$$

a robust $e_{ss}(t) = 0$ would be obtained for *both* step and ramp inputs in light of Eqs. (7.3.18) and (7.3.19). The coefficient q_2 then can be selected so that $q(s)$ cancels the closed-loop pole closest to the origin, thus minimizing the $y(t)$ tail.

Since the roots of

$$\delta(s) = a(s)k(s) + c(s)h(s) = (s^3 + 11s^2 + 10s)(s^2 + 17s) + 1000(s^2 + 1.87s + 0.289)$$

$$= (s + 0.173)(s + 1.944)(s + 2.396 \pm j5.923)(s + 21.092)$$

define the closed-loop poles in this case, a choice of q_2 such that

$$q(s = -0.173) = q_2(-0.173)^2 + (1.87)(-0.173) + 0.289 = 0$$

namely,

$$q_2 = \frac{(1.87)(0.173) - 0.289}{(-0.173)^2} = 1.153$$

will imply a resulting

$$q(s) = 1.153s^2 + 1.87s + 0.289 = 1.153(s + 0.173)(s + 1.449)$$

which cancels the tail-producing pole at $s = -0.173$, thus improving the settling time of the closed-loop system.

A 2 DOF implementation of this limiting case lag-lead, or practical PID compensator, was depicted in Figure 8.13. The unit step response, as depicted in Figure 8.26, is characterized by a 52% overshoot, a 2% settling time t_s of 1.87 seconds, and a $y(t = 5) = 0.9994$, that is, a tail error of only 0.06% at $t = 5$ seconds. Values of q_2 below 1.153 will decrease the overshoot but increase the settling time. This particular compensator appears to do a good job of ensuring the desired performance goals by means of "classical" loopshaping techniques. However, as we will show in the next chapter, a "modern" LQR optimal design can be used to further improve both the loop and the response performance of the DC servomotor.

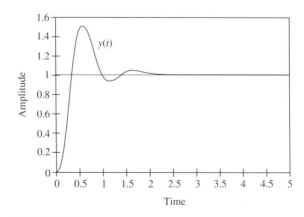

• **FIGURE 8.26**
The Unit Step Response of a 2 DOF PID Controller

8.3   SUMMARY

We have now discussed the main classical techniques that are used to control dynamic systems, namely PID and lag-lead compensators. PI controllers were shown to increase the loop gain at the lower frequencies and to decrease the loop gain at the higher frequencies, without compromising the stability margins, relative to plant parameter variations. Therefore, PI compensation improves many of the loop performance goals of a closed-loop system, such as the steady-state tracking accuracy, although it generally produces a more sluggish response because of the reduction in the loop gain bandwidth.

On the other hand, PD controllers were shown to increase the phase lead of the loop gain at the medium and high frequencies, thus improving the stability margins of the closed-loop system. They also can increase the loop gain bandwidth of the system, thus producing a faster response. However, such an increase in the bandwidth can produce an undesirable amplification of any high frequency noise present within the loop. Ideal PD compensation can also cause stability problems due to unmodeled high frequency dynamics. Therefore, the derivative terms of practical PD compensators often are implemented with a roll-off pole that maintains a fixed magnitude at the higher frequencies. Moreover, the derivative term can be implemented by a minor loop around $G(s)$ to prevent plant input saturation.

PID controllers were shown to combine the benefits of PI and PD designs. They also imply added design flexibility by providing three arbitrarily adjustable terms. This form of "three term" compensation can be used in situations in which a nominal plant transfer function $G(s)$ is unknown. The derivative term of a practical PID compensator also can be implemented by a minor loop around the plant, to prevent plant input saturation, with a PI compensator in the forward path.

Closed-loop designs that employ PID controllers are usually restrictive 1 DOF unity feedback in configuration, although they need not be so. In particular, a practical 2 DOF PID compensator was presented that provides four adjustable terms. Such a design retains the loop performance of a 1 DOF PID controller, but allows adjustments to the response performance of the closed-loop system.

The other main classical control techniques were then presented, namely, lag, lead, or a combination of the two, which is termed lag-lead compensation. Lag compensation was shown to be analogous to PI compensation, with a low frequency roll-off pole; lead compensation was shown to be analogous to practical PD compensation; and lag-lead compensation was shown to be analogous to practical PID compensation, with a low frequency roll-off pole. When a plant is characterized by desirable low frequency and desirable high frequency loop gain properties, loop-shaping lag-lead compensators can be used to attenuate the amplitude of the loop gain at the mid-range frequencies to improve the stability margins of the compensated system, without affecting its low and high frequency loop gain characteristics.

PROBLEMS

8-1. Consider the proportional liquid level control system of Example 8.1.12, where the loop gain

$$L(s) = K_P G(s) = K_P \frac{5}{s^2 + 6s + 5} = \frac{5K_P}{(s+1)(s+5)}$$

(a) Plot a root locus for this system for $0 \leq K_P \leq \infty$. For what positive and negative values of the gain K_P will the closed-loop system be stable?

(b) Sketch a polar plot of $L(s)$, and determine the effect on both the GM and the ΦM of an increasing gain K_P.

(c) Show that when $K_P = 10$, the gain crossover frequency $\omega_g = 6.2$, and $\angle L(j\omega_g = j6.2) = -132°$, so that the ΦM $= 48°$.

8-2. Assume that

$$L(s) = K_P \frac{5}{s^2 + 4s - 5} = \frac{5K_P}{(s-1)(s+5)}$$

in the previous problem.

(a) For what values of K_P will the closed-loop system be stable?

(b) Sketch a polar plot of $L(s)$, and discuss the closed-loop stability properties of the system as a function of the gain K_P.

8-3. Consider the PID liquid level control system of Example 8.1.21 where

$$H_{PID}(s) = \frac{\alpha}{5} \frac{(s+1)(s+5)}{s}$$

cancels the two nominal plant $G(s)$ poles at $s = -1$ and -5. Show that such a controller actually "freezes" these two poles at corresponding closed-loop locations by rendering them uncontrollable by the reference input.

8-4. Consider the control of the liquid level system of Example 8.1.12 via a proportional gain $K_P = 10$, which was shown to imply a nominal $e_{ss}(t) = 9.1\%$ for step changes in the reference input. If $c_0 = 5$ does not change, determine how large a_0 can become before $e_{ss}(t)$ exceeds 10%.

Determine a K_P that will ensure a nominal $e_{ss}(t) \leq 5\%$ despite such a change in a_0.

8-5. An X-Y plotter is a widely used electromechanical device whose pen must follow independent, time-varying, X-axis and Y-axis input signals quickly and accurately. Assume that the DC servomotor and pen carriage associated with both axes can be modeled by the nominal, open-loop transfer function

$$G(s) = \frac{y(s)}{u(s)} = \frac{100}{s(s+5)(s+100)}$$

and that an optical encoder (sensor) provides low-noise output position and velocity feedback signals to a microprocessor-based PD control system, as depicted for the Y-axis in Figure 8.27.

(a) Determine the output response transfer function $T(s) = \frac{y(s)}{r(s)}$ as a function of the control system gains K_P and K_D.

(b) Determine the input response transfer function $\frac{u(s)}{r(s)}$ as a function of K_P and K_D.

(c) If $|u(t)|$ must not exceed 2600 to prevent plant input saturation and $|r(t)| \leq 10$, determine appropriate values for K_P and K_D that will ensure the fastest possible step response characterized by a critically damped pair of dominant poles, hence no overshoot. It may be noted that in this case, $\max_t |u(t)| = u(0)$ for step changes in $r(t)$.

(d) Determine the resulting steady-state error for ramp changes in $r(t)$.

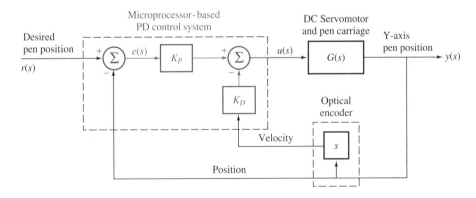

• **FIGURE 8.27**
A PD Controller for an X-Y Plotter

8-6. Consider the liquid level system of Example 8.1.12, as defined by the nominal transfer function

$$G(s) = \frac{5}{s^2 + 6s + 5} = \frac{c(s) = c}{a(s)}$$

Show that a 2 DOF

$$k(s) = s(s + k_1) = p_r(s)(s + k_1) \quad \text{and} \quad h(s) = h_2 s^2 + h_1 s + h_0$$

pair can be chosen to obtain any stable, monic

$$\delta(s) = a(s)k(s) + c(s)h(s) = \hat{\delta}(s)\hat{q}(s)$$

of degree 4, with $\deg[\hat{\delta}(s)] = \deg[\hat{q}(s)] = 2$. Also show that if the corresponding $q(s) = \alpha\hat{q}(s)$, with $\alpha = \hat{\delta}(0)/c$, then the resulting output response of the system will be characterized by a robust $e_{ss}(t) = 0$ for step changes in the reference input, as defined by the transfer function

$$T(s) = \frac{y(s)}{r(s)} = \frac{\alpha c \hat{q}(s)}{\hat{\delta}(s)\hat{q}(s)} = \frac{\alpha c = \omega_n^2}{\hat{\delta}(s) = s^2 + 2\zeta\omega_n s + \omega_n^2}$$

Note that such a design is equivalent to the practical 2 DOF PID compensator depicted in Figure 8.13.

8-7. Show that the lack of an exact cancellation of the $T(s)$ pole-zero pair defined by $(0.98s + 1)$ and $(1.2s + 1)$ in Example 8.1.26 results in the presence of the modal term $+0.51e^{-1.02t}$ in the unit step output response of the system.

8-8.* Recall the practical 2 DOF PID compensator of Example 8.1.35, where

$$\frac{y(s)}{r(s)} = \frac{5q(s)}{(s + 1)(s + 2.19)(s + 5)(s + 22.81)}$$

and

$$\frac{u(s)}{r(s)} = \frac{q(s)}{(s + 2.19)(s + 22.81)}$$

Assume a $q(s) = 50/2.19(s + 1)(s + 2.19)$, which cancels the two slowest closed-loop poles.

(a) Compare the unit step $y(t)$ response of the resulting closed-loop system with those depicted in Figure 8.14.

(b) Compare the unit step $u(t)$ response of the resulting closed-loop system with those depicted in Figure 8.15.

(c) Verify that $\max_t |u(t)| = u(0) = 50/2.19 = 22.83$, as noted earlier.

8-9. Consider the DC servomotor of Example 8.2.4, as defined by the nominal transfer function

$$G(s) = \frac{10K_a}{s(s+1)(s+10)}$$

Show that a system defined by such a loop gain $L(s) = G(s)$, with $K_a = 100$, would be unstable if closed under unity feedback, because at the phase crossover frequency $\omega_\phi = \sqrt{10} = 3.16$, $|G(j3.16)| = 1000/110 = 9.09 > 1$.

8-10.* Consider a motor driven antenna that must automatically track a slowly moving light source, as depicted in Figure 8.28, where voltage signals v_1 and v_2 from photodiodes positioned at either end of the antenna vary in intensity depending on the pointing angle θ of the antenna, as depicted in Figure 8.29(a). The angle θ_0, which depends on the position of the light source, defines the value of θ at which the two voltage intensities are equal, that is, $v_1 = v_2$, and the antenna is pointing directly at the source, so that $\rho = \theta_0 - \theta = 0°$, as desired.

For small angular deviations about any given θ_0, a linearized model of Figure 8.29(a), as depicted in Figure 8.29(b), implies the relationships:

$$v_1 = -g\theta + v_{10} \qquad \text{and} \qquad v_2 = g\theta + v_{20}$$

Therefore, when $v_1 = v_2$, it follows that $2g\theta_0 = v_{10} - v_{20}$, so that

$$v_1 - v_2 = 2g(\theta_0 - \theta) = 2g\rho$$

This latter relationship forms the basis of the unity feedback, error-driven, antenna positioning control scheme for maintaining a $\rho = 0°$, as depicted in the block diagram of Figure 8.30.

Since $v_2 = g\theta + v_{20}$, and the light source is slowly moving, v_{20} can be treated as a constant disturbance signal at the "output" v_2, which

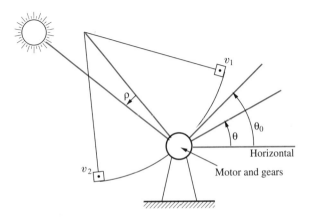

● **FIGURE 8.28**
An Antenna Tracking a Light Source

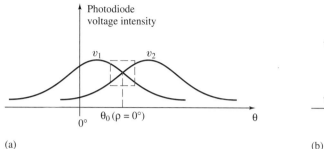

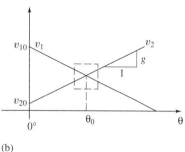

(a)

(b)

• FIGURE 8.29
Photodiode Voltage Intensity versus Antenna Position

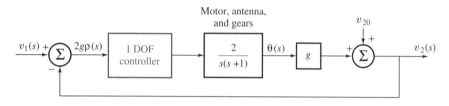

• FIGURE 8.30
A 1 DOF Controller for Light Source Antenna Tracking

is compared with the reference input v_1 to produce the angular pointing error $2g\rho$. Moreover, since the nominal plant is type 1, the constant "disturbance" v_{20} will be completely rejected at the output, and a robust

$$e_{ss}(t) = 2g\rho_{ss}(t) = 0 \implies \theta_{ss}(t) = \theta_0$$

will be ensured provided the closed-loop system remains stable. Assume that $g = 0.3$ for the remainder of this problem.

(a) Determine the values of K, T, and α that define a 1 DOF lead compensator $H_{le}(s)$ for this system, as defined by Eq. (8.2.9), which will imply a nominal, second-order, closed-loop system whose step response is characterized by an undamped natural frequency $\omega_n = 3$ and a damping ratio $\zeta = 0.7$.

(b) Derive three different, first-order, 2 DOF designs that yield the same nominal closed-loop transfer function as in Part (a).

(c) Now assume that a tachometer mounted on the drive shaft of the antenna produces a voltage proportional to the angular velocity $\dot{\theta}(t)$ of the antenna. Determine appropriate numerical values for minor loop, derivative feedback $K_D s$ around the plant, and a corresponding forward path proportional gain K_P that together imply the same nominal closed-loop transfer function as obtained in Parts (a) and (b).

(d) Discuss some of the differences among the various designs.

8-11.* Consider the lag compensated DC servomotor of Example 8.2.4, where

$$G(s) = \frac{1000}{s^3 + 11s^2 + 10s} \quad \text{and} \quad H_{la}(s) = \frac{0.01(s + 0.1)}{s + 0.001} = \frac{h(s)}{k(s)}$$

so that

$$\frac{y(s)}{r(s)} = \frac{G(s)H_{la}(s)}{1 + G(s)H_{la}(s)} = \frac{10(s + 0.1)}{(s + 0.111)(s + 0.391 \pm j.861)(s + 10.108)}$$

Recall that the unit step output response is characterized by an $e^{-0.111t}$ modal tail due to the uncancelled closed-loop pole at $s = -0.111$. Now consider the alternative, 2 DOF lag implementation depicted in Figure 8.31, where $q(s) = 0.009s + 0.001 = 0.009(s + 0.111)$ cancels the tail-producing pole. Determine the output response transfer function $T(s) = \frac{y(s)}{r(s)}$ in this 2 DOF case, and compare the unit step response to that depicted in Figure 8.18. Verify a maximum overshoot $\approx 23\%$ and a 2% settling time of 9 seconds in the 2 DOF case, as compared with respective values of $\approx 35\%$ and 14 seconds in the 1 DOF case.

8-12. Verify that the transfer function of the RC network depicted in Figure 8.32 is defined by

$$\frac{u(s)}{e(s)} = \frac{(R_1C_1s + 1)(R_2C_2s + 1)}{R_1R_2C_1C_2s^2 + (R_1C_1 + R_1C_2 + R_2C_2)s + 1}$$

Now show that the RC network can be classified as a lag-lead network

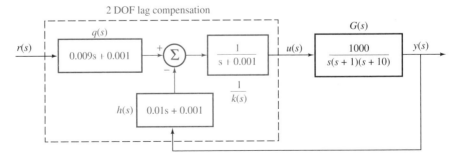

• **FIGURE 8.31**
2 DOF Tail-Cancelling Lag Compensation

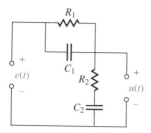

• **FIGURE 8.32**
An RC Lag-Lead Network

in view of the general lag-lead transfer function defined by Eq. (8.2.23)

$$H_{ll}(s) = K \underbrace{\left(\frac{T_1 s + 1}{\alpha_1 T_1 s + 1} \right)}_{\text{lag}} \underbrace{\left(\frac{T_2 s + 1}{\alpha_2 T_2 s + 1} \right)}_{\text{lead}}$$

In particular, express R_1, R_2, C_1, and C_2 as functions of the $H_{ll}(s)$ parameters when $K = \alpha_1 \alpha_2 = 1$.

8-13.* An astronaut inside a space station visually controls the position of an external robot arm, as depicted in Figure 8.33a. A block diagram of

• **FIGURE 8.33a**
Courtesy of NASA

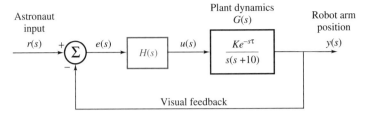

● FIGURE 8.33b
Remote Visual Control of a Robot Arm

the resulting 1 DOF unity feedback control system, which is characterized by a human reflex time delay of τ seconds, is depicted in Figure 8.33b.

(a) If $\tau = 0.5$ second, determine the values of the gain $K > 0$ for which the closed-loop system will be unstable. Establish, in particular, that if $H(s) = 1$, a $K = 30$ will imply an unstable closed-loop system.

(b) If a $K = 30$ is desired, however, to ensure adequate low frequency loop gain magnitudes, determine the value ω^* at which

$$|G(j\omega^*)| = \left| \frac{30e^{-j\omega^*\tau}}{j\omega^*(j\omega^* + 10)} \right| = +10 \text{ dB}$$

as in 8.2.27, to define a stabilizing lag-lead compensator

$$H_{ll}(s) = \frac{s^2 + 0.348\omega^*s + 0.01\omega^{*2}}{s^2 + 3.163\omega^*s + 0.01\omega^{*2}} = \frac{h(s)}{k(s)}$$

as given by Eq. (8.2.29).

(c) Determine the GM and the ΦM associated with the employment of this $H_{ll}(s) = H(s)$, as in Figure 8.33b.

8-14. The linearized, nominal pitch $p(t)$ dynamics of the private jet aircraft depicted in Figure 8.34 can be defined by the differential equation

$$\ddot{p}(t) + k_v\dot{p}(t) + 21p(t) = 1.5\delta\dot{a}(t) + 6\delta a(t)$$

where the input control signal $\delta a(t)$ represents a differential aileron displacement.

(a) Sketch a root locus of the open-loop poles of this system as k_v varies from 1.5 to 4.8.

(b) Determine a simple proportional controller K_P (as a function of k_v) that will imply a fixed, k_v dependent, critically damped pair of closed-loop poles using 1 DOF unity feedback compensation.

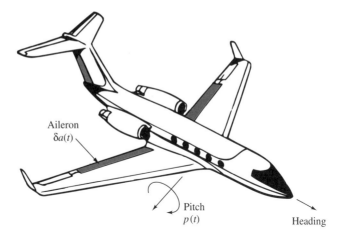

Aileron
$\delta a(t)$

Pitch
$p(t)$

Heading

● **FIGURE 8.34**
Pitch Control of a Private Jet Aircraft

(c) Show that either a lag compensator or a lead compensator can be used to obtain any desired, critically damped pair of closed-loop poles using 1 DOF unity feedback compensation.

8-15. * For both open-loop systems defined below, with input $u(t)$ and output $y(t)$, determine appropriate 1 DOF unity feedback lag-lead compensators $H_{ll}(s)$. In each case, determine the resulting GM and ΦM, the steady-state error for both a unit step and a unit ramp input, and the damping ratio associated with any dominant pair of closed-loop poles.

(a) $G(s) = \dfrac{y(s)}{u(s)} = \dfrac{100}{(s+1)(s+10)}$

(b) $D^2 y(t) = u(t)$

9

MODERN CONTROL TECHNIQUES

9.1   LINEAR STATE VARIABLE FEEDBACK

The popularity of state-space representations like this

$$\dot{\mathbf{x}}(t) = A\mathbf{x}(t) + B\mathbf{u}(t); \qquad \mathbf{y}(t) = C\mathbf{x}(t) + E\mathbf{u}(t) \qquad (9.1.1)$$

for describing the behavior of linear dynamic systems is due in large part to the determination by Kalman [39] that a **linear state variable feedback (LSVF) control law**, which is defined by the relationship

$$\mathbf{u}(t) \overset{\text{def}}{=} F\mathbf{x}(t) + \mathbf{r}(t) \qquad (9.1.2)$$

minimizes the linear quadratic regulator (LQR) performance index defined by Eq. (7.1.4), as we will show later. Numerous subsequent investigations have established that LSVF can be used to achieve a variety of other design goals as well, in both the SISO and MIMO cases.

In the SISO case, F in Eq. (9.1.2) is a real, constant, $(1 \times n)$ gain vector,

$$F = [f_1, \ f_2, \ \ldots, \ f_n]$$

each element f_i of which multiplies a corresponding state variable $x_i(t)$, with $r(t)$ an external reference input. The substitution of Eq. (9.1.2) for

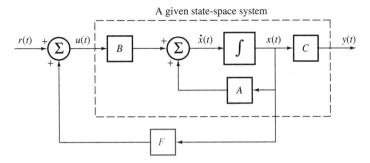

A given state-space system

● **FIGURE 9.1**
A LSVF Closed-Loop State-Space System

$u(t)$ in Eq. (9.1.1), with $E = 0$,[1] implies the *closed* (under LSVF) *loop state-space system*

$$\dot{\mathbf{x}}(t) = (A + BF)\mathbf{x}(t) + Br(t); \qquad y(t) = C\mathbf{x}(t) \qquad (9.1.3)$$

which is depicted in Figure 9.1.

In light of Eq. (3.1.8), the *closed-loop transfer function* of the system is given by

$$T(s) = \frac{y(s)}{r(s)} = C(sI - A - BF)^{-1}B \qquad (9.1.4)$$

whose n *closed-loop poles* are defined by the roots of

$$|sI - A - BF| \stackrel{\text{def}}{=} \delta^F(s) \qquad (9.1.5)$$

which also correspond to the n eigenvalues of the *closed-loop state matrix* $A + BF$.

Pole Placement for State-Space Systems

We will now show that if a state-space system is completely state controllable, then all n of its closed-loop poles can be arbitrarily positioned in the complex s-plane by an appropriate choice of the LSVF gain vector F.[2]

To prove this fact, we note that if a single-input state-space system is completely state controllable, it can be transformed to the controllable canonical form defined by Eq. (2.7.18) via the nonsingular transformation

---

[1]For convenience, we will assume this to be the case, since the vast majority of physical systems are characterized by strictly proper transfer functions.

[2]This fact holds for differential operator systems as well, as we will show. It also holds in the multi-input case [62], although MIMO cases will not be considered here.

matrix

$$Q_c \stackrel{\text{def}}{=} \begin{bmatrix} q_c \\ q_c A \\ \vdots \\ q_c A^{n-1} \end{bmatrix} \tag{9.1.6}$$

where

$$q_c \stackrel{\text{def}}{=} [0 \ 0 \ \ldots \ 0 \ 1] \mathcal{C}^{-1} \tag{9.1.7}$$

the last row of the inverse of the controllability matrix $\mathcal{C} = [B, AB, \ldots, A^{n-1}B]$, as defined by Eq. (2.6.5).

In particular, if

$$\hat{\mathbf{x}}(t) \stackrel{\text{def}}{=} Q_c \mathbf{x}(t) \tag{9.1.8}$$

with Q_c defined by Eqs. (9.1.6) and (9.1.7), it follows that

$$\underbrace{\begin{bmatrix} \dot{\hat{x}}_1(t) \\ \dot{\hat{x}}_2(t) \\ \vdots \\ \dot{\hat{x}}_n(t) \end{bmatrix}}_{\dot{\hat{\mathbf{x}}}(t)} = \underbrace{\begin{bmatrix} 0 & 1 & 0 & \cdots & 0 \\ 0 & 0 & 1 & \cdots & 0 \\ \vdots & \vdots & & \ddots & \vdots \\ -a_0 & -a_1 & & \cdots & -a_{n-1} \end{bmatrix}}_{\hat{A} = Q_c A Q_c^{-1}} \underbrace{\begin{bmatrix} \hat{x}_1(t) \\ \hat{x}_2(t) \\ \vdots \\ \hat{x}_n(t) \end{bmatrix}}_{\hat{\mathbf{x}}(t)} + \underbrace{\begin{bmatrix} 0 \\ \vdots \\ 0 \\ 1 \end{bmatrix}}_{\hat{B} = Q_c B} u(t);$$

$$y(t) = [\hat{C} = C Q_c^{-1}] \hat{\mathbf{x}}(t) \tag{9.1.9}$$

with

$$|sI - \hat{A}| = |sI - Q_c A Q_c^{-1}| = |Q_c||sI - A||Q_c^{-1}| = |sI - A|$$

$$= a(s) = s^n + a_{n-1}s^{n-1} + \cdots + a_1 s + a_0$$

$$= s^n + \underbrace{[a_0 \ a_1 \ \ldots \ a_{n-1}]}_{\stackrel{\text{def}}{=} \hat{a}} \begin{bmatrix} 1 \\ s \\ \vdots \\ s^{n-1} \end{bmatrix} \tag{9.1.10}$$

(See Problem 9-2.)

Therefore, if

$$\delta^F(s) = s^n + \delta^F_{n-1}s^{n-1} + \cdots + \delta^F_1 s + \delta^F_0$$

$$= s^n + \underbrace{[\delta^F_0 \ \delta^F_1 \ \cdots \ \delta^F_{n-1}]}_{\stackrel{\text{def}}{=} \hat{\delta}^F} \begin{bmatrix} 1 \\ s \\ \vdots \\ s^{n-1} \end{bmatrix} \tag{9.1.11}$$

is a polynomial with n desired roots, then the unique LSVF control law

$$u(t) = \hat{F}\hat{\mathbf{x}}(t) + r(t) = Fx(t) + r(t) \tag{9.1.12}$$

with

$$\hat{F} = \hat{a} - \hat{\delta}^F = F Q_c^{-1} \tag{9.1.13}$$

implies that

$$\hat{A} + \hat{B}\hat{F} = \begin{bmatrix} 0 & 1 & 0 & \cdots & 0 \\ 0 & 0 & 1 & \cdots & 0 \\ \vdots & \vdots & & \ddots & \vdots \\ -\delta_0^F & -\delta_1^F & \cdots & & -\delta_{n-1}^F \end{bmatrix}$$

so that

$$|sI - \hat{A} - \hat{B}\hat{F}| = |Q_c||sI - A - BF||Q_c^{-1}| = |sI - A - BF| = \delta^F(s) \tag{9.1.14}$$

As a consequence, the LSVF closed-loop state-space system defined by

$$\dot{\mathbf{x}}(t) = (A + BF)\mathbf{x}(t) + Br(t) \tag{9.1.15}$$

with

$$F = \underbrace{[a_0 - \delta_0^F, \ a_1 - \delta_1^F, \ \ldots, \ a_{n-1} - \delta_{n-1}^F]}_{\hat{a} - \hat{\delta}^F} \underbrace{\begin{bmatrix} q_c \\ q_c A \\ \vdots \\ q_c A^{n-1} \end{bmatrix}}_{Q_c} \tag{9.1.16}$$

is characterized by the desired set of closed-loop poles, as given by the roots of $\delta^F(s)$.

It may be noted that Eq. (9.1.16) also can be written as

$$F = \underbrace{a_0 q_c + a_1 q_c A + \cdots + a_{n-1} q_c A^{n-1}}_{q_c[a_0 I + a_1 A + \ldots + a_{n-1} A^{n-1}]} - \hat{\delta}^F Q_c = -q_c A^n - \hat{\delta}^F Q_c \tag{9.1.17}$$

independent of the coefficients of $a(s)$, since the Cayley-Hamilton Theorem (see Appendix A) implies that

$$a_0 I + a_1 A + \ldots + a_{n-1} A^{n-1} = -A^n \tag{9.1.18}$$

. . . .

EXAMPLE 9.1.19 To illustrate the preceding discussion, consider a dynamic system defined by the state-space pair

$$A = \begin{bmatrix} 0 & 0 & -5 \\ 1 & 0 & -1 \\ 0 & 1 & -4 \end{bmatrix}; \quad B = \begin{bmatrix} 1 \\ 2 \\ 0 \end{bmatrix}, \quad \text{whose} \quad \mathcal{C} = [B, \ AB, \ A^2 B] = \begin{bmatrix} 1 & 0 & -10 \\ 2 & 1 & -2 \\ 0 & 2 & -7 \end{bmatrix}$$

is nonsingular; that is, $|\mathcal{C}| = -7 - 40 + 4 = -43$, with the last row of $\mathcal{C}^{-1}$, namely,

$$q_c = \frac{[4, \ -2, \ 1]}{-43} = [-0.093, \ 0.0465, \ -0.0233]$$

as defined by Eq. (9.1.7). Since A is a right-column companion matrix in this example,

$$|sI - A| = a(s) = s^3 + 4s^2 + s + 5 = s^3 + \underbrace{[5, \quad 1, \quad 4]}_{\hat{a}} \begin{bmatrix} 1 \\ s \\ s^2 \end{bmatrix}$$

Therefore, the open-loop poles of the system, as given by the roots of $a(s)$, lie at -4.06 and $-0.029 \pm j1.11$ in the complex s-plane, so that the open-loop system is nearly unstable.

The LSVF gain vector F that positions the closed-loop poles of the system at the more desirable $s = -1$ and $-2 \pm j$ locations, so that

$$\delta^F(s) = |sI - A - BF| = (s+1)(s+2\pm j) = s^3 + \underbrace{[5, \quad 9, \quad 5]}_{\hat{\delta}^F} \begin{bmatrix} 1 \\ s \\ s^2 \end{bmatrix}$$

is given uniquely by Eq. (9.1.16),

$$F = \underbrace{[0, \ -8, \ -1]}_{\hat{a} - \hat{\delta}^F} \underbrace{\begin{bmatrix} -0.093 & 0.0465 & -0.0233 \\ 0.0465 & -0.0233 & 0.5116 \\ -0.0233 & 0.5116 & -2.256 \end{bmatrix}}_{Q_c} = [-0.349, \ -0.326, \ -1.837]$$

Finally, it can be verified (see Problem 9-3) that Q_c transforms the given $\{A, B\}$ pair to controllable companion form, so that

$$Q_c A Q_c^{-1} = \hat{A} = \begin{bmatrix} 0 & 1 & 0 \\ 0 & 0 & 1 \\ -5 & -1 & -4 \end{bmatrix} \quad \text{and} \quad Q_c B = \hat{B} = \begin{bmatrix} 0 \\ 0 \\ 1 \end{bmatrix}$$

Therefore, the LSVF gain vector $\hat{F} = \hat{a} - \hat{\delta}^F = [0, \ -8, \ -1]$, as given by Eq. (9.1.13), implies that

$$\hat{A} + \hat{B}\hat{F} = \begin{bmatrix} 0 & 1 & 0 \\ 0 & 0 & 1 \\ -5 & -9 & -5 \end{bmatrix}$$

so that

$$|sI - \hat{A} - \hat{B}\hat{F}| = |sI - A - BF| = s^3 + 5s^2 + 9s + 5 = \delta^F(s)$$

as desired.

....

Pole Placement for Differential Operator Systems

Let us now consider a differential operator system in controllable canonical form, as defined by Eq. (2.7.17),

$$\underbrace{(D^n + a_{n-1}D^{n-1} + \cdots + a_1 D + a_0)}_{a(D)} z(t) = \underbrace{b(D)}_{1} u(t); \quad y(t) = c(D)z(t)$$

$$(9.1.20)$$

Referring to Eq. (2.7.22), this differential operator system is equivalent to the controllable canonical state-space system defined by Eq. (9.1.9), with

$$
\begin{bmatrix} 1 \\ D \\ \vdots \\ D^{n-1} \end{bmatrix} z(t) = \begin{bmatrix} \hat{x}_1(t) \\ \hat{x}_2(t) \\ \vdots \\ \hat{x}_n(t) \end{bmatrix} = \hat{\mathbf{x}}(t) \tag{9.1.21}
$$

Therefore, the state-space LSVF control law given by Eq. (9.1.12) is equivalent to the **differential operator LSVF control law**

$$
u(t) = \hat{F}\hat{\mathbf{x}}(t) + r(t) = \underbrace{[\hat{a} - \hat{\delta}^F]}_{\hat{F}} \begin{bmatrix} 1 \\ D \\ \vdots \\ D^{n-1} \end{bmatrix} z(t) + r(t) \overset{\text{def}}{=} f(D)z(t) + r(t)
$$

$$\tag{9.1.22}$$

Equations (9.1.10), (9.1.11), and (9.1.22) now imply that

$$
a(D) - f(D) = D^n + [\hat{a} - \hat{a} + \hat{\delta}^F] \begin{bmatrix} 1 \\ D \\ \vdots \\ D^{n-1} \end{bmatrix} = \delta^F(D) \tag{9.1.23}
$$

so that the substitution of Eq. (9.1.22) for $u(t)$ in Eq. (9.1.20) implies an equivalent *closed* (under LSVF) *loop differential operator system* defined by

$$
\underbrace{[a(D) - f(D)]}_{\delta^F(D)} z(t) = r(t); \qquad y(t) = c(D)z(t) \tag{9.1.24}
$$

which is depicted in Figure 9.2.

Referring to Eq. (3.1.25), the closed-loop transfer function of the system is given by

$$
T(s) = \frac{y(s)}{r(s)} = \frac{c(s)}{a(s) - f(s)} = \frac{c(s)}{\delta^F(s)} \tag{9.1.25}
$$

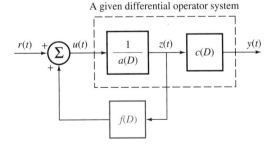

● **FIGURE 9.2**
A LSVF Closed-Loop Differential Operator System

It is of interest to note that the numerator polynomial $c(s)$ of the closed-loop system is equal to that of the open-loop system. This establishes the fact that the *zeros of a system are not affected by LSVF compensation*, although they can be "cancelled" by corresponding zeros (closed-loop poles) of $\delta^F(s)$, thereby rendering them unobservable, since $\delta^F(D)$ and $c(D)$ in Eq. (9.1.24) will not be coprime.

. . . .

EXAMPLE 9.1.26 Consider a system defined by the controllable canonical differential operator representation

$$\underbrace{(D^3 + 4D^2 + D + 5)}_{a(D)} z(t) = u(t) \qquad (9.1.27)$$

which is equivalent to both of the equivalent state-space systems defined in Example 9.1.19. The desired

$$\delta^F(D) = D^3 + 5D^2 + 9D + 5 = a(D) - f(D)$$

of that example can be obtained here by employing the differential operator LSVF control law Eq. (9.1.22), namely,

$$u(t) = \underbrace{[0, \ -8, \ -1]}_{\hat{F} = \hat{a} - \hat{\delta}^F} \begin{bmatrix} 1 \\ D \\ D^2 \end{bmatrix} z(t) + r(t) = \underbrace{(-D^2 - 8D)}_{f(D)} z(t) + r(t) \qquad (9.1.28)$$

In particular, the substitution of Eq. (9.1.28) for $u(t)$ in Eq. (9.1.27) implies that

$$(D^3 + 4D^2 + D + 5)z(t) = (-D^2 - 8D)z(t) + r(t)$$

or that

$$\underbrace{(D^3 + 5D^2 + 9D + 5)}_{\delta^F(D)} z(t) = r(t)$$

. . . .

Stabilizability

We have now shown how all n poles of a controllable (state-space or differential operator) system can be arbitrarily positioned in the complex s-plane by means of LSVF. Given such flexibility, we would obviously want to place the poles "far enough" into the stable half-plane $Re(s) < 0$ to ensure closed-loop stability despite any plant parameter variations. Other factors can influence their placement as well, as we will show later.

When a system is not completely state controllable, it is impossible to alter the uncontrollable poles (or modes) of the system, as we noted in Section 2.6. However, it is possible to arbitrarily position all of the controllable poles via LSVF. This can be shown by employing a minimal realization of the system with an observable output; that is, since the pole-placement effect of LSVF is independent of the defined output $y(t)$, we can redefine the output, if necessary, to obtain an observable (but uncontrollable) system. If a minimal realization of the resulting system is then placed in the controllable companion state-space form defined by Eq. (9.1.9), it follows that all of the poles of the minimal realization, that is, all of the controllable poles of the system, can be arbitrarily assigned using LSVF.

In light of this observation, we will say that a given dynamic system is **stabilizable** if all of its uncontrollable poles are stable, since its remaining controllable poles can be arbitrarily positioned in the stable half-plane $Re(s) < 0$. Therefore, a controllable system is always stabilizable, but a stabilizable system need not be controllable.

9.2   OBSERVERS

In the previous section, we showed how LSVF could be used to arbitrarily position all of the controllable poles of a linear system. Note that this requires the ability to directly measure or *observe* all n *internal* state variables, as illustrated in Figure 9.1. In most cases, however, it is not possible to directly measure the entire n-dimensional state. For example, in the SISO case, $y(t)$ is the only "state variable" that is available for feedback purposes. In such cases, a LSVF control law could not be implemented without employing $n - 1$ additional sensors, assuming their existence; for example, in many cases, it is physically impossible to measure certain states because appropriate sensors do not exist.

In 1964, D. Luenberger [43] introduced the notion of an **observer**, which is a dynamic system whose state variables $\bar{x}_i(t)$ approach those of the actual system $x_i(t)$ "as fast as desired." To construct an observer, Luenberger employed a pole-placement technique that is the *dual* of that introduced in the previous section for arbitrary pole placement by LSVF. He further illustrated that the subsequent employment of the observer state (rather that the actual system state) for feedback control produces a closed-loop system with many of the characteristics that would be obtained if actual LSVF were used. We will now illustrate the construction of observers and the subsequent utilization of observer feedback in the SISO case for both state-space and differential operator systems.

State-Space Observers

We will first show that if a given state-space system

$$\dot{\mathbf{x}}(t) = A\mathbf{x}(t) + Bu(t); \qquad y(t) = C\mathbf{x}(t) \tag{9.2.1}$$

is completely state observable, then all n eigenvalues of $|sI - A - HC|$ can be arbitrarily positioned in the complex s-plane by an appropriate choice of the $(n \times 1)$ gain vector H.

To prove this fact, by *duality*, we note that if a given, single-output state-space system is completely state observable, it can be transformed to the observable canonical form defined by Eqs. (2.7.2) and (2.7.10) by means of the nonsingular transformation matrix

$$Q_o \overset{\text{def}}{=} [q_o, \ Aq_o, \ \ldots, \ A^{n-1}q_o]^{-1} \tag{9.2.2}$$

where

$$q_o \overset{\text{def}}{=} \mathcal{O}^{-1} \begin{bmatrix} 0 \\ 0 \\ \vdots \\ 0 \\ 1 \end{bmatrix} \tag{9.2.3}$$

the last column of the inverse of the observability matrix

$$\mathcal{O} = \begin{bmatrix} C \\ CA \\ \vdots \\ CA^{n-1} \end{bmatrix}$$

as given by Eq. (2.6.9).

In particular, if

$$\hat{\mathbf{x}}(t) \overset{\text{def}}{=} Q_o \mathbf{x}(t) \tag{9.2.4}$$

with Q_o given by Eqs. (9.2.2) and (9.2.3), it follows that

$$\underbrace{\begin{bmatrix} \dot{\hat{x}}_1(t) \\ \dot{\hat{x}}_2(t) \\ \vdots \\ \dot{\hat{x}}_n(t) \end{bmatrix}}_{\dot{\hat{\mathbf{x}}}(t)} = \underbrace{\begin{bmatrix} 0 & 0 & 0 & \cdots & -a_0 \\ 1 & 0 & 0 & \cdots & -a_1 \\ 0 & 1 & 0 & \cdots & \\ \vdots & \vdots & \vdots & \ddots & \vdots \\ 0 & 0 & \cdots & 1 & -a_{n-1} \end{bmatrix}}_{\hat{A} = Q_o A Q_o^{-1}} \underbrace{\begin{bmatrix} \hat{x}_1(t) \\ \hat{x}_2(t) \\ \vdots \\ \hat{x}_n(t) \end{bmatrix}}_{\hat{\mathbf{x}}(t)} + \underbrace{\hat{B}}_{Q_o B} u(t)$$

$$y(t) = \underbrace{[0, \ 0 \ \ldots, \ 0, \ 1]}_{\hat{C} = C Q_o^{-1}} \hat{\mathbf{x}}(t) \tag{9.2.5}$$

with

$$|sI - \hat{A}| = |sI - Q_o A Q_o^{-1}| = |Q_o||sI - A||Q_o^{-1}| = |sI - A|$$

$$= a(s) = s^n + \underbrace{[a_0 \ a_1 \ \ldots \ a_{n-1}]}_{\hat{a}} \begin{bmatrix} 1 \\ s \\ \vdots \\ s^{n-1} \end{bmatrix} \tag{9.2.6}$$

Therefore, if

$$\delta^H(s) = s^n + \underbrace{[\delta_0^H \ \delta_1^H \ \cdots \ \delta_{n-1}^H]}_{\overset{\text{def}}{=} \ \hat{\delta}^H} \begin{bmatrix} 1 \\ s \\ \vdots \\ s^{n-1} \end{bmatrix} \qquad (9.2.7)$$

is a polynomial with n desired roots, then the unique choice of

$$\hat{H} = [\hat{a} - \hat{\delta}^H]^T = Q_o H \qquad (9.2.8)$$

implies that

$$\hat{A} + \hat{H}\hat{C} = \begin{bmatrix} 0 & 0 & 0 & \cdots & -\delta_0^H \\ 1 & 0 & 0 & \cdots & -\delta_1^H \\ 0 & 1 & 0 & \cdots & \\ \vdots & \vdots & & \ddots & \vdots \\ 0 & 0 & \cdots & 1 & -\delta_{n-1}^H \end{bmatrix} \qquad (9.2.9)$$

so that

$$|sI - \hat{A} - \hat{H}\hat{C}| = |Q_o||sI - A - H||Q_o^{-1}| = |sI - A - HC| = \delta^H(s) \qquad (9.2.10)$$

with

$$H = Q_o^{-1}\hat{H} = \underbrace{[q_o, \ Aq_o, \ \ldots, \ A^{n-1}q_o]}_{Q_o^{-1}} [\hat{a} - \hat{\delta}^H]^T \qquad (9.2.11)$$

Note that this observer pole-placement result is completely independent of the system input $u(t)$.

. . . .

EXAMPLE 9.2.12 To illustrate the preceding discussion, consider a dynamic system defined by the state-space pair

$$A = \begin{bmatrix} 0 & 1 & 0 \\ 0 & 0 & 1 \\ -5 & -1 & -4 \end{bmatrix}; \qquad C = [1, \ 2, \ 0]$$

which is *dual* to the $\{A, B\}$ pair of Example 9.1.19. Therefore, its observability matrix

$$\mathcal{O} = \begin{bmatrix} C \\ CA \\ CA^2 \end{bmatrix} = \begin{bmatrix} 1 & 2 & 0 \\ 0 & 1 & 2 \\ -10 & -2 & -7 \end{bmatrix}$$

is nonsingular; that is, $|\mathcal{O}| = -7 - 40 + 4 = -43$, with the last column of $\mathcal{O}^{-1}$, namely,

$$q_o = \begin{bmatrix} 4 \\ -2 \\ 1 \end{bmatrix} \times \frac{1}{-43} = \begin{bmatrix} -0.093 \\ 0.0465 \\ -0.0233 \end{bmatrix}$$

in light of Eq. (9.2.3). Since A is a bottom row companion matrix in this example,

$$|sI - A| = a(s) = s^3 + 4s^2 + s + 5 = s^3 + \underbrace{[5, \quad 1, \quad 4]}_{\hat{a}} \begin{bmatrix} 1 \\ s \\ s^2 \end{bmatrix}$$

as in Example 9.1.19.

The gain vector H that positions the closed-loop poles of the system at desired $s = -1$ and $-2 \pm j$ locations, so that

$$\delta^H(s) = |sI - A - HC| = (s+1)(s+2 \pm j) = s^3 + \underbrace{[5, \quad 9, \quad 5]}_{\hat{\delta}^H} \begin{bmatrix} 1 \\ s \\ s^2 \end{bmatrix}$$

is given uniquely by Eq. (9.2.11),

$$H = \underbrace{\begin{bmatrix} -0.093 & 0.0465 & -0.0233 \\ 0.0465 & -0.0233 & 0.5116 \\ -0.0233 & 0.5116 & -2.256 \end{bmatrix}}_{[q_0, \ Aq_0, \ A^2q_0]} \underbrace{\begin{bmatrix} 0 \\ -8 \\ -1 \end{bmatrix}}_{[\hat{a} - \hat{\delta}^H]^T} = \begin{bmatrix} -0.349 \\ -0.326 \\ -1.837 \end{bmatrix}$$

Finally, it can be verified that

$$Q_o = [q_o, \ Aq_o, \ A^2q_o]^{-1} = \begin{bmatrix} -9 & 4 & 1 \\ 4 & 9 & 2 \\ 1 & 2 & 0 \end{bmatrix}$$

transforms the given $\{A, C\}$ pair to observable canonical form, so that

$$Q_o A Q_o^{-1} = \hat{A} = \begin{bmatrix} 0 & 0 & -5 \\ 1 & 0 & -1 \\ 0 & 1 & -4 \end{bmatrix} \quad \text{and} \quad C Q_o^{-1} = \hat{C} = [0, \ 0, \ 1]$$

(See Problem 9-4.)

Therefore, the gain vector $\hat{H} = [\hat{a} - \hat{\delta}^H]^T = [0, \ -8, \ -1]^T$ given by Eq. (9.2.8) implies that

$$\hat{A} + \hat{H}\hat{C} = \begin{bmatrix} 0 & 0 & -5 \\ 1 & 0 & -9 \\ 0 & 1 & -5 \end{bmatrix}$$

so that

$$|sI - \hat{A} - \hat{H}\hat{C}| = |sI - A - HC| = s^3 + 5s^2 + 9s + 5 = \delta^H(s)$$

as desired.

A **state-space observer** of the entire state of an observable system defined by Eq. (9.2.1) is now defined as any nth-order, stable, state-space system

$$\dot{\bar{\mathbf{x}}}(t) = (A + HC)\bar{\mathbf{x}}(t) + Bu(t) - Hy(t) \qquad (9.2.13)$$

In particular, since

$$[\dot{\mathbf{x}}(t) - \dot{\bar{\mathbf{x}}}(t)] = A\mathbf{x}(t) + Bu(t) - (A + HC)\bar{\mathbf{x}}(t) - Bu(t) + HC\mathbf{x}(t)$$

$$= (A + HC)[\mathbf{x}(t) - \bar{\mathbf{x}}(t)], \qquad (9.2.14)$$

Eq. (2.5.2) implies that

$$[\mathbf{x}(t) - \bar{\mathbf{x}}(t)] = e^{(A+HC)(t-t_0)}[\mathbf{x}(t_0) - \bar{\mathbf{x}}(t_0)] \qquad (9.2.15)$$

Therefore, since all of the eigenvalues of $A + HC$, which correspond to the poles (or modes) of the observer, lie in the stable half-plane $Re(s) < 0$, Eq. (9.2.15) implies that the state of an observer will *exponentially approach* the state of the given system. Moreover, the further into the left-half s-plane the observer poles are positioned, the faster $\bar{\mathbf{x}}(t)$ will approach $\mathbf{x}(t)$.

Detectability

We have now shown how all n poles of an observable state-space system can be arbitrarily positioned in the complex s-plane by means of the gain vector H. When a system is not completely state observable, it is impossible to alter the unobservable poles of the system using H. However, it is possible to arbitrarily position all of the observable poles by means of H. This follows directly by *duality* with respect to the stabilizability results presented in the previous section.

Using this observation, we will say that a given dynamic system is **detectable** if all of its unobservable poles are stable, since its remaining poles can be arbitrarily positioned in the stable half-plane $Re(s) < 0$ via H. Therefore, an observable system is always detectable, but a detectable system may not be observable. It might be noted that the state space system defined in Example 2.6.11 is not detectable because it has an unstable, unobservable pole at $s = +1$.

State-Space Observer State Feedback

It is now of interest to determine the effect of employing a **linear observer state feedback control law**, namely,

$$u(t) \overset{\text{def}}{=} F\bar{\mathbf{x}}(t) + r(t) \qquad (9.2.16)$$

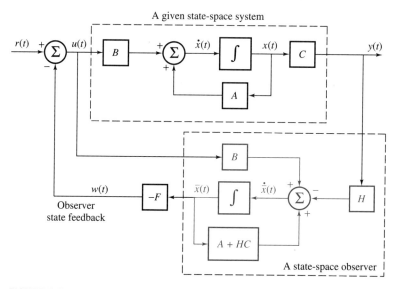

● **FIGURE 9.3**
Linear Observer State Feedback

instead of the actual LSVF control law defined by Eq. (9.1.2). The result-ing $2n$th-order, closed-loop system is depicted in Figure 9.3. Note that the defined *observer output*

$$w(t) \stackrel{\text{def}}{=} -F\bar{\mathbf{x}}(t) \tag{9.2.17}$$

is consistent with Eq. (9.2.16) in light of the negative feedback sign at the left-most summation junction, so that

$$u(t) = r(t) - \underbrace{[-F\bar{\mathbf{x}}(t)]}_{w(t)} = F\bar{\mathbf{x}}(t) + r(t)$$

If we define the state of this $2n$th-order system as

$$\underline{\mathbf{x}}(t) \stackrel{\text{def}}{=} \begin{bmatrix} \mathbf{x}(t) \\ \mathbf{x}(t) - \bar{\mathbf{x}}(t) \end{bmatrix} \tag{9.2.18}$$

noting that in light of Eq. (9.2.16),

$$\dot{\mathbf{x}}(t) = A\mathbf{x}(t) + Bu(t) = A\mathbf{x}(t) + BF\bar{\mathbf{x}}(t) + Br(t) + \underbrace{BF\mathbf{x}(t) - BF\mathbf{x}(t)}_{0},$$

Eq. (9.2.14) then implies that the dynamic behavior of the overall closed-loop system can be defined by the 2*n*th-order state-space representation

$$\underbrace{\begin{bmatrix} \dot{\mathbf{x}}(t) \\ \dot{\mathbf{x}}(t) - \dot{\bar{\mathbf{x}}}(t) \end{bmatrix}}_{\dot{\underline{\mathbf{x}}}(t)} = \underbrace{\begin{bmatrix} A + BF & -BF \\ 0 & A + HC \end{bmatrix}}_{\underline{A}} \underbrace{\begin{bmatrix} \mathbf{x}(t) \\ \mathbf{x}(t) - \bar{\mathbf{x}}(t) \end{bmatrix}}_{\underline{\mathbf{x}}(t)} + \underbrace{\begin{bmatrix} B \\ 0 \end{bmatrix}}_{\underline{B}} r(t)$$

$$y(t) = \underbrace{[C, \quad 0]}_{\underline{C}} \underline{\mathbf{x}}(t) \tag{9.2.19}$$

In view of Eq. (9.2.19), the 2*n* closed-loop poles of the system are given by the roots of

$$|sI - \underline{A}| = |sI - A - BF| \times |sI - A - HC| = \delta^F(s)\delta^H(s) \tag{9.2.20}$$

the *n* desired LSVF poles and the *n* selected observer poles. Note that Eq. (9.2.19) and Eq. (9.2.14) imply that the observer poles, as given by the roots of $|sI - A - HC| = \delta^H(s)$, are uncontrollable, as they clearly must be, since $\mathbf{x}(t) - \bar{\mathbf{x}}(t) \to 0$ as $t \to \infty$, independent of any input signals. Moreover, the closed-loop transfer function

$$T(s) = \frac{y(s)}{r(s)} = [C, \quad 0] \begin{bmatrix} sI - A - BF & BF \\ 0 & sI - A - HC \end{bmatrix}^{-1} \begin{bmatrix} B \\ 0 \end{bmatrix}$$

$$= \frac{C(sI - A - BF)^+ \delta^H(s)B}{\delta^F(s)\delta^H(s)} = C(sI - A - BF)^{-1}B \tag{9.2.21}$$

after the *n* (stable) observer pole-zero pairs, as given by the roots of $\delta^H(s)$, are cancelled. In light of Eq. (9.1.4), it should be noted that this $T(s)$ is exactly the same $T(s)$ that would be obtained using actual LSVF.

....

EXAMPLE 9.2.22 Consider the following state-space system in controllable canonical form:

$$\dot{\mathbf{x}}(t) = \underbrace{\begin{bmatrix} 0 & 1 & 0 \\ 0 & 0 & 1 \\ -5 & -1 & -4 \end{bmatrix}}_{A} \mathbf{x}(t) + \underbrace{\begin{bmatrix} 0 \\ 0 \\ 1 \end{bmatrix}}_{B} u(t); \qquad y(t) = \underbrace{[1, \quad 2, \quad 0]}_{C} \mathbf{x}(t)$$

whose state matrix A and output matrix C are the same as those in Example 9.2.12, so that

$$|sI - A| = a(s) = s^3 + 4s^2 + s + 5 = s^3 + \underbrace{[5, \quad 1, \quad 4]}_{\hat{a}} \begin{bmatrix} 1 \\ s \\ s^2 \end{bmatrix}$$

If we wish to position the LSVF closed-loop poles of the system at -3 and $-1 \pm j$ in the complex *s*-plane, Eq. (9.1.11) implies that

$$\delta^F(s) = (s+3)(s+1 \pm j) = s^3 + 5s^2 + 8s + 6 = s^3 + \underbrace{[6, \quad 8, \quad 5]}_{\hat{\delta}^F} \begin{bmatrix} 1 \\ s \\ s^2 \end{bmatrix}$$

Since the given system is in controllable canonical form, $Q_c = I = Q_c^{-1}$ in this case. Therefore, Eq. (9.1.13) implies that

$$\hat{F} = [\hat{a} - \hat{\delta}^F] = F = [5, \quad 1, \quad 4] - [6, \quad 8, \quad 5] = [-1, \ -7, \ -1]$$

If we wish to position the observer poles at $s = -1$ and $-2 \pm j$, as in Example 9.2.12, the unique observer gain vector

$$H = \begin{bmatrix} -0.349 \\ -0.326 \\ -1.837 \end{bmatrix}$$

of that example is the appropriate one to use. The determination of F and H completes our design. The resulting $(2n = 6)$th-order closed-loop system will be characterized by the prime or *minimal transfer function*

$$T(s) = C(sI - A - BF)^{-1}B = \frac{2s + 1}{s^3 + 5s^2 + 8s + 6}$$

after the uncontrollable observer poles defined by the roots of $|sI - A - HC| = \delta^H(s) = s^3 + 5s^2 + 9s + 5$ have been cancelled.

....

Differential Operator Observers

Note that a state-space observer is simply a dynamic system that is "driven" by both the given system input $u(t)$ and the given system output $y(t)$ to produce its own output $w(t) = -F\bar{\mathbf{x}}(t)$, which exponentially approaches a desired LSVF control law $-F\mathbf{x}(t) = -\hat{F}\hat{\mathbf{x}}(t) = -f(D)z(t)$, in light of Eqs. (9.1.12), (9.1.22), and (9.2.15).

To extend this concept to differential operator representations, consider the differential operator system depicted in Figure 9.4, whose open-loop (in solid lines) dynamic behavior is defined by the given system or *plant*

$$a(D)z(t) = u(t); \qquad y(t) = c(D)z(t) \tag{9.2.23}$$

and the **differential operator observer**

$$\hat{q}(D)w(t) = m(D)u(t) + h(D)y(t) \tag{9.2.24}$$

with

$$a(D) = D^n + a_{n-1}D^{n-1} + \cdots + a_1 D + a_0 \tag{9.2.25}$$

a monic polynomial of degree n, which defines the *order* of the plant, and $c(D)$ is a polynomial of degree strictly less than n. Note that the substitution of Eq. (9.2.23) into Eq. (9.2.24) implies that

$$\hat{q}(D)w(t) = [a(D)m(D) + c(D)h(D)]z(t) \tag{9.2.26}$$

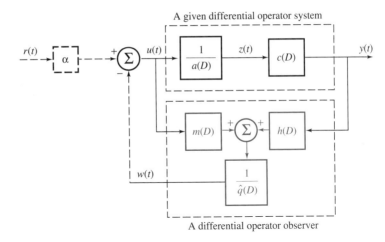

• FIGURE 9.4
A Differential Operator Observer

To explicitly determine an appropriate $m(D)$ and $h(D)$, we now recall the pole placement algorithm of Appendix B, which was first presented in Section 7.2 by means of Eqs. (7.2.6) through (7.2.8). If we assume that the plant polynomials $a(D)$ and $c(D)$ are coprime, so that the system defined by Eq. (9.2.23) is observable, then a monic $(n-1)$-degree polynomial

$$k(D) = D^{n-1} + k_{n-2}D^{n-2} + \cdots + k_1 D + k_0 \qquad (9.2.27)$$

and a corresponding $(n-1)$-degree polynomial

$$h(D) = h_{n-1}D^{n-1} + h_{n-2}D^{n-2} + \cdots + h_1 D + h_0 \qquad (9.2.28)$$

can be chosen to obtain any desired $\delta(D)$ of degree $2n-1$. We will therefore choose $k(D)$ and $h(D)$ so that

$$a(D)k(D) + c(D)h(D) = \delta(D) = \underbrace{[a(D) - f(D)]}_{\delta^F(D)} \hat{q}(D) \qquad (9.2.29)$$

in light of Eq. (9.1.23), with both

$$\delta^F(D) = a(D) - f(D) = D^n + \delta_{n-1}^F D^{n-1} + \cdots + \delta_1^F D + \delta_0^F \quad (9.2.30)$$

and

$$\hat{q}(D) = D^{n-1} + \hat{q}_{n-2}D^{n-2} + \cdots + \hat{q}_1 D + \hat{q}_0 \qquad (9.2.31)$$

monic, arbitrary, and stable.

If the $(n-2)$-degree polynomial

$$m(D) \overset{\text{def}}{=} k(D) - \hat{q}(D) \qquad (9.2.32)$$

so that $k(D) = m(D) + \hat{q}(D)$, Eq. (9.2.29) implies that

$$a(D)m(D) + c(D)h(D) = -\hat{q}(D)f(D) \qquad (9.2.33)$$

The substitution of Eq. (9.2.33) into Eq. (9.2.26) therefore implies that

$$\hat{q}(D)[w(t) + f(D)z(t)] = 0 \qquad (9.2.34)$$

or that $w(t)$ approaches $-f(D)z(t)$ as fast as the modal response defined by the roots of $\hat{q}(D)$, which define the observer poles. Note that the *order* of this differential operator observer is $n - 1$, and not n, as in the state-space case considered earlier, although a state-space observer of order $n - 1$ does exist. However, the explicit determination of such an observer requires additional computations that can be avoided by using this differential operator approach.

It may be noted that the monic polynomial $k(D)$ defined by Eq. (9.2.27) also can be of degree n, with $h(D)$ given by Eq. (9.2.28). Using Eq. (9.2.29), $\delta(D) = \delta^F(D)\hat{q}(D)$ will then be of degree $2n$. If $\delta^F(D)$ is defined by Eq. (9.2.30), $\deg[\hat{q}(D)] = n$, which implies an observer of order n, as in the state-space case. Such a full nth-order observer can imply a lower loop bandwidth and, as a consequence, may be employed when there is significant measurement noise or unmodeled high frequency dynamics.

Differential Operator Observer State Feedback

If we now "close the loop," as depicted by the dotted lines in Figure 9.4, so that $u(t) = a(D)z(t) = \alpha r(t) - w(t)$, hence that

$$\hat{q}(D)a(D)z(t) = \alpha\hat{q}(D)r(t) - \hat{q}(D)w(t),$$

Eq. (9.2.26) implies that

$$a(D)\hat{q}(D)z(t) = \alpha\hat{q}(D)r(t) - [a(D)m(D) + c(D)h(D)]z(t)$$

Therefore, the closed-loop system can be defined by

$$a(D)\underbrace{[\hat{q}(D) + m(D)]}_{k(D)}z(t) = -h(D)y(t) + \underbrace{\alpha\hat{q}(D)}_{q(D)}r(t) \; ;$$

$$y(t) = c(D)z(t) \qquad (9.2.35)$$

It is of interest to note that Eq. (9.2.35) represents the dynamic behavior of the 2 DOF differential operator observer feedback system depicted in Figure 9.5, in which the zeros of the polynomial $q(D) = \alpha\hat{q}(D)$ correspond to $(n-1)$ of the closed-loop poles. This 2 DOF differential operator system is completely analogous to its Laplace transformed "counterpart," namely, the "2 DOF Design Procedure" of Section 7.2. In particular, Eq. (7.2.8) corresponds to the Laplace transform of Eq. (9.2.29), with $\mathcal{L}[\delta^F(D)] = \delta^F(s)$ and $\mathcal{L}[\hat{q}(D)] = \hat{q}(s)$ analogous to $\hat{\delta}(s)$ and $\hat{q}(s)$ of Eq. (7.2.8), respectively. Therefore, the 2 DOF design procedure introduced in Section 7.2 is equivalent to (Laplace transformed) observer state feedback.

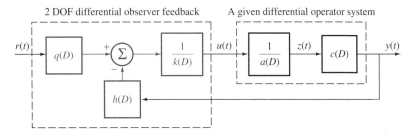

- **FIGURE 9.5**
2 DOF Differential Operator Observer Feedback

Using Eq. (9.2.35), the dynamic behavior of the closed-loop systems of Figures 9.4 and 9.5 can be represented in differential operator form as

$$\underbrace{[a(D)k(D) + c(D)h(D)]}_{\delta(D) = \delta^F(D)\hat{q}(D)} z(t) = \underbrace{\alpha\hat{q}(D)}_{q(D)} r(t); \qquad y(t) = c(D)z(t)$$

$$(9.2.36)$$

Therefore, the closed-loop poles of the system are given by the n roots of $\delta^F(D) = a(D) - f(D)$, which represent the desired LSVF poles, and the $n - 1$ roots of $\hat{q}(D)$, which represent the uncontrollable but stable poles of the differential operator observer. Moreover, the closed-loop transfer function

$$T(s) = \frac{y(s)}{r(s)} = \frac{\alpha c(s)\hat{q}(s)}{\delta^F(s)\hat{q}(s)} = \frac{\alpha c(s)}{\delta^F(s)} \qquad (9.2.37)$$

with $\alpha = 1$, is equivalent to that given by Eq. (9.1.25), that is, the same $T(s)$ that would be obtained using the actual differential operator LSVF control law $f(D)z(t)$, rather than $w(t)$, in the feedback path.

. . . .

EXAMPLE 9.2.38 Consider a differential operator system in controllable canonical form, which is equivalent to the state-space system of Example 9.2.22, namely

$$\underbrace{(D^3 + 4D^2 + D + 5)}_{a(D)} z(t) = u(t); \qquad y(t) = \underbrace{(2D + 1)}_{c(D)} z(t)$$

Note that the system also is observable, since $a(D)$ and $c(D)$ are coprime.

If we decide to position the closed-loop poles of the system at -3 and $-1 \pm j$ in the complex s-plane, as in Example 9.2.22, it follows that

$$\delta^F(D) = (D + 3)(D + 1 \pm j) = D^3 + 5D^2 + 8D + 6$$

and if we position the poles of a differential operator observer (of order $n - 1 = 2$) at $-2 \pm j$, it follows that

$$\hat{q}(D) = (D + 2 \pm j) = D^2 + 4D + 5$$

Therefore, an appropriate pair $\{k(D), h(D)\}$ of compensator polynomials can be determined using Eq. (9.2.29). In this particular case, this implies solving the equation

$$\underbrace{(D^3 + 4D^2 + D + 5)}_{a(D)} \underbrace{(D^2 + k_1 D + k_0)}_{k(D)} + \underbrace{(2D + 1)}_{c(D)} \underbrace{(h_2 D^2 + h_1 D + h_0)}_{h(D)}$$

$$= \underbrace{(D^3 + 5D^2 + 8D + 6)}_{\delta^F(D)} \underbrace{(D^2 + 4D + 5)}_{\hat{q}(D)} = D^5 + 9D^4 + 33D^3 + 63D^2 + 64D + 30$$

for the five unknown coefficients of $k(D)$ and $h(D)$.

It can be shown (see Problem 9-6) that this is equivalent to solving the following matrix/vector equation:

$$\begin{bmatrix} 5 & 0 & 0 & 1 & 0 & 0 \\ 1 & 5 & 0 & 2 & 1 & 0 \\ 4 & 1 & 5 & 0 & 2 & 1 \\ 1 & 4 & 1 & 0 & 0 & 2 \\ 0 & 1 & 4 & 0 & 0 & 0 \\ 0 & 0 & 1 & 0 & 0 & 0 \end{bmatrix} \begin{bmatrix} k_0 \\ k_1 \\ 1 \\ h_0 \\ h_1 \\ h_2 \end{bmatrix} = \begin{bmatrix} 30 \\ 64 \\ 63 \\ 33 \\ 9 \\ 1 \end{bmatrix}$$

which is given by means of MATLAB [46] by $k_1 = 5$, $k_0 = 4.14$, $h_2 = 3.93$, $h_1 = 16.256$, and $h_0 = 9.302$. Therefore, the polynomial pair

$$k(D) = D^2 + 5D + 4.14 \quad \text{and} \quad h(D) = 3.93D^2 + 16.256D + 9.302$$

associated with the "implicit observer" of Figure 9.5, or the "explicit observer" of Figure 9.4, with

$$m(D) = k(D) - \hat{q}(D) = D - 0.86$$

and $\alpha = 1$, both imply a closed-loop system characterized by the desired transfer function

$$T(s) = \frac{c(s)\hat{q}(s)}{\delta^F(s)\hat{q}(s)} = \frac{(2s+1)(s^2 + 4s + 5)}{(s^3 + 5s^2 + 8s + 6)(s^2 + 4s + 5)} = \frac{2s+1}{s^3 + 5s^2 + 8s + 6}$$

The Separation Principle

We have now shown how the controllable poles of a closed-loop system can be arbitrarily positioned in the stable half-plane $Re(s) < 0$ using LSVF, and that if the poles also are observable, how an appropriate LSVF control law can be physically implemented using an observer whose poles also can be arbitrarily positioned in the stable half-plane. Moreover, the selection of an appropriate LSVF gain vector F that arbitrarily assigns the closed-loop poles is completely independent of, or "separate" from, the selection of the gain vector H that assigns the observer poles. This rather simple but useful observation is known as the **separation principle** [37].

Lower Order and Multiple Output Observers

In certain situations, a $k(D)$ of degree less than $n-1$, and a corresponding $h(D)$, with $\deg[h(D)] \leq \deg[k(D)] < n-1$, also can yield a desired monic

$$\delta(D) = a(D)k(D) + c(D)h(D) = \delta^F(D)\hat{q}(D) \qquad (9.2.39)$$

as in Eq. (9.2.29), with $\delta^F(D)$ defined by Eq. (9.2.30) and $\deg[\hat{q}(D)] = \deg[k(D)] < n - 1$. Moreover, the subsequent implementation of a 2 DOF compensator defined by such a $k(D)$, $h(D)$, and $q(D) = \alpha\hat{q}(D)$ is equivalent to differential operator observer state feedback. Since the loop performance of such a system is dependent only on $k(D)$ and $h(D)$, and not $q(D)$, it follows that virtually any proper loop compensator[3] is analogous to the "loop portion" of an observer state feedback compensator.

The availability of more than one measured output $y(t)$ can also imply an observer of order less than $n - 1$. Generally speaking, direct access to p independent outputs will imply an observer of order $r \leq n - p$, as we will now illustrate. Suppose there are $p > 1$ independent outputs, namely, $y_1(t) = c_1(D)z(t)$, $y_2(t) = c_2(D)z(t)$, ..., $y_p(t) = c_p(D)z(t)$. If all $p \leq n$ of these outputs are employed in a 2 DOF multi-output (MO) feedback configuration, as depicted in Figure 9.6, with $h(D) = [h_1(D), h_2(D), \ldots, h_p(D)]$, a p-vector of polynomials in D, the

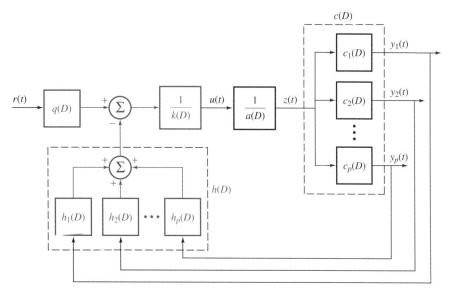

- **FIGURE 9.6**
2 DOF MO Differential Operator Observer Feedback

---

[3]Such as a PID or a lag-lead compensator.

dynamic behavior of the resulting closed-loop system would be defined by the relationship

$$a(D)k(D)z(t)$$

$$= -[h_1(D)c_1(D) + h_2(D)c_2(D) + \cdots h_p(D)c_p(D)]z(t) + q(D)r(t)$$

or

$$\underbrace{[a(D)k(D) + c_1(D)h_1(D) + c_2(D)h_2(D) + \cdots + c_p(D)h_p(D)]}_{\delta(D)}z(t)$$

$$= q(D)r(t) \qquad (9.2.40)$$

Therefore, if $k(D)$ is an arbitrary monic polynomial of degree r, that is, if

$$k(D) = D^r + k_{r-1}D^{r-1} + \cdots k_1 D + k_0$$

and if each element of $h(D)$ is an arbitrary polynomial of degree r, then provided r is "large enough," the r arbitrary coefficients of $k(D)$ and the $(r+1) \times p$ arbitrary coefficients of $h(D)$ can be used to obtain any arbitrary $\delta(D)$ of degree $n+r$, as defined by Eq. (9.2.40). Since $r + (r+1) \times p \geq n + r$, an $r = n - p$ will always be large enough. As a consequence,

$$\frac{n-p}{p} \leq r \leq n - p \qquad (9.2.41)$$

Therefore, we can choose

$$\delta(D) = \underbrace{[a(D) - f(D)]}_{\delta^F(D)}\hat{q}(D) \qquad (9.2.42)$$

as in Eq. (9.2.29), with $\delta^F(D)$ defined by Eq. (9.2.30) and $\hat{q}(D)$ an arbitrary monic polynomial of degree r. If $q(D) = \alpha\hat{q}(D)$, it then follows that

$$\frac{y_i(s)}{r(s)} = \frac{\alpha c_i(s)\hat{q}(s)}{\delta^F(s)\hat{q}(s)} = \frac{\alpha c_i(s)}{\delta^F(s)} \quad \text{for} \quad i = 1, 2, \ldots, p \qquad (9.2.43)$$

. . . .

EXAMPLE 9.2.44 To illustrate the preceding discussion, consider the differential operator system of Example 9.2.38, with a second available output defined by $\ddot{z}(t) - z(t)$, so that

$$\underbrace{(D^3 + 4D^2 + D + 5)}_{a(D)}z(t) = u(t); \qquad \mathbf{y}(t) = \begin{bmatrix} y_1(t) \\ y_2(t) \end{bmatrix} = \underbrace{\begin{bmatrix} c_1(D) \\ c_2(D) \end{bmatrix}}_{c(D)}z(t) = \begin{bmatrix} 2D+1 \\ D^2-1 \end{bmatrix}z(t)$$

Since $\frac{n-p}{p} = \frac{1}{2}$ in this case, a monic $k(D) = D + k_0$ of degree $r = 1$, together with a $h_1(D) =$

$h_1 D + h_0$ and a scalar $h_2(D) = h_2,$[4] will imply any desired

$$\delta(D) = \underbrace{(D^3 + 4D^2 + D + 5)}_{a(D)} \underbrace{(D + k_0)}_{k(D)} + \underbrace{(2D + 1)}_{c_1(D)} \underbrace{(h_1 D + h_0)}_{h_1(D)} + \underbrace{(D^2 - 1)}_{c_2(D)} \underbrace{(h_2)}_{h_2(D)}$$

$$= D^4 + (4 + k_0)D^3 + (1 + 4k_0 + 2h_1 + h_2)D^2 + (5 + k_0 + 2h_0 + h_1)D + 5k_0 + h_0 - h_2$$

of degree $n + r = 3 + 1 = 4$, as in Eq. (9.2.40).

In light of Eq. (9.2.42) and Example 9.2.38, a choice of

$$\delta(D) = \underbrace{(D^3 + 5D^2 + 8D + 6)}_{\delta^F(D)} \underbrace{(D + 2)}_{\hat{q}(D)} = D^4 + 7D^3 + 18D^2 + 22D + 12$$

can be obtained if $k_0 = 3$, $h_0 = 8.67$, $h_1 = -3.33$, and $h_2 = 11.67$ or, equivalently, if

$$k(D) = D + 3, \qquad h_1(D) = -3.33D + 8.67, \quad \text{and} \quad h_2(D) = 11.67$$

A 2 DOF MO implementation of these compensator polynomials, with $q(s) = \alpha\hat{q}(s) = \hat{q}(s) = (s + 2)$, would then imply a closed-loop system characterized by the same (first output) transfer function obtained in Example 9.2.38, namely,

$$\frac{y_1(s)}{r(s)} = \frac{c_1(s)\hat{q}(s)}{\delta^F(s)\hat{q}(s)} = \frac{(2s + 1)(s + 2)}{(s^3 + 5s^2 + 8s + 6)(s + 2)} = \frac{2s + 1}{s^3 + 5s^2 + 8s + 6}$$

9.3 OPTIMAL ERROR-DEPENDENT POLE PLACEMENT

As we have shown, system poles that are both controllable and observable can be arbitrarily assigned using linear feedback from an observer, whose poles also can be arbitrarily assigned. We will now consider the appropriate placement of both the closed-loop system poles and the observer poles.

In certain cases, such as aircraft controller design, desired "handling qualities" can be used to determine appropriate closed-loop pole locations [10] [56]. Also, in the case of either a second-order system or a "dominant poles" system, a desired rise time, a maximum overshoot, and a minimum settling time can be used to define corresponding closed-loop pole locations. In most cases, however, choosing appropriate locations for all of the system poles is not as straightforward.

Generally speaking, the further into the left-half s-plane the poles are positioned, the higher the loop bandwidth and the faster the output response. However, if sensor noise or unmodeled high frequency dynamics

[4]Other choices for $h_1(D)$ and $h_2(D)$ are possible as well—see Problem 9-9.

are prevalent, high loop bandwidths should be avoided. Furthermore, large control signals are usually required to place the poles farther into the left-half s-plane or to move poles away from nearby zeros, and there generally are practical limits on the allowable amplitude of such signals. For example, the plant input $u(t)$ often must be held below certain values to avoid saturation, although a system that never saturates is probably overdesigned.

The effect of control system gain on the closed-loop pole locations can be seen in both Eqs. (9.1.16) and (9.2.11), where $[\hat{a} - \hat{\delta}^F]$ and $[\hat{a} - \hat{\delta}^H]^T$ represent a measure of the "distance" that the poles must be moved. The greater this distance, the larger the required gains, as given by the elements of both the state feedback vector F and the observer gain vector H. Large control signals can be avoided by keeping the closed-loop poles "close to" the open-loop poles. However, this may not be desirable or even possible, for example, when the open-loop plant is unstable.

Some Error-Dependent Performance Indices

There are a number of factors that should be considered when placing the closed-loop poles, and design "trade-offs" must usually be made relative to the desired and often conflicting performance goals. Nonetheless, certain systematic techniques have been developed for pole-placement, many of which are based on the minimization of an appropriate performance index, such as the LQR performance index defined by Eq. (7.1.4).

Other performance indices also can be employed. In particular, consider the various functions of the error

$$e(t) = r(t) - y(t) \tag{9.3.1}$$

between the reference input $r(t)$ and the plant output $y(t)$ that are depicted in Figure 9.7. A desire to zero this error "as quickly as possible" may be achieved by minimizing any one of the following four performance indices:

- the *integral* of the *square error* (ISE) performance index,

$$J_S \overset{\text{def}}{=} \int_0^\infty e^2(t)\, dt$$

- the *integral* of the *time* multiplied by the *square error* (ITSE) performance index,

$$J_{TS} \overset{\text{def}}{=} \int_0^\infty t e^2(t)\, dt$$

- the *integral* of the *absolute error* (IAE) performance index, and

$$J_A \overset{\text{def}}{=} \int_0^\infty |e(t)|\, dt$$

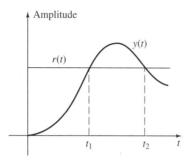
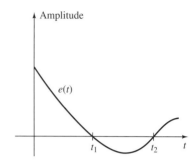
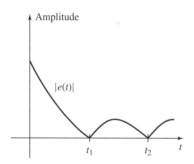
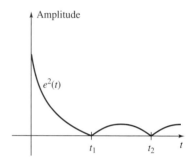

- **FIGURE 9.7**
Functions of the Error $e(t) = r(t) - y(t)$

- the *integral* of the *time* multiplied by the *absolute error* (ITAE)
performance index,

$$J_{TA} \overset{\text{def}}{=} \int_0^\infty t|e(t)|\, dt$$

Figure 9.8 depicts the values of these four performance indices for unit
step inputs, as functions of the damping ratio ζ, when the closed-loop
transfer function

$$T(s) = \frac{y(s)}{r(s)} = \frac{1}{s^2 + 2\zeta s + 1} \qquad (9.3.2)$$

It may be noted that the time $t \geq 0$ acts as a variable *weight* in the
ITSE and ITAE cases, minimizing the contribution to the performance
index of the unavoidably larger values of $e(t)$ at the lower values of t,
while zeroing $e(t)$ more rapidly at the higher values of t. Clearly, the
steady-state value of the error

$$e_{ss}(t) = \lim_{t \to \infty} [r(t) - y(t)]$$

must equal 0 in all cases in order to ensure finite values for the performance
indices.

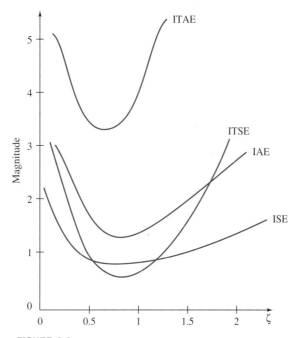

● **FIGURE 9.8**
The Four Error Performance Indices

The ITAE Performance Index

Note that all four indices imply an *optimal* value of ζ between 0.5 and 1 and that the ITAE performance index is the most sensitive to changes in ζ. For this reason, the ITAE performance index has been applied in other cases as well.

Consider a zeroless[5] nth-order system, as defined by the open-loop transfer function

$$G(s) = \frac{y(s)}{u(s)} = \frac{c(s)}{a(s)} = \frac{c}{s^n + a_{n-1}s^{n-1} + \cdots + a_1 s + a_0} \qquad (9.3.3)$$

The optimal closed-loop values for $\delta^F(s)$, as defined by Eq. (9.1.5), that minimize $J_{ITAE} = \int_0^\infty t|e(t)|\,dt$ for step inputs have been determined by Graham and Lathrop [29] for $n = 1$ through 6, respectively, as follows:

$$s + \sqrt{a_0}$$

$$s^2 + 1.4\sqrt{a_0}s + a_0$$

$$s^3 + 1.75\sqrt{a_0}s^2 + 2.15a_0 s + a_0\sqrt{a_0}$$

---

[5]If the given plant has only stable zeros, they can be "cancelled" by corresponding closed-loop poles, thus yielding a zeroless system of lower order, whose remaining poles then can be placed arbitrarily.

$$s^4 + 2.1\sqrt{a_0}s^3 + 3.4a_0 s^2 + 2.7a_0\sqrt{a_o}s + a_0^2$$

$$s^5 + 2.8\sqrt{a_0}s^4 + 5a_0 s^3 + 5.5a_0\sqrt{a_0}s^2 + 3.4a_0^2 s + a_0^2\sqrt{a_0}$$

and

$$s^6 + 3.25\sqrt{a_0}s^5 + 6.6a_0 s^4 + 8.6a_0\sqrt{a_0}s^3 + 7.45a_0^2 s^2 + 3.95a_0^2\sqrt{a_0}s + a_0^3$$
$$(9.3.4)$$

Figure 9.9 displays the unit step responses of these six ITAE optimal systems, as characterized by the closed-loop transfer function

$$T(s) = \frac{y(s)}{r(s)} = \frac{\alpha c = (\sqrt{a_0})^n}{\delta^F(s)} \qquad (9.3.5)$$

Note that the numerator of $T(s)$ must equal $(\sqrt{a_0})^n = \delta^F(0)$ to obtain a final steady-state value of $y(t) = r(t)$, hence an $e_{ss}(t) = 0$. In particular, for the unit step input case depicted in Figure 9.9, the final value theorem implies that

$$\lim_{t\to\infty} y(t) = \lim_{s\to 0} sy(s) = \lim_{s\to 0} \frac{sT(s)}{s} = \frac{\alpha c}{\delta^F(0)} = \frac{(\sqrt{a_0})^n}{(\sqrt{a_0})^n} = 1 \quad (9.3.6)$$

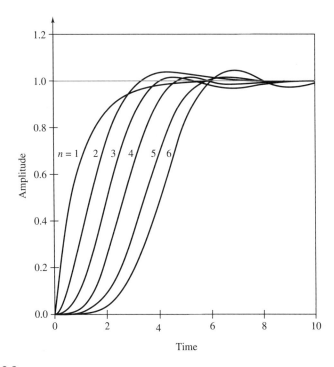

- **FIGURE 9.9**
Unit Step Responses of Optimal ITAE Systems

The closed-loop plant pole positions associated with an optimal ITAE system do not necessarily represent the "best" possible locations relative to all of the performance goals that we may attempt to attain, and alternative choices do exist, as we will show in the next section. Moreover, nothing has been said, thus far, regarding the "optimal" observer pole locations. This question also will be addressed in the next section.

EXAMPLE 9.3.7 To illustrate pole-placement by means of the ITAE performance index, consider a fourth-order system defined by the differential equation

$$\frac{d^4 y(t)}{dt} + \frac{d^3 y(t)}{dt} + 8\frac{dy(t)}{dt} + 8y(t) = \frac{du(t)}{dt} + 2u(t)$$

or the equivalent differential operator representation

$$(D^4 + D^3 + 8D + 8)y(t) = (D+1)(D+2)(D^2 - 2D + 4)y(t) = (D+2)u(t)$$

which is both stabilizable and detectable, since the only common factor $D+2$ implies an uncontrollable but stable mode, e^{-2t}.

In this example, we will first "cancel" the $D+2$ factor and base our subsequent design on the minimal, or prime transfer function

$$G(s) = \frac{y(s)}{u(s)} = \frac{1}{(s+1)(s-1\pm j\sqrt{3})} = \frac{1}{s^3 - s^2 + 2s + 4} = \frac{c}{s^3 + a_2 s^2 + a_1 s + a_0}$$

as defined by Eq. (9.3.3), when $n = 3$. The optimal ITAE closed-loop transfer function is given by Eqs. (9.3.4) and (9.3.5),

$$T(s) = \frac{y(s)}{r(s)} = \frac{(\sqrt{a_0})^3}{\delta^F(s)} = \frac{8}{s^3 + 3.5s^2 + 8.6s + 8} \qquad (9.3.8)$$

with closed-loop poles at $s = -1.416$ and $-1.042 \pm j2.136$.

We can obtain this $T(s)$ in a variety of ways, in view of the results presented in Section 9.2. We can employ a third-order state-space observer by first defining a minimal, third-order state-space realization of $G(s)$. Alternatively, we can derive a second-order differential operator observer. If we choose the differential operator observer, it can be implemented using either the "explicit" observer feedback configuration of Figure 9.4 or the "implicit" 2 DOF configuration of Figure 9.5.

In this case, we will choose to employ a Laplace transformed differential operator observer, based on the implicit 2 DOF configuration of Figure 9.5. If we arbitrarily position the observer poles at $s = -1 \pm j$, so that

$$\hat{q}(s) = s^2 + 2s + 2$$

we then can solve the Laplace transformed polynomial Eq. (9.2.29), namely

$$\underbrace{(s^3 - s^2 + 2s + 4)}_{a(s)} \underbrace{(s^2 + k_1 s + k_0)}_{k(s)} + \underbrace{(1)}_{c(s)} \underbrace{(h_2 s^2 + h_1 s + h_0)}_{h(s)}$$

$$= \delta(s) = \underbrace{(s^2 + 2s + 2)}_{\hat{q}(s)} \underbrace{(s^3 + 3.5s^2 + 8.6s + 8)}_{\delta^F(s)}$$

or

$$s^5 + (k_1 - 1)s^4 + (k_0 - k_1 + 2)s^3 + (2k_1 - k_0 + 4 + h_2)s^2 + (2k_0 + 4k_1 + h_1)s + 4k_0 + h_0$$

$$= s^5 + 5.5s^4 + 17.6s^3 + 32.2s^2 + 33.2s + 16$$

for $k(s)$ and $h(s)$. In this case, we can sequentially determine that

$$k_1 = 5.5 + 1 = 6.5, \qquad k_0 = 17.6 - 2 + 6.5 = 22.1$$

$$h_2 = 32.2 - 4 + 22.1 - 2 \times 6.5 = 37.3, \qquad h_1 = 33.2 - 4 \times 6.5 - 2 \times 22.1 = -37$$

$$h_0 = 16 - 4 \times 22.1 = -72.4$$

so that

$$k(s) = s^2 + 6.5s + 22.1 \quad \text{and} \quad h(s) = 37.3s^2 - 37s - 72.4$$

Finally, to obtain a numerator of $T(s)$ equal to $(\sqrt{a_0})^3 = 2^3 = 8$, we will multiply $\hat{q}(s)$ by $8 = \alpha$, as in Eq. (7.2.11), thereby obtaining the closed-loop system depicted in Figure 9.10,[6] which is characterized by the ITAE optimal $T(s)$ defined by Eq. (9.3.8).

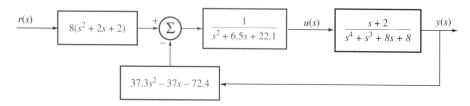

- **FIGURE 9.10**
  A Non-Robust $e_{ss}(t) = 0$ ITAE Optimal Closed-Loop System

It is of interest to note that this particular closed-loop design implies a *non-robust* $e_{ss}(t) = 0$ for step inputs. In particular, Eq. (5.1.19) implies that

$$\frac{e(s)}{r(s)} = \frac{a(s)k(s) + c(s)[h(s) - q(s)]}{\underbrace{a(s)k(s) + c(s)h(s)}_{\delta(s)}} = \frac{a(s)k(s) + h(s) - q(s)}{a(s)k(s) + h(s)}$$

since $c(s) = c = 1$ in this case. Therefore, for a unit step reference input $r(s) = \frac{1}{s}$,

$$e_{ss}(t) = \lim_{s \to 0} \frac{se(s)}{s} = \frac{a_0 k_0 + h_0 - \alpha \hat{q}_0}{a_0 k_0 + h_0} \tag{9.3.9}$$

with a *fixed* $k_0 = 22.1$, $h_0 = -72.4$, and $\alpha \hat{q}_0 = 8 \times 2 = 16$. Although the nominal $a_0 = 4$ does imply that

$$a_0 k_0 + h_0 - \alpha \hat{q}_0 = 4 \times 22.1 - 72.4 - 16 = 0$$

---

[6]It may be noted that both a differential operator realization and a state-space realization of the second order, 2 DOF compensator depicted in Figure 9.10 are given in Problem 3-18.

and hence a nominal $e_{ss}(t) = 0$, any changes in a_0 due to plant parameter variations or modeling inaccuracies will imply an $e_{ss}(t) = 22.1a_0 - 88.4 \neq 0$.

To obtain a *robust* $e_{ss}(t) = 0$ for step changes in the input, we can now invoke the internal model principle (IMP) of Section 7.3. Using Eq. (7.3.11), we can ensure a *type 1* system by choosing

$$k(s) = \hat{k}p_r(s) = (s^2 + \hat{k}_1 s + \hat{k}_0)s = s^3 + k_2 s^2 + k_1 s$$

since $p_r(s) = s$ for step inputs, with the corresponding $h(s)$ and $\hat{q}(s)$ also of degree 3. Therefore, in order to ensure a robust $e_{ss}(t) = 0$ in this case, the compensator order will be increased from 2 to 3.

If we arbitrarily position the observer poles at $s = -3$ and $-1 \pm j$, so that

$$\hat{q}(s) = s^3 + 5s^2 + 8s + 6$$

we then can solve Eq. (9.2.29),

$$\underbrace{(s^3 - s^2 + 2s + 4)}_{a(s)} \underbrace{(s^3 + k_2 s^2 + k_1 s)}_{k(s)} + \underbrace{(1)}_{c(s)} \underbrace{(h_3 s^3 + h_2 s^2 + h_1 s + h_0)}_{h(s)}$$

$$= \underbrace{(s^3 + 5s^2 + 8s + 6)}_{\hat{q}(s)} \underbrace{(s^3 + 3.5s^2 + 8.6s + 8)}_{\delta^F(s)}$$

or

$$s^6 + (k_2 - 1)s^5 + (k_1 - k_2 + 2)s^4 + (2k_2 - k_1 + h_3 + 4)s^3 + (2k_1 + 4k_2 + h_2)s^2 + (4k_1 + h_1)s + h_0$$

$$= s^6 + 8.5s^5 + 34.1s^4 + 85s^3 + 129.8s^2 + 115.6s + 48$$

for $k(s)$ and $h(s)$. As in the case of the second-order observer, we can sequentially determine that

$$k_2 = 8.5 + 1 = 9.5, \qquad k_1 = 34.1 - 2 + 9.5 = 41.6, \qquad k_0 = 0$$

$$h_3 = 85 - 4 + 41.6 - 2 \times 9.5 = 103.6, \qquad h_2 = 129.8 - 4 \times 9.5 - 2 \times 41.6 = 8.6$$

$$h_1 = 115.6 - 4 \times 41.6 = -50.8 \quad \text{and} \quad h_0 = 48$$

so that

$$k(s) = s^3 + 9.5s^2 + 41.6s \quad \text{and} \quad h(s) = 103.6s^3 + 8.6s^2 - 50.8s + 48$$

To obtain a numerator of $T(s)$ equal to $(\sqrt{a_0})^3 = 2^3 = 8$, we will multiply $\hat{q}(s)$ by $8 = \alpha$, as in the second-order case, thereby obtaining the closed-loop system depicted in Figure 9.11, which also is characterized by the ITAE optimal $T(s)$ defined by Eq. (9.3.8). However, in this particular case a robust $e_{ss}(t) = 0$ is ensured for step changes in the reference input; that is, in light of Eq. (9.3.9), the fixed $k_0 = 0$, $h_0 = 48$, and $\alpha \hat{q}_0 = 48$ imply that

$$a_0 k_0 + h_0 - \alpha \hat{q}_0 = h_0 - \alpha \hat{q}_0 = 48 - 8 \times 6 = 0$$

despite any changes in the value of a_0.

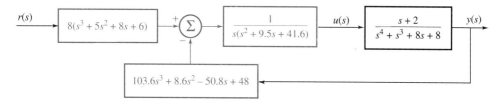

- **FIGURE 9.11**
  A Robust $e_{ss}(t) = 0$ ITAE Optimal Closed-Loop System

9.4   OPTIMAL LQR COMPENSATION

One of the most useful and commonly employed performance indices is the linear quadratic regulator (LQR) performance index defined by Eq. (7.1.4), namely, .

$$J_{LQR} = \int_0^\infty \{\rho y^2(t) + u^2(t)\}\, dt \qquad (9.4.1)$$

Kalman [39] was the first to show that a LSVF feedback control law minimizes a LQR performance index. His state-space solution employed a so-called **matrix Riccati equation** (see Problem 9-16). We will employ a differential operator approach here, based on a "spectral factorization," to obtain an equivalent result. As we will show, a LQR performance index can be used to obtain improved closed-loop performance and zero-error tracking, as well as to design an optimal observer.

Spectral Factorization

Consider a system defined by the minimal, strictly proper, rational transfer function

$$G(s) = \frac{y(s)}{u(s)} = \frac{c(s)}{a(s)} = \frac{c_m s^m + \cdots + c_1 s + c_0}{s^n + \cdots + a_1 s + a_0}, \quad \text{with} \quad m < n \qquad (9.4.2)$$

Since the polynomials $a(s)$ and $c(s)$ have real coefficients, it follows that both

$$a(j\omega)a(-j\omega) = |a(j\omega)|^2 \geq 0$$

and $\quad c(j\omega)c(-j\omega) = |c(j\omega)|^2 \geq 0 \quad \forall \text{ real } \omega \qquad (9.4.3)$

We next note that for any real *weighting factor* $\rho > 0$,

$$\Delta(s) \overset{\text{def}}{=} a(s)a(-s) + \rho c(s)c(-s) \qquad (9.4.4)$$

is an *even polynomial*, in the sense that the coefficients of all of its odd s power terms are zero, so that

$$\Delta(s) = (-1)^n s^{2n} + \Delta_{2n-2} s^{2n-2} + \cdots + \Delta_2 s^2 + \Delta_0 \qquad (9.4.5)$$

Therefore, if λ_j is a root of $\Delta(s)$ then $-\lambda_j$ is also a root of $\Delta(s)$. Moreover, since $a(s)$ and $c(s)$ have no common roots, Eqs. (9.4.3) and (9.4.4) imply that

$$\Delta(j\omega) = |a(j\omega)|^2 + \rho|c(j\omega)|^2 > 0 \quad \forall \text{ real } \omega \tag{9.4.6}$$

so that $\Delta(s)$ has no roots on the $j\omega$-axis. Consequently, $\Delta(s)$ can be expressed by a **spectral factorization**, as in Chang [13], namely

$$\Delta(s) = [\Delta(s)]^+[\Delta(s)]^- \tag{9.4.7}$$

Here all n roots of

$$[\Delta(s)]^+ \stackrel{\text{def}}{=} s^n + \Delta_{n-1}^+ s^{n-1} + \cdots + \Delta_1^+ s + \Delta_0^+ \tag{9.4.8}$$

the *left half-plane spectral factor of* $\Delta(s)$, lie in the stable half-plane $Re(s) < 0$, and all n roots of $[\Delta(s)]^- = [\Delta(-s)]^+$, the *right half-plane spectral factor of* $\Delta(s)$, lie in the unstable half-plane $Re(s) > 0$ at the $j\omega$-axis mirror images of the roots of $[\Delta(s)]^+$.

Since Eq. (9.4.2) implies the minimal, differential operator realization

$$a(D)z(t) = u(t); \qquad y(t) = c(D)z(t) \tag{9.4.9}$$

in controllable canonical form, $\Delta(D)$ and $[\Delta(D)]^+$ are defined by Eqs. (9.4.4) through (9.4.8) by simply replacing the Laplace operator s by the differential operator D.

LQR Optimal LSVF Control

Consider a system defined by the transfer function Eq. (9.4.2) or, equivalently, the differential operator representation Eq. (9.4.9). The substitution of Eq. (9.4.9) into Eq. (9.4.1) implies that

$$J_{LQR} = \int_0^\infty \{\rho[c(D)z(t)]^2 + [a(D)z(t)]^2\}\, dt$$

$$= \int_0^\infty \{\rho c^2(D) + a^2(D) \underbrace{-([\Delta(D)]^+)^2 + ([\Delta(D)]^+)^2}_{0}\}z^2(t)\, dt \tag{9.4.10}$$

We next note that the integral

$$\int_0^\infty \{\rho c^2(D) + a^2(D) - ([\Delta(D)]^+)^2\}z^2(t)\, dt$$

is independent of the $z(t)$ path [12] and, therefore, is dependent only on $z(t)$ and its derivatives at the initial time $t = 0$. As a consequence, J_{LQR} is minimized if $+ ([\Delta(D)]^+)^2 z^2(t) = 0$ or, equivalently, if $[\Delta(D)]^+ z(t) = 0$. Note, however, that the substitution of $\{a(D) - [\Delta(D)]^+\}z(t)$ for $u(t)$ in Eq. (9.4.9) implies that $[\Delta(D)]^+ z(t) = 0$.

In light of this observation, it follows that if $a(D)$ and $c(D)$ are coprime, then the unique LQR optimal LSVF control law

$$u^*(t) = \{a(D) - \underbrace{[a(D)a(-D) + \rho c(D)c(-D)]^+}_{[\Delta(D)]^+ \overset{\text{def}}{=} \delta^{F^*}(D)}\}z(t) \overset{\text{def}}{=} f^*(D)z(t)$$

(9.4.11)

minimizes J_{LQR} for any arbitrary set of nonzero initial conditions on $z(t)$ and its derivatives. Moreover, the closed-loop poles of the resulting LQR optimal system are given by the roots of

$$[\Delta(s)]^+ = a(s) - f^*(s) = \delta^{F^*}(s)$$

(9.4.12)

An LQR optimal LSVF control law implies the "best" system output response with respect to arbitrary, nonzero initial conditions on its state when the reference input $r(t) = 0$, which is optimal output regulation. However, the LQR optimal feedback control law defined by Eq. (9.4.11) can very often be used to obtain a desirable tracking response as well, as we will show.

The Root-Square Locus

Referring to the preceding discussions, the LQR optimal closed-loop poles of a system defined by Eq. (9.4.2) are given by the n roots of $\delta^{F^*}(s)$, which correspond to the n stable roots of the spectral factorization

$$\Delta(s) = a(s)a(-s) + \rho c(s)c(-s) = \underbrace{\delta^{F^*}(s)}_{[\Delta(s)]^+} \underbrace{\delta^{F^*}(-s)}_{[\Delta(s)]^-}$$

(9.4.13)

An s-plane plot of all $2n$ roots of $\Delta(s)$, as ρ varies from 0 to ∞, represents a special root locus (as defined in Section 5.3), which is termed a **root-square locus**. A root-square locus plot of $\Delta(s)$ is symmetric with respect to both the real and imaginary axes, and since $\Delta(j\omega) > 0$, in light of Eq. (9.4.6), such a plot never crosses the $j\omega$-axis.

A root-square locus plot can provide significant insight relative to the design trade-offs between "large" output excursions and the control effort required to constrain such excursions. In particular, as $\rho \to 0$, Eq. (9.4.1) implies a minimization of the "expensive" control $u(t)$ at the expense of potentially large $y(t)$ excursions that result from nonzero initial conditions. This can be verified using Eq. (9.4.4), where

$$\delta^{F^*}(s) = [\Delta(s)]^+ \to [a(s)a(-s)]^+ \quad \text{as} \quad \rho \to 0$$

thereby implying that all of the stable roots of $a(s)$ will move only slightly, while any unstable roots of $a(s)$ at $s = +\sigma_i \pm j\omega_i$ will be reflected across the $j\omega$-axis near $s = -\sigma_i \pm j\omega_i$.

Conversely, as $\rho \to \infty$, Eq. (9.4.1) implies a maximization of the "inexpensive" control $u(t)$ that prevents large $y(t)$ excursions, hence a resultant movement of $n - m$ closed-loop poles farther into the stable half-plane, thereby producing a faster response of $y(t) \to 0$. Again, this can be seen in Eq. (9.4.4) where

$$[\Delta(s)] = [\Delta(s)]^+[\Delta(s)]^- \to [(-1)^n s^{2n} + \rho c(s)c(-s)] \quad \text{as} \quad \rho \to \infty$$

thereby implying that $m = \deg[c(s)]$ of the optimal closed-loop poles will move toward the stable zeros of $[c(s)c(-s)]^+$. Since

$$\Delta(s) \approx (-1)^n s^{2n} + \rho(-1)^m c_m^2 s^{2m} = s^{2m}[(-1)^n s^{2(n-m)} + (-1)^m \rho c_m^2] \tag{9.4.14}$$

for large values of s, the remaining $n - m$ roots of $[\Delta(s)]^+$ will correspond to the stable zeros of

$$s^{2(n-m)} = (-1)^{n-m+1} \rho c_m^2 \tag{9.4.15}$$

These zeros lie on a circle of radius $(\sqrt{\rho} c_m)^{\frac{1}{n-m}} \to \infty$ as $\rho \to \infty$, in the complex s-plane, at asymptotic radial directions that are defined by $\frac{k\pi}{n-m}$, if $n - m$ is odd, and $\frac{(2k+1)\pi}{2(n-m)}$, if $n - m$ is even, for $k = 0, 1, \ldots, 2(n-m) - 1$. The resulting stable pole distribution, which is termed a **Butterworth configuration** [38], is depicted in Figure 9.12 for $n - m = 1$ through 4.

Optimal Observers

By *duality*, spectral factorization also can be used to determine **LQR optimal observers**. If any noise $v(t)$ enters the system at the input $u(t)$, then it can be shown that the optimal[7] observer poles are given by the n stable roots of $\delta^{H^*}(s)$, which is defined as the left half-plane spectral factor of

$$\bar{\Delta}(s) = a(s)a(-s) + \sigma c(s)c(-s) = [\bar{\Delta}(s)]^+[\bar{\Delta}(s)]^- \stackrel{\text{def}}{=} \delta^{H^*}(s)\delta^{H^*}(-s) \tag{9.4.16}$$

The adjustable weighting factor $\sigma > 0$ represents the "spectral intensity" of the input noise $v(t)$ relative to that of the measurement noise $\eta(t)$.[8] However, as noted in Reference [26],

> ... such characteristics are rarely known with sufficient accuracy to justify the claim of true optimality.

---

[7] In the sense of minimizing the "mean-square error" associated with the "expected" value of the observed state [26].

[8] A formal development of this result extends beyond the scope of this text. However, it may be noted that an optimal observer whose pole locations are determined by the "spectral properties" of the exogenous noise signals is termed a **Kalman filter** [26].

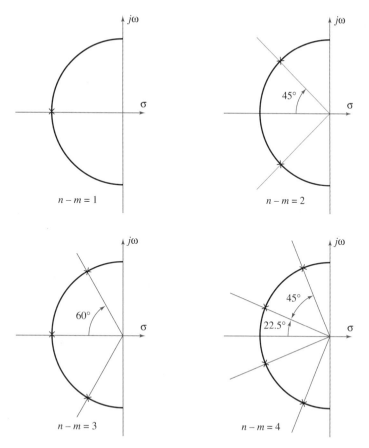

● **FIGURE 9.12**
Some Butterworth Pole Configurations

Therefore, for our purposes, σ will represent the "intensity" of $\eta(t)$, irrespective of any actual input noise $v(t)$. In particular, as $\sigma \to 0$ (the high measurement noise case), the optimal observer poles will approach the stable roots of $a(s)$ and the $j\omega$-axis mirror images of the unstable roots of $a(s)$. Conversely, as $\sigma \to \infty$ (the low measurement noise case), $m = \deg[c(s)]$ of the optimal observer poles will move toward the stable zeros of $[c(s)c(-s)]^+$, while the remaining $n - m$ optimal observer poles will approach infinity in a Butterworth configuration on a semicircle of radius $(\sqrt{\sigma}c_m)^{\frac{1}{n-m}}$. The resulting loop bandwidth of the system will be increased, as expected, in the low measurement noise case.

It is of interest to note that the same spectral factorization equation with a different weighting factor, Eq. (9.4.13) or Eq. (9.4.16), can be used to determine both the optimal LSVF closed-loop pole locations and the corresponding, *dual* optimal observer pole locations. Therefore, LQR optimization by means of spectral factorization implies that a single root-square locus plot will display all $2n$ optimal closed-loop poles.

Factors such as unmodeled high frequency dynamics, output disturbances, and input saturation can also be considered when a LQR design is employed, as we will show. The virtually limitless number of "optimal" closed-loop pole configurations associated with a root-square locus plot can be used for trade-off studies that reflect a final design compromise among all of the desired performance goals.

LQR Optimal Design

We will now illustrate a LQR optimal design that is based on the determination of two sets of n closed-loop poles. The first set of LQR optimal system poles is given by the n roots of $[\Delta(s)]^+ = \delta^{F^*}(s)$, in light of Eq. (9.4.13), and the second set of LQR optimal observer poles is given by the n roots of $[\bar{\Delta}(s)]^+ = \delta^{H^*}(s)$, in light of Eq. (9.4.16). In both cases, the variation of a single parameter (ρ or σ) implies all of the optimal pole locations. If an observer of order $n-1$ is employed, $n-1$ roots of $\delta^{H^*}(s)$ can be used to define $\hat{q}(s)$, whose roots represent the observer poles.

Since the output response transfer function $T(s)$ is independent of the observer, in view of the separation principle, we will determine $\delta^{F^*}(s)$ first by varying ρ in Eq. (9.4.13) until an acceptable LQR optimal transfer function

$$T(s) = \frac{y(s)}{r(s)} = \frac{\alpha c(s)}{\delta^{F^*}(s)} \qquad (9.4.17)$$

analogous to Eq. (9.2.37) is obtained.

Once $\delta^{F^*}(s)$ has been selected, each choice of σ in Eq. (9.4.16) implies a corresponding $\delta^{H^*}(s)$, hence an appropriate $\hat{q}(s)$. We then can solve the polynomial Diophantine Eq. (9.2.29), namely,

$$a(s)k(s) + c(s)h(s) = \delta^{F^*}(s)\hat{q}(s) \qquad (9.4.18)$$

for

$$k(s) = s^{n-1} + k_{n-2}s^{n-2} + \cdots + k_1 s + k_0 \qquad (9.4.19)$$

and

$$h(s) = h_{n-1}s^{n-1} + \cdots + h_1 s + h_0 \qquad (9.4.20)$$

that, together with

$$q(s) = \alpha\hat{q}(s) \qquad (9.4.21)$$

as in Eq. (9.2.36), define a 2 DOF LQR optimal observer, analogous to that depicted in differential operator form in Figure 9.5.

The nominal loop gain

$$L(s) = G(s)H(s) = \frac{c(s)h(s)}{a(s)k(s)}$$ (9.4.22)

and the corresponding sensitivity function

$$S(s) = \frac{1}{1 + L(s)} = \frac{a(s)k(s)}{\delta^{F^*}(s)\hat{q}(s)}$$ (9.4.23)

can then be evaluated and altered (via σ), if necessary, until "acceptable" loop performance is obtained.

. . . .

EXAMPLE 9.4.24 Recall the linearized state-space equations of motion of the orbiting satellite, which was introduced in Chapter 2 with Examples 2.4.2, 2.4.7, and 2.6.14,

$$\underbrace{\begin{bmatrix} \dot{x}_1(t) \\ \dot{x}_2(t) \\ \dot{x}_3(t) \\ \dot{x}_4(t) \end{bmatrix}}_{\dot{\mathbf{x}}(t)} = \underbrace{\begin{bmatrix} 0 & 1 & 0 & 0 \\ 3\omega^2 & 0 & 0 & 2d\omega \\ 0 & 0 & 0 & 1 \\ 0 & \frac{-2\omega}{d} & 0 & 0 \end{bmatrix}}_{A} \underbrace{\begin{bmatrix} x_1(t) \\ x_2(t) \\ x_3(t) \\ x_4(t) \end{bmatrix}}_{\mathbf{x}(t)} + \underbrace{\begin{bmatrix} 0 & 0 \\ 1 & 0 \\ 0 & 0 \\ 0 & \frac{1}{d^2} \end{bmatrix}}_{[B_1 \ B_2]} \underbrace{\begin{bmatrix} u_1(t) \\ u_2(t) \end{bmatrix}}_{\mathbf{u}(t)}$$

with output

$$\underbrace{\begin{bmatrix} y_1(t) \\ y_2(t) \end{bmatrix}}_{\mathbf{y}(t)} = \underbrace{\begin{bmatrix} 1 & 0 & 0 & 0 \\ 0 & 0 & 1 & 0 \end{bmatrix}}_{[C_1 \ C_2]^T} \mathbf{x}(t)$$

Also recall that the complete state can be controlled by the tangential thruster $u_2(t)$ alone, if the radial thruster $u_1(t)$ were to fail, and that the entire state can be observed by $y_2(t)$ alone, if the $y_1(t)$ sensor were to fail. Therefore, it is useful to design a SISO feedback control system from $y_2(t)$ to $u_2(t)$, which could be activated in the event of a $u_1(t)$ or a $y_1(t)$ failure. The primary goal of such a control system would be to maintain a nominal orbiting trajectory of radius d and angular velocity ω, as noted in Example 2.4.7.

If (say) $\omega = 1$ and $d = 2$, it follows that

$$G_{22}(s) = \frac{y_2(s)}{u_2(s)} = C_2(sI - A)^{-1}B_2$$

$$= \frac{0.25(s^2 - 3)}{s^2(s^2 + 1)} = \frac{0.25(s + 1.732)(s - 1.732)}{s^2(s + j)(s - j)} \overset{\text{def}}{=} \frac{c_2(s)}{a(s)}$$

an open-loop transfer function with a stable zero at $s = -1.732$, an unstable zero at $s = +1.732$, and four marginally stable poles, two at $s = 0$ and the other two at $s = \pm j$. If $y_2(s)$ were fed back to $u_2(s)$ by a proportional gain K_P, the resulting closed-loop poles would vary as in the root locus

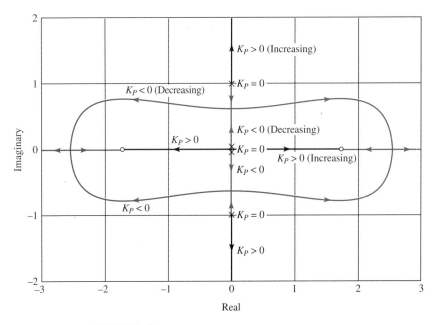

● **FIGURE 9.13**
A $G_{22}(s)$ Root Locus Plot for the Orbiting Satellite

plot of Figure 9.13, and the resulting closed-loop system would not be stable for any positive (black) or negative (blue) value of K_P.

To improve both the loop and the response performance, an optimal LQR design will now be employed.[9] We first determine the optimal system poles by means of a root-square locus plot of

$$\Delta(s) = a(s)a(-s) + \rho c_2(s)c_2(-s) = (s^4 + s^2)^2 + \rho(0.25s^2 - 0.75)^2$$

$$= s^8 + 2s^6 + s^4 + \rho(0.0625s^4 - 0.375s^2 + 0.5625)$$

as defined by Eq. (9.4.13) and depicted in Figure 9.14, noting that the imaginary poles at $s = \pm j$ approach a $n - m = 2$ Butterworth configuration as $\rho \to \infty$. If the weighting factor $\rho = 10$ in Eq. (9.4.1),

$$[\Delta(s)]^+ = \delta^{F^*}(s) = (s + 0.549 \pm j1.413)(s + 0.93 \pm j.41)$$

$$= s^4 + 2.958s^3 + 5.373s^2 + 5.409s + 2.374$$

which implies an acceptable 2 DOF unit step response, as we will show later.

Optimal observer poles, as defined by the roots of $\delta^{H^*}(s)$, can now be determined using Eq. (9.4.16), or

$$a(s)a(-s) + \sigma c_2(s)c_2(-s) = \delta^{H^*}(s)\delta^{H^*}(-s)$$

---

[9]Note that it is not clear how a PID or a lag-lead compensator might be employed in this case.

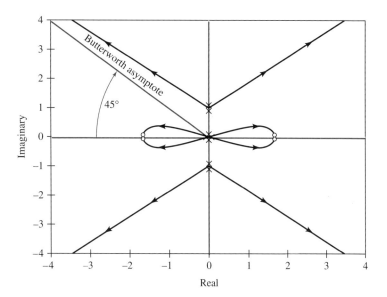

• **FIGURE 9.14**
A Root-Square Locus Plot for the Orbiting Satellite

whose roots also are depicted in the Figure 9.14 root-square locus plot. A choice of $\sigma = 1$ implies that

$$\delta^{H^*}(s) = (s + 0.304 \pm j1.161)(s + 0.623 \pm j0.365)$$

The elimination of the imaginary part of the slower complex pair yields a resultant, third-degree

$$\hat{q}(s) = (s + 0.304)(s + 0.623 \pm j.365) = s^3 + 1.55s^2 + 0.9s + 0.1585$$

We next employ Eq. (9.4.18) to determine $k(s)$ and $h(s)$,

$$\underbrace{(s^4 + s^2)}_{a(s)}\underbrace{(s^3 + k_2 s^2 + k_1 s + k_0)}_{k(s)} + \underbrace{(0.25s^2 - 0.75)}_{c_2(s)}\underbrace{(h_3 s^3 + h_2 s^2 + h_1 s + h_0)}_{h(s)}$$

$$= \underbrace{(s^4 + 2.958s^3 + 5.373s^2 + 5.409s + 2.374)}_{\delta^{F^*}(s)}\underbrace{(s^3 + 1.55s^2 + 0.9s + 0.1585)}_{\hat{q}(s)}$$

$$= s^7 + 4.508s^6 + 10.858s^5 + 16.558s^4 + 16.062s^3 + 9.399s^2 + 2.994s + 0.376$$

which can be represented in vector/matrix form by means of Eq. (B7) of Appendix B, that is,

$$\underbrace{\begin{bmatrix} 0 & 0 & 0 & 0 & -0.75 & 0 & 0 & 0 \\ 0 & 0 & 0 & 0 & 0 & -0.75 & 0 & 0 \\ 1 & 0 & 0 & 0 & 0.25 & 0 & -0.75 & 0 \\ 0 & 1 & 0 & 0 & 0 & 0.25 & 0 & -0.75 \\ 1 & 0 & 1 & 0 & 0 & 0 & 0.25 & 0 \\ 0 & 1 & 0 & 1 & 0 & 0 & 0 & 0.25 \\ 0 & 0 & 1 & 0 & 0 & 0 & 0 & 0 \\ 0 & 0 & 0 & 1 & 0 & 0 & 0 & 0 \end{bmatrix}}_{\mathcal{S}} \underbrace{\begin{bmatrix} k_0 \\ k_1 \\ k_2 \\ 1 \\ h_0 \\ h_1 \\ h_2 \\ h_3 \end{bmatrix}}_{W} = \underbrace{\begin{bmatrix} 0.376 \\ 2.994 \\ 9.399 \\ 16.062 \\ 16.558 \\ 10.858 \\ 4.508 \\ 1 \end{bmatrix}}_{\bar{\delta}}$$

Solving this vector/matrix equation using MATLAB for the coefficient vector, we determine that

$$k(s) = s^3 + 4.508s^2 + 11.659s + 11.419 = (s + 1.642)(s + 1.433 \pm j2.214)$$

and

$$h(s) = -7.202s^3 + 2.526s^2 - 3.99s - 0.501 = -7.202(s + 0.115)(s - 0.233 \pm j.744)$$

A negative root locus (since $h(s)$ is characterized by a negative gain in this case) of this LQR optimal system is depicted in Figure 9.15. The small black boxes indicate all $2n - 1 = 7$ closed-loop pole locations at the LQR optimal $h(s)$ gain of -7.202.

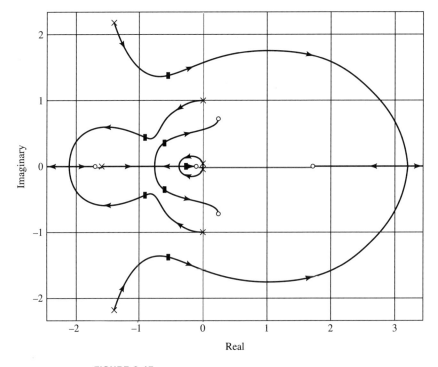

• **FIGURE 9.15**
A Root Locus Plot of the LQR Optimal Systems

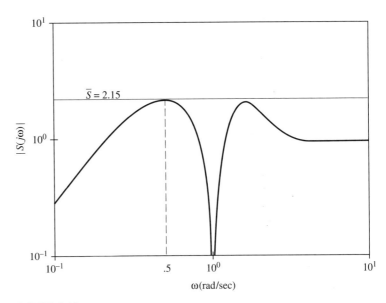

• **FIGURE 9.16**
A Bode Plot of $|S(j\omega)|$

A Bode plot of the magnitude of the sensitivity function

$$S(s) = \frac{1}{1 + L(s)} = \frac{a(s)k(s)}{a(s)k(s) + c_2(s)h(s)}$$

as depicted in Figure 9.16, reflects a $\max_\omega |S(j\omega)| = \bar{S} = 2.15$ at $\omega = 0.5$, which implies acceptable stability margins in this somewhat "difficult to control" case. The Bode diagram of the loop gain

$$L(s) = \frac{c_2(s)h(s)}{a(s)k(s)} = \frac{(0.25s^2 - 0.75)(-7.202s^3 + 2.526s^2 - 3.99s - 0.501)}{(s^4 + s^2)(s^3 + 4.508s^2 + 11.659s + 11.419)}$$

depicted in Figure 9.17, and the corresponding Nyquist plot of Figure 9.18, display a gain margin GM $= 1/0.515 = 1.942$ at a phase crossover frequency $\omega_\phi = 0.578$, and a phase margin ΦM $= 36.08°$ at a gain crossover frequency $\omega_g = 0.303$, which are consistent with the $\bar{S} = 2.15$ of Figure 9.16. Moreover, we will assume that the $|L(j\omega)|$ plot reflects appropriate magnitudes at both the lower and the higher frequencies to ensure acceptable loop goal performance.

The 2 DOF control configuration of Figure 9.19 now is employed, with $q(s) = \alpha\hat{q}(s) = -3.161\hat{q}(s)$. As a consequence, $q(0) = h(0)$, as in Eq. (7.3.18), which will ensure a robust $e_{ss}(t) = 0$ for step inputs. The resulting $y_2(t)$ unit step response, as defined by the transfer function

$$\frac{y_2(s)}{r(s)} = \frac{c_2(s)q(s)}{a(s)k(s) + c_2(s)h(s)} = \frac{c_2(s)\alpha\hat{q}(s)}{\delta^{F^*}(s)\hat{q}(s)} = \frac{\alpha c_2(s)}{\delta^{F^*}(s)}$$

$$= \frac{-0.79(s + 1.732)(s - 1.732)}{s^4 + 2.958s^3 + 5.373s^2 + 5.409s + 2.374}$$

is depicted by the solid line in Figure 9.20. It is characterized by a maximum overshoot M_p of only 6.8% at $t_p = 4$, with a 2% settling time $t_s = 6.7$.

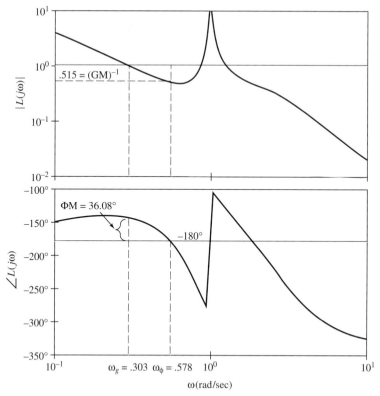

• **FIGURE 9.17**
A Bode Diagram of the LQR Optimal $L(s)$

The $y_2(t)$ unit step response resulting from a 1 DOF, unity feedback implementation, as defined by the transfer function

$$\frac{y_2(s)}{r(s)} = \frac{c_2(s)h(s)}{a(s)k(s) + c_2(s)h(s) = \delta^{F^*}(s)\hat{q}(s)}$$

$$= \frac{-1.8(s + 1.732)(s - 1.732)(s + 0.115)(s - 0.233 \pm j.744)}{(s^4 + 2.958s^3 + 5.373s^2 + 5.409s + 2.374)(s^3 + 1.55s^2 + 0.9s + 0.1585)}$$

is depicted by the dashed line for comparison purposes. It may be noted that the initial negative $y_2(t)$ responses of Figure 9.20 characterize stable systems that have an odd number of unstable (nonminimum phase) zeros [54].

Finally, noting that $c_1(s) = s$ in this example, the 2 DOF unit step $y_1(t)$ response, as defined by the transfer function

$$\frac{y_1(s)}{r(s)} = \frac{c_1(s)q(s) = \alpha s\hat{q}(s)}{\delta^{F^*}(s)\hat{q}(s)} = \frac{\alpha s}{\delta^{F^*}(s)} = \frac{-3.161s}{s^4 + 2.958s^3 + 5.373s^2 + 5.409s + 2.374}$$

is depicted by the dotted line in Figure 9.20. We therefore conclude that this 2 DOF LQR optimal design will ensure acceptable loop and response performance in the event of a $u_1(t)$ or $y_1(t)$ failure.

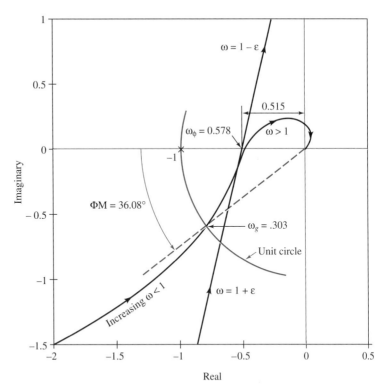

A Nyquist Plot of the LQR Optimal $L(s)$

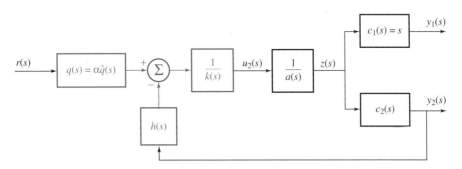

● **FIGURE 9.19**
A 2 DOF LQR Optimal Controller for the Orbiting Satellite

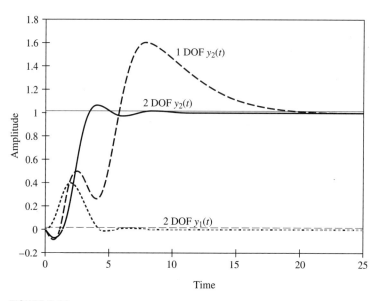

- **FIGURE 9.20**
  A Unit Step Response of a 2 DOF $y_2(t)$ and $y_1(t)$, and a 1 DOF $y_2(t)$

Robust Zero-Error LQR Tracking

We will now determine conditions under which it is possible to obtain *both* an LQR optimal closed-loop transfer function *and* a robust $e_{ss}(t) = 0$ due to the application of a nondiminishing reference input signal, as defined by Eq. (7.2.2), namely

$$r(s) = \mathcal{L}[r(t)] = \frac{m_r(s)}{p_r(s)} \qquad (9.4.25)$$

with all $r = \deg[p_r(s)]$ roots of $p_r(s)$ on the $j\omega$-axis.

To ensure a robust $e_{ss}(t) = 0$, recall that Eqs. (7.3.16) and (7.3.17) both must hold, that is,

$$a(s)k(s) = \bar{m}_e(s)p_r(s) \quad \text{and} \quad h(s) - q(s) = \hat{h}(s)p_r(s)$$

Therefore, we first define $\bar{a}(s)$ as the unique monic least common denominator of $a(s)$ and $p_r(s)$, so that

$$\bar{a}(s) = a(s)\bar{p}_r(s) = \tilde{a}(s)p_r(s) \qquad (9.4.26)$$

with both

$$\bar{n} \overset{\text{def}}{=} \deg[\bar{a}(s)] \geq n \quad \text{and} \quad \bar{r} \overset{\text{def}}{=} \deg[\bar{p}_r(s)] \leq r \qquad (9.4.27)$$

as low as possible. Assuming that $c(s)$ and $\bar{a}(s)$ are coprime, we next solve the "extended" spectral factorization

$$\bar{a}(s)\bar{a}(-s) + \rho c(s)c(-s) = \bar{\delta}^{F^*}(s)\bar{\delta}^{F^*}(-s) \qquad (9.4.28)$$

varying the weighting factor ρ to obtain a desired set of $\bar{n} = n + \bar{r}$ closed-loop poles. Alternatively, we can define $\delta^{F^*}(s)$ using Eq. (9.4.13), and set

$$\bar{\delta}^{F^*}(s) = \delta^{F^*}(s)\delta_r(s) \qquad (9.4.29)$$

for any arbitrary stable polynomial $\delta_r(s)$ of degree $\bar{r}$.

In either case, if the monic polynomial $\hat{q}(s)$ is then defined by $n - 1$ roots of $\delta^{H^*}(s)$, as given by Eq. (9.4.16), the equation

$$\underbrace{a(s)\bar{p}_r(s)}_{\bar{a}(s)}\bar{k}(s) + c(s)h(s) = \bar{\delta}^{F^*}(s)\hat{q}(s) \qquad (9.4.30)$$

has a unique solution, as given by the polynomials

$$\bar{k}(s) = s^{n-1} + \bar{k}_{n-2}s^{n-2} + \cdots + \bar{k}_1 s + \bar{k}_0 \qquad (9.4.31)$$

and

$$h(s) = h_{\bar{n}-1}s^{\bar{n}-1} + h_{\bar{n}-2}s^{\bar{n}-2} + \cdots + h_1 s + h_0 \qquad (9.4.32)$$

Once $h(s)$ has been determined, the polynomial Diophantine equation

$$\hat{q}(s)\bar{q}(s) + p_r(s)\hat{h}(s) = h(s) \qquad (9.4.33)$$

can be solved for the nonunique polynomials $\bar{q}(s)$ and $\hat{h}(s)$, of degrees $\bar{r}$ and $\bar{n} - r - 1$, respectively.

The polynomials

$$h(s) = h(s), \qquad k(s) = \bar{p}_r(s)\bar{k}(s) \quad \text{and} \quad q(s) = \hat{q}(s)\bar{q}(s) \qquad (9.4.34)$$

will then imply a proper 2 DOF compensator characterized by a robust $e_{ss}(t) = 0$, since

$$a(s)k(s) = a(s)\bar{p}_r(s)\bar{k}(s) = \tilde{a}(s)\bar{k}(s)p_r(s) \qquad (9.4.35)$$

which implies Eq. (7.3.16), and Eq. (9.4.33) implies Eq. (7.3.17). The closed-loop system will be characterized by the output response transfer function

$$T(s) = \frac{y(s)}{r(s)} = \frac{c(s)q(s)}{a(s)k(s) + c(s)h(s)} = \frac{c(s)\bar{q}(s)\hat{q}(s)}{\bar{\delta}^{F^*}(s)\hat{q}(s)} = \frac{c(s)\bar{q}(s)}{\bar{\delta}^{F^*}(s)} \qquad (9.4.36)$$

which is similar to the LQR optimal $T(s)$ defined by Eq. (9.4.17).

We finally note that $\hat{q}(s)$ can be replaced in Eqs. (9.4.33) and (9.4.34) by a monic $\tilde{q}(s)$, which is defined by *any* $n - 1$ roots of $\bar{\delta}^{F^*}(s)\hat{q}(s)$. The resulting

$$q(s) = \tilde{q}(s)\bar{q}(s) \qquad (9.4.37)$$

would not affect the loop performance of the system. However, it could imply a faster response performance than the $q(s) = \hat{q}(s)\bar{q}(s)$ of Eq. (9.4.34) if, for example, $\tilde{q}(s)$ were defined by the slowest $n - 1$ roots of $\bar{\delta}^{F^*}(s)\hat{q}(s)$, as we now will illustrate.

. . . .

EXAMPLE 9.4.38 Consider the "generic" DC servomotor defined by the transfer function

$$G(s) = \frac{K_a}{s(s+1)(0.1s+1)} = \frac{10K_a}{s^3 + 11s^2 + 10s} = \frac{c(s)}{a(s)}$$

as in Eq. (8.2.2), with K_a an adjustable amplifier gain. In light of Example 8.2.37, we will determine an LQR optimal controller for the $K_a = 100$ servomotor that will enable it to track both step and ramp inputs with a robust $e_{ss}(t) = 0$, so that $p_r(s) = s^2$ in Eq. (9.4.25).

Since $a(s) = s^3 + 11s^2 + 10s = s(s^2 + 11s + 10)$, it follows that the monic

$$\bar{a}(s) = s^4 + 11s^3 + 10s^2 = \underbrace{(s^3 + 11s^2 + 10s)}_{a(s)}\underbrace{s}_{\bar{p}_r(s)} = \underbrace{(s^2 + 11s + 10)}_{\tilde{a}(s)}\underbrace{s^2}_{p_r(s)}$$

is the least common denominator of $a(s)$ and $p_r(s)$, with $\bar{n} = \deg[\bar{a}(s)] = 4$, and $\bar{r} = \deg[\bar{p}_r(s)] = 1$.

The "extended" spectral factorization defined by Eq. (9.4.28), with $\rho = 0.01$, now implies that

$$\bar{a}(s)\bar{a}(-s) + \rho c(s)c(-s) = (s^4 + 11s^3 + 10s^2)(s^4 - 11s^3 + 10s^2) + 10{,}000 = \bar{\delta}^{F^*}(s)\bar{\delta}^{F^*}(-s)$$

with a LQR optimal

$$\bar{\delta}^{F^*}(s) = \underbrace{s^4 + 14.446s^3 + 53.838s^2 + 103.786s + 100.024}_{(s+10)(s+2.254)(s+1.096 \pm j1.799)}$$

A choice of $\sigma = 0.1$ in Eq. (9.4.16) then implies a

$$\delta^{H^*}(s) = (s + 10.42)(s + 3.65 \pm j4.15)$$

(see Example 9.4.43), so that

$$\hat{q}(s) = (s + 3.65 \pm j4.15) = s^2 + 7.29s + 30.5$$

We now solve the polynomial Eq. (9.4.30) for $\bar{k}(s)$ and $h(s)$:

$$\underbrace{(s^4 + 11s^3 + 10s^2)}_{\bar{a}(s)}\underbrace{(s^2 + \bar{k}_1 s + \bar{k}_0)}_{\bar{k}(s)} + \underbrace{1000}_{c(s)}\underbrace{(h_3 s^3 + h_2 s^2 + h_1 s + h_0)}_{h(s)}$$

$$= s^6 + (11 + \bar{k}_1)s^5 + (\bar{k}_0 + 11\bar{k}_1 + 10)s^4 + (11\bar{k}_0 + 10\bar{k}_1 + 1000h_3)s^3$$
$$+ (10\bar{k}_0 + 1000h_2)s^2 + 1000h_1 s + 1000h_0$$

$$= \underbrace{s^6 + 21.736s^5 + 189.65s^4 + 936.87s^3 + 2498.7s^2 + 3894.7s + 3050.7}_{\bar{\delta}^{F^*}(s)\hat{q}(s)}$$

In this case, we can sequentially determine that $h_0 = 3.051$, $h_1 = 3.895$, $\bar{k}_1 = 10.736$, $\bar{k}_0 = 61.554$, $h_2 = 1.883$, and $h_3 = 0.1524$, so that

$$\bar{k}(s) = \underbrace{s^2 + 10.736s + 61.554}_{(s + 5.368 \pm j5.722)} \quad \text{and} \quad h(s) = \underbrace{0.1524s^3 + 1.883s^2 + 3.895s + 3.051}_{0.1524(s+10)(s + 1.178 \pm j.784)}$$

In light of Eq. (9.4.37), $\tilde{q}(s) = (s + 1.096 \pm j1.799) = s^2 + 2.192s + 4.438$ will be defined by the slowest $n - 1 = 2$ roots of $\bar{\delta}^{F^*}(s)\hat{q}(s)$, with $\bar{q}(s) = \bar{q}_1 s + \bar{q}_0$ of degree $\bar{r} = 1$. Therefore,

$$q(s) = \tilde{q}(s)\bar{q}(s) = (s^2 + 2.192s + 4.438)(\bar{q}_1 s + \bar{q}_0)$$

will satisfy Eq. (9.4.33), that is,

$$\underbrace{(s^2 + 2.192s + 4.438)(\bar{q}_1 s + \bar{q}_0) + s^2 \hat{h}(s)}_{q(s) = q_3 s^3 + q_2 s^2 + q_1 s + q_0} = \underbrace{0.1524s^3 + 1.883s^2 + 3.895s + 3.051}_{h(s)}$$

provided

$$q_0 = 4.438\bar{q}_0 = 3.051 = h_0 \quad \Longrightarrow \quad \bar{q}_0 = 0.687$$

and

$$q_1 = 4.438\bar{q}_1 + 2.192\bar{q}_0 = 3.895 = h_1 \quad \Longrightarrow \quad \bar{q}_1 = 0.538$$

as in Eqs. (7.3.18) and (7.3.19), respectively. This choice for $\bar{q}(s) = 0.538s + 0.687$, hence $q(s)$, together with the $k(s) = \bar{p}_r(s)\bar{k}(s) = s\bar{k}(s)$ and $h(s)$ defined above, as depicted in Figure 9.21, will imply a resulting 2 DOF output response transfer function

$$T(s) = \frac{y(s)}{r(s)} = \frac{c(s)q(s)}{\delta^{F^*}(s)\hat{q}(s)} = \frac{1000(0.538s + 0.687) = 538(s + 1.277)}{\underbrace{s^4 + 19.554s^3 + 142.54s^2 + 538.84s + 688.48}_{(s + 10)(s + 2.254)(s + 3.65 \pm j4.15)}}$$

that is characterized by a robust $e_{ss}(t) = 0$ for both step and ramp inputs.

Figure 9.22 compares the unit step output and plant input responses of this 2 DOF LQR optimal design (the solid lines) with those of the 2 DOF practical PID design of Example 8.2.37 (the dotted lines). The LQR optimal design clearly displays a faster response with less overshoot than the PID design, as well as a lower plant input. In particular, the LQR optimal $max_t\ u(t) = u(0) = 0.538$, as compared with a PID $max_t\ u(t) = u(0) = 1.15$. Moreover, $y(t = 5) = 1.0001$ in the optimal case, which implies a tail error of only 0.01% at $t = 5$ seconds, as compared to a 0.06% tail error at $t = 5$ seconds in the PID case. Plant parameter variations will also have less effect on the tail error in the LQR optimal case, because the slowest "cancelled" optimal poles at $s = -1.096 \pm j1.799$ are farther from the $j\omega$-axis than the slowest "cancelled" PID pole at $s = -0.173$.

Figure 9.23 compares the magnitudes of the loop gains and the sensitivity functions in the two cases. The loop gain magnitude of the LQR optimal design (the solid line) clearly displays both better disturbance rejection and noise attenuation properties than the PID design (the dotted line). Finally, the maximum sensitivity function magnitudes, both of which are ≈ 1.95, imply essentially the same acceptable stability margins in the two cases.

Clearly, the LQR optimal design is superior to the PID design. However, since its dynamic order is greater (by one) than the PID compensator, a control system engineer would have to decide whether the increase in order would justify its actual implementation.

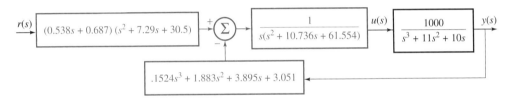

● FIGURE 9.21

2 DOF LQR Optimal Control of the DC Servomotor

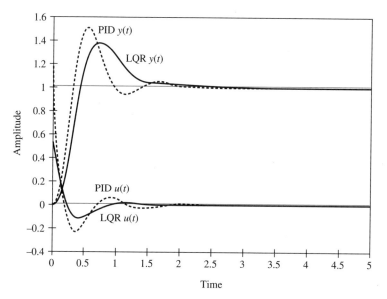

● **FIGURE 9.22**
LQR Optimal and PID Step Response Comparisons

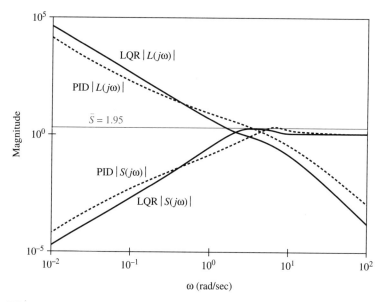

● **FIGURE 9.23**
LQR Optimal and PID $|S(j\omega)|$ and $|L(j\omega)|$ Comparisons

Robust Zero-Error LQR Step Tracking

In the case of step changes in reference input $r(t)$, it generally is possible to obtain the exact LQR optimal closed-loop transfer function defined by Eq. (9.4.17), as well as a robust $e_{ss}(t) = 0$.

If the plant $G(s)$ is type 0, so that $p_r(s) = \bar{p}_r(s) = s$, a choice of

$$\bar{\delta}^{F^*}(s) = \delta^{F^*}(s) \underbrace{(s + \gamma)}_{\delta_r(s)} \tag{9.4.39}$$

as in Eq. (9.4.29), for any arbitrary $\gamma > 0$, then will imply a unique solution $\{\bar{k}(s), h(s)\}$ to Eq. (9.4.30). A subsequent choice of

$$\bar{q}(s) = \alpha(s + \gamma), \qquad \text{with} \qquad \alpha = \frac{h(0) = h_0}{\gamma \hat{q}(0)} \tag{9.4.40}$$

so that

$$q(s) = \bar{q}(s)\hat{q}(s) = \underbrace{\frac{h_0}{\gamma \hat{q}(0)}}_{\alpha}(s + \gamma)\hat{q}(s)$$

in Eq. (9.4.34) will imply that $q(0) = q_0 = h_0$, which will ensure a robust $e_{ss}(t) = 0$ for step inputs in view of Eq. (7.3.18). The resulting 2 DOF output response transfer function will be identical to the LQR optimal $T(s)$ defined by Eq. (9.4.17), since

$$T(s) = \frac{y(s)}{r(s)} = \frac{c(s)q(s)}{a(s)s\bar{k}(s) + c(s)h(s)} = \frac{\alpha c(s)(s + \gamma)\hat{q}(s)}{\delta^{F^*}(s)(s + \gamma)\hat{q}(s)} = \frac{\alpha c(s)}{\delta^{F^*}(s)} \tag{9.4.41}$$

If $G(s)$ is type 1 or higher, $\bar{p}_r(s) = 1$ in Eq. (9.4.26) and Eq. (9.4.30), with $\bar{n} = n$, $\bar{a}(s) = a(s)$, and $\bar{k}(s) = k(s)$ in Eqs. (9.4.31) and (9.4.34). Therefore, $\bar{\delta}^{F^*}(s) = \delta^{F^*}(s)$ in Eq. (9.4.30), and a choice of

$$\bar{q}(s) = \alpha = \frac{h(0) = h_0}{\hat{q}(0)} \tag{9.4.42}$$

in Eq. (9.4.34) will imply a robust $e_{ss}(t) = 0$ for step inputs, as well as the LQR optimal $T(s)$ defined by Eq. (9.4.17).

EXAMPLE 9.4.43 Consider the type 1 DC servomotor of Example 9.4.38, as defined by the $K_a = 100$ transfer function

$$G(s) = \frac{c(s)}{a(s)} = \frac{1000}{s^3 + 11s^2 + 10s} = \frac{10^3}{s(s + 1)(s + 10)}$$

We now will assume that we need only track step (positional) changes in the reference input $r(t)$ with zero steady-state errors.

A root-square locus plot of the zeros of

$$\Delta(s) = a(s)a(-s) + \rho c(s)c(-s) = (s^3 + 11s^2 + 10s)(-s^3 + 11s^2 - 10s) + \rho 10^6$$

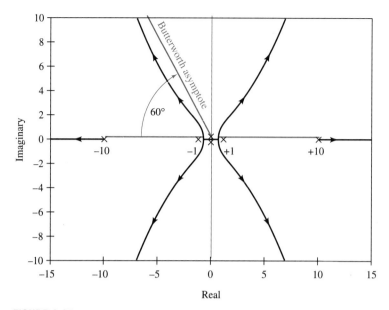

- **FIGURE 9.24**
  A Root-Square Locus Plot

as a function of $\rho \geq 0$ is depicted in Figure 9.24. If $\rho = 1$, which implies equal weighting between large output excursions and the control effort required to constrain such excursions,

$$\Delta(s) = (s^3 + 11s^2 + 10s)(-s^3 + 11s^2 - 10s) + 10^6$$
$$= -s^6 + 101s^4 - 100s^2 + 10^6 = -(s \pm 12.12)(s \pm 5.46 \pm j7.26)$$

or a desired

$$\delta^{F^*}(s) = (s + 12.12)(s + 5.46 \pm j7.26) = s^3 + 23.03s^2 + 214.76s + 1000$$

If the magnitude of the loop gain $|G(j\omega)| = |L(j\omega)|$ of the uncompensated plant should not be increased at the higher frequencies, as in Example 8.2.30, we can base our choice of σ on the requirement that

$$\lim_{\omega \to \infty} \left\{ |H(s = j\omega)| = \left| \frac{h(j\omega)}{k(j\omega)} \right| \right\} = 1$$

or that $|L(j\omega)| = |H(j\omega)G(j\omega)| \to |G(j\omega)|$ as $\omega \to \infty$. Since $\bar{n} = n$ in this case, it follows that $h(s) = h_2 s^2 + h_1 s + h_0$ and $k(s) = \bar{k}(s) = s^2 + k_1 s + k_0$. Therefore, a choice of σ such that

$$\lim_{\omega \to \infty} \left| \frac{h(j\omega)}{k(j\omega)} \right| = \frac{-h_2 \omega^2}{-\omega^2} = h_2 = 1$$

will ensure sufficient attenuation of the measurement noise, as well as robust stability with respect to unmodeled high frequency dynamics.

If $\sigma = 0.1,$[10] Eq. (9.4.16) implies that

$$\bar{\Delta}(s) = -s^6 + 101s^4 - 100s^2 + 10^5 = \underbrace{(s + 10.42)(s + 3.65 \pm j4.15)}_{[\Delta(s)]^+ = \delta^{H^*}(s)} \underbrace{[\Delta(s)]^-}_{\delta^{H^*}(-s)}$$

so that

$$\hat{q}(s) = (s + 3.65 \pm j4.15) = s^2 + 7.29s + 30.5$$

We can now solve the polynomial Eq. (9.4.18) for $k(s)$ and $h(s)$,

$$\underbrace{(s^3 + 11s^2 + 10s)}_{a(s)} \underbrace{(s^2 + k_1 s + k_0)}_{k(s)} + \underbrace{1000}_{c(s)} \underbrace{(h_2 s^2 + h_1 s + h_0)}_{h(s)}$$

$$= s^5 + (k_1 + 11)s^4 + (k_0 + 11k_1 + 10)s^3 + (11k_0 + 10k_1 + 10^3 h_2)s^2$$
$$+ (10k_0 + 10^3 h_1)s + 10^3 h_0$$

$$= \underbrace{s^5 + 30.32s^4 + 413.15s^3 + 3268s^2 + 13840s + 30{,}500}_{\delta^{F^*}(s)\hat{q}(s)}$$

As in Example 9.4.38, we can sequentially determine that $k_1 = 19.32$, $k_0 = 190.63$, $h_2 = 0.98 \approx 1$, $h_1 = 11.93$, and $h_0 = 30.5$, so that

$$k(s) = \underbrace{s^2 + 19.32s + 190.63}_{(s + 9.66 \pm j9.86)} \quad \text{and} \quad h(s) = \frac{0.98s^2 + 11.93s + 30.5}{0.98(s + 3.65)(s + 8.52)}$$

with

$$q(s) = \alpha\hat{q}(s) = \frac{h(0)}{\hat{q}(0)}\hat{q}(s) = \underbrace{\frac{30.5}{30.5}}_{\alpha = 1}\hat{q}(s) = s^2 + 7.29s + 30.5 = \hat{q}(s)$$

These compensator polynomials imply the 2 DOF LQR optimal closed-loop system depicted in Figure 9.25, which is characterized by a robust $e_{ss}(t) = 0$ for step changes in the reference input

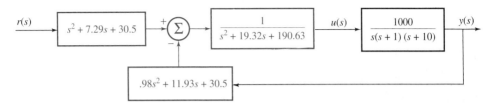

● **FIGURE 9.25**
An LQR Optimal Closed-Loop Design

---

[10]This particular value of 0.1 for σ, which was obtained after three iterations, implies an $h_2 \approx 1$, as we will show.

$r(t)$, the loop gain transfer function

$$L(s) = \frac{c(s)h(s)}{a(s)k(s)} = \frac{1000(0.98s^2 + 11.93s + 30.5)}{s(s^2 + 11s + 10)(s^2 + 19.32s + 190.63)}$$

and the output response transfer function

$$T(s) = \frac{y(s)}{r(s)} = \frac{c(s)q(s)}{a(s)k(s) + c(s)h(s)} = \frac{\alpha c(s)}{\delta^{F^*}(s)} = \frac{1000}{s^3 + 23.03s^2 + 214.76s + 1000}$$

A Bode diagram of both the magnitude and phase of the LQR optimal loop compensator

$$H(s = j\omega) = \frac{h(j\omega)}{k(j\omega)} = \frac{30.5 - 0.98\omega^2 + j11.93\omega}{190.63 - \omega^2 + j19.32\omega}$$

is depicted in Figure 9.26. In light of these plots, we may characterize $H(s)$ as a *double order lead* characterized by complex poles. This is typical of LQR designs, where lag compensation is rarely encountered, because a lag compensator implies a pole close to the origin, hence a slowly decaying "tail" mode, which could excessively increase the $\int_0^\infty \rho y^2(t)\, dt$ contribution to J_{LQR}.

Figure 9.27 displays Bode plots of the loop gain magnitude $|L(j\omega)|$ (the solid line) and the sensitivity function magnitude $|S(j\omega)|$ (the dashed line) associated with this LQR optimal compensator. A $\max_\omega |S(j\omega)| = \bar{S} = 1.893$ at $\omega = 7.9$ implies acceptable stability margins in this case. A Bode plot of $|G(j\omega)|$ (the dotted line) also is depicted to illustrate that $\lim_{\omega \to \infty} \{|L(j\omega)| = |H(j\omega)G(j\omega)|\} \to |G(j\omega)|$, as desired. Finally, Figure 9.28 compares the unit step output and plant input responses of this LQR optimal system (the solid lines) with those associated with the lead compensator of Example 8.2.19 (the dotted lines). Clearly, the optimal output response is superior. However, it does require a higher order (by one) compensator as well as a larger plant input.

9.5   LOW LOOP GAIN (LLG)

The *response performance* of an LQR optimal, closed-loop system can be defined by the output response transfer function

$$T(s) = \frac{y(s)}{r(s)} = \frac{\alpha c(s)}{\delta^{F^*}(s)} \tag{9.5.1}$$

in light of Eq. (9.4.17), completely independent of an observer. However, an LQR optimal observer, as defined by Eq. (9.4.18),

$$a(s)k(s) + c(s)h(s) = \delta^{F^*}(s)\hat{q}(s) \tag{9.5.2}$$

with $\hat{q}(s)$ determined by the weighting factor σ in Eq. (9.4.16), can be used to obtain $T(s)$ by means of a 2 DOF LQR optimal design.

If the given system, as defined by $G(s)$, is minimum phase, σ can be selected somewhat arbitrarily. However, if $G(s)$ has either unstable zeros or unstable poles, the two limiting cases, namely, when $\sigma \to 0$ and when $\sigma \to \infty$, can imply a desirable loop performance as well, as we will show. This section will focus on **low loop gain** (LLG) compensation,

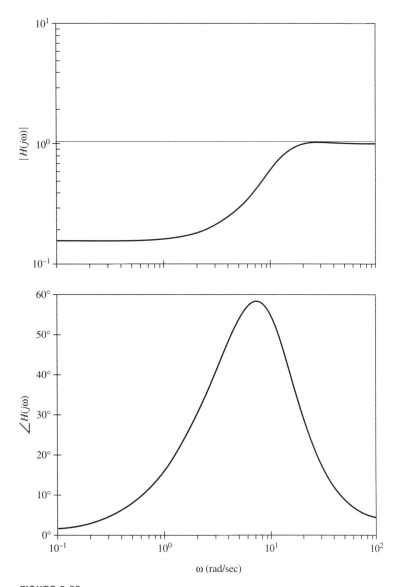

• **FIGURE 9.26**
A Bode Diagram of the LQR Optimal $H(j\omega)$

which is a special case of LQR optimal control, defined when $\sigma = 0$, that can be employed for *stable* systems.

Consider a system defined by the minimal, strictly proper, rational transfer function

$$G(s) = \frac{y(s)}{u(s)} = \frac{c(s)}{a(s)} = \frac{c_m s^m + \cdots + c_1 s + c_0}{s^n + \cdots + a_1 s + a_0} \qquad (9.5.3)$$

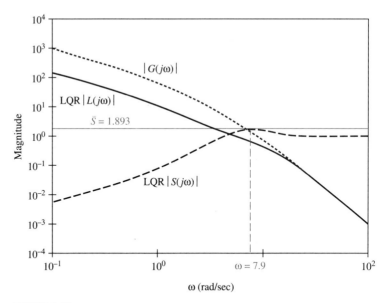

● **FIGURE 9.27**
LQR Optimal Loop Gain and Sensitivity Function Magnitude Plots

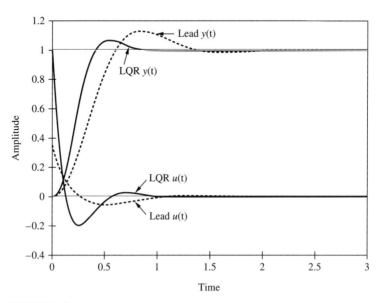

● **FIGURE 9.28**
Unit Step Output and Plant Input Comparisons

with all n roots of $a(s)$ in the stable half-plane $Re(s) < 0$. Assume that an LQR optimal $\delta^{F^*}(s)$ has been determined using Eq. (9.4.13). If $\sigma = 0$ in Eq. (9.4.16), it follows that

$$[\bar{\Delta}(s)]^+ = \delta^{H^*}(s) = a(s) \tag{9.5.4}$$

We will therefore define an **LLG optimal observer** by setting $\hat{q}(s) = a(s)^{11}$ in Eq. (9.5.2) and solving the resulting equation for nth degree polynomials $k(s)$ and $h(s)$; that is, the substitution of $a(s)$ for $\hat{q}(s)$ in Eq. (9.5.2) implies that

$$c(s)h(s) = a(s)[\delta^{F^*}(s) - k(s)] \tag{9.5.5}$$

and since $a(s)$ and $c(s)$ are coprime, it follows that

$$h(s) = h_n s^n + h_{n-1} s^{n-1} + \cdots + h_1 s + h_0 = h_n a(s) \tag{9.5.6}$$

for some scalar h_n. Therefore, in view of Eq. (9.5.5),

$$k(s) = s^n + k_{n-1} s^{n-1} + \cdots + k_1 s + k_0 = \delta^{F^*}(s) - h_n c(s) \tag{9.5.7}$$

If $q(s) = \alpha \hat{q}(s) = \hat{\alpha} h_n a(s)$, with $\hat{\alpha} = \alpha/h_n$, the polynomial $h_n a(s)$ can be placed within the loop "above" $k(s) = \delta^{F^*}(s) - h_n c(s)$, thereby implying the unity feedback, LLG optimal observer configuration depicted in Figure 9.29. Note that the scalar h_n directly determines the magnitude of the loop gain

$$L(s) = \frac{h_n a(s) c(s)}{a(s)[\delta^{F^*}(s) - h_n c(s)]} = \frac{h_n c(s)}{\delta^{F^*}(s) - h_n c(s)} \tag{9.5.8}$$

Moreover, h_n can be arbitrarily assigned, since the n coefficients of $k(s)$ will ensure Eq. (9.5.7) independent of $h_n c(s)$.

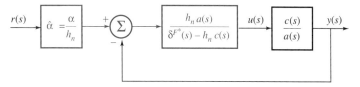

- **FIGURE 9.29**
A Unity Feedback LLG Optimal Observer Configuration

As $h_n \to 0$, the closed-loop system of Figure 9.29 will approach the open-loop system of Figure 9.30, where the plant poles, as defined by the roots of $a(s)$, are cancelled by the compensator zeros and replaced by optimal (open-loop) poles, as defined by the roots of $\delta^{F^*}(s)$. Note that the

---

[11] Any nominal roots of $a(s)$ "close to" the $j\omega$-axis, which may cross the axis due to parameter variations, should be excluded from $\hat{q}(s)$.

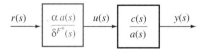

● **FIGURE 9.30**
The $h_n \to 0$ Limiting LLG Case

Figure 9.30 system is characterized by a sensitivity function $S(s = j\omega) = 1$ (in which $s = j\omega$) at all frequencies, that is, using Eq. (5.1.16),

$$\lim_{h_n \to 0} S(s) = \lim_{h_n \to 0} \left\{ \frac{a(s)k(s)}{a(s)k(s) + h_n a(s)c(s)} \right\} \to \frac{a(s)k(s)}{a(s)k(s)} = 1 \quad (9.5.9)$$

which is characteristic of an open-loop system. However, one would rarely set $h_n = 0$, because none of the benefits associated with feedback compensation would be obtained.

LLG/IMP Robust Zero-Error Tracking

The preceding discussion now serves to motivate a LLG closed-loop design procedure for stable systems that invokes the internal model principle (IMP) to ensure robust zero-error tracking in addition to low loop sensitivity, hence desirable stability margins. The resulting LLG/IMP design also implies robust stability with respect to unmodeled high frequency dynamics, as well as desirable noise attenuation properties.

As we will show, a LLG/IMP design can be particularly effective in nonminimum phase cases, which are characterized by transport lag or plant zeros in the unstable half-plane $Re(s) > 0$. In such cases, a low loop gain prevents the stable plant poles from moving "too close" to the unstable plant zeros. The LLG/IMP design is based on the robust zero error LQR tracking results presented in Section 9.4, when $\sigma = 0$, so that $\delta^{H^*}(s) = a(s)$, as in Eq. (9.5.4).

To begin, let us assume that the reference input $r(t)$ is a known, nondiminishing signal,

$$r(s) = \mathcal{L}[r(t)] = \frac{m_r(s)}{p_r(s)}$$

as in Eq. (9.4.25), with all r roots of $p_r(s)$ on the $j\omega$-axis, so that $a(s)$ and $p_r(s)$ are coprime. If $p_r(s)$ and $c(s)$ are also coprime, we can employ the "extended" spectral factorization Eq. (9.4.28), namely,

$$\underbrace{a(s)p_r(s)}_{\bar{a}(s)} \underbrace{a(-s)p_r(-s)}_{\bar{a}(-s)} + \rho c(s)c(-s) = \bar{\delta}^{F^*}(s)\bar{\delta}^{F^*}(-s) \quad (9.5.10)$$

varying the weighting factor ρ until a desired monic $\bar{\delta}^{F^*}(s)$ of degree $n+r$ is obtained.

We now replace $\delta^{F^*}(s)$ by $\bar{\delta}^{F^*}(s)$ and $\hat{q}(s)$ by $a(s)$ in Eq. (9.5.2), noting that $a(s)$ must divide $h(s)$, so that

$$a(s)\underbrace{p_r(s)\bar{k}(s)}_{k(s)} + c(s)\underbrace{a(s)\bar{h}(s)}_{h(s)} = \bar{\delta}^{F^*}(s)a(s) \tag{9.5.11}$$

As a consequence,

$$p_r(s)\bar{k}(s) + c(s)\bar{h}(s) = \bar{\delta}^{F^*}(s) \tag{9.5.12}$$

has a unique solution, as defined by a monic

$$\bar{k}(s) = s^n + \bar{k}_{n-1}s^{n-1} + \cdots + \bar{k}_1 s + \bar{k}_0 \tag{9.5.13}$$

of degree n, and an

$$\bar{h}(s) = \bar{h}_{r-1}s^{r-1} + \ldots + \bar{h}_1 s + \bar{h}_0 \tag{9.5.14}$$

of degree $r - 1$.

Once $h(s) = a(s)\bar{h}(s)$ has been determined, Eq. (7.3.17) then implies that the polynomial equation

$$a(s)\bar{q}(s) + p_r(s)\hat{h}(s) = a(s)\bar{h}(s) = h(s) \tag{9.5.15}$$

can be solved for the nonunique polynomials $\bar{q}(s)$ and $\hat{h}(s)$, of degrees r and n, respectively. Since $p_r(s)$ and $a(s)$ are coprime, however,

$$\hat{h}(s) = \hat{h}_n s^n + \cdots + \hat{h}_1 s + \hat{h}_0 = \hat{h}_n a(s) \tag{9.5.16}$$

so that

$$\bar{q}(s) = \bar{h}(s) - \hat{h}_n p_r(s) \tag{9.5.17}$$

with $\hat{h}_n$ the only adjustable parameter.

A subsequent choice of the 2 DOF compensator polynomials

$$k(s) = p_r(s)\bar{k}(s), \qquad h(s) = a(s)\bar{h}(s),$$

$$\text{and} \quad q(s) = \delta^{H^*}(s)\bar{q}(s) = a(s)\bar{q}(s) \tag{9.5.18}$$

will then imply a proper 2 DOF compensator characterized by a robust $e_{ss}(t) = 0$, since

$$a(s)k(s) = a(s)p_r(s)\bar{k}(s) \tag{9.5.19}$$

will imply Eq. (7.3.16), and Eq. (9.5.17) will imply Eq. (7.3.17).

Since $a(s) = \delta^{H^*}(s)$ is common to both $q(s)$ and $h(s)$ it can be placed within the loop "above" $k(s) = p_r(s)\bar{k}(s)$, thereby implying the closed-loop compensator depicted in Figure 9.31. The resulting closed-loop system will be characterized by the output response transfer function

$$T(s) = \frac{y(s)}{r(s)} = \frac{c(s)\bar{q}(s)}{p_r(s)\bar{k}(s) + c(s)\bar{h}(s)} = \frac{c(s)\bar{q}(s)}{\bar{\delta}^{F^*}(s)} \tag{9.5.20}$$

which is similar to the LQR optimal $T(s)$ defined by Eq. (9.4.17).

Moreover, the magnitude of the sensitivity function

$$S(s) = \frac{p_r(s)\bar{k}(s)}{p_r(s)\bar{k}(s) + c(s)\bar{h}(s)} = \frac{p_r(s)\bar{k}(s)}{\bar{\delta}^{F^*}(s)} \tag{9.5.21}$$

will be zero at those frequencies ω where $p_r(j\omega) = 0$[12] and will not get too "large" provided $c(s)\bar{h}(s)$ is "small." We can ensure such a low loop gain (LLG) condition by choosing a "small" ρ in Eq. (9.5.10) that, referring to Eq. (9.5.12), would imply a

$$\bar{\delta}^{F^*}(s) \approx p_r(s)a(s) \approx p_r(s)\bar{k}(s), \qquad \text{with} \qquad c(s)\bar{h}(s) \approx 0$$

In general, design trade-offs via ρ can be made between the response performance of the system, as defined by the $T(s)$ of Eq. (9.5.20), and the loop performance of the system, as defined by the $S(s)$ of Eq. (9.5.21).

LLG/IMP Robust Zero-Error Step Tracking

In the case of step changes in the reference input, that is, when $p_r(s) = s$ and $c(0) = c_0 \neq 0$, it is possible to obtain both a robust $e_{ss}(t) = 0$, and the desired LQR optimal $T(s)$ defined by Eq. (9.5.1). In light of Eq. (9.4.29), we first set $\bar{\delta}^{F^*}(s) = \delta^{F^*}(s)(s+\gamma)$ in Eq. (9.5.12), for any arbitrary $\gamma > 0$, and solve the resulting equation for $\bar{k}(s)$, as given by Eq. (9.5.13), and a corresponding $\bar{h}(s) = \bar{h}_0$, as defined by Eq. (9.5.14),

$$s \underbrace{[s^n + \bar{k}_{n-1}s^{n-1} + \cdots + \bar{k}_1 s + \bar{k}_0]}_{\bar{k}(s)} + \bar{h}_0 c(s) = \delta^{F^*}(s)\underbrace{(s+\gamma)}_{\delta_r(s)} = \bar{\delta}^{F^*}(s)$$

$$\tag{9.5.22}$$

We can then set

$$\bar{q}(s) = \bar{h}(s) - \hat{h}_n p_r(s) = \bar{h}_0 - \hat{h}_n s = \bar{q}_0 + \bar{q}_1 s = \alpha(s+\gamma) = \frac{\bar{h}_0}{\gamma}(s+\gamma)$$

$$\tag{9.5.23}$$

using Eq. (9.5.17), so that $\bar{q}_0 = \bar{h}_0$. Since $h(s) = \bar{h}_0 a(s)$ and $q(s) = \bar{q}(s)a(s)$ in this case, it follows that

$$q(0) = a(0)\bar{q}_0 = \bar{h}_0 a(0) = h(0) \tag{9.5.24}$$

which will ensure a robust $e_{ss}(t) = 0$ for step changes in the input.

The resulting 2 DOF closed-loop design of Figure 9.31 can then be represented by the equivalent unity feedback design of Figure 9.32 by placing $\bar{h}(s) = \bar{h}_0$ in the forward path "above" $k(s) = s\bar{k}(s)$. The output

---

[12]Therefore, if certain of the roots of $p_r(s)$ correspond to output disturbance modes, this LLG/IMP design will imply a *complete disturbance rejection* of these modes.

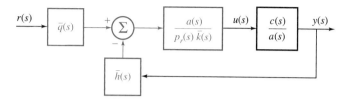

- **FIGURE 9.31**
  2 DOF LLG/IMP Compensation

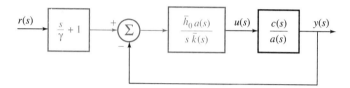

- **FIGURE 9.32**
  2 DOF Unity Feedback LLG/IMP Compensation for Step Inputs

response transfer function will be identical to that defined by Eq. (9.4.17),

$$T(s) = \frac{y(s)}{r(s)} = \underbrace{\frac{\bar{h}_0}{\gamma}}_{\alpha} \frac{c(s)(s+\gamma)}{s\bar{k}(s) + \bar{h}_0 c(s)} = \frac{\alpha c(s)(s+\gamma)}{\delta^{F^*}(s)(s+\gamma)} = \frac{\alpha c(s)}{\delta^{F^*}(s)}$$

$$(9.5.25)$$

Moreover, the sensitivity function

$$S(s) = \frac{s\bar{k}(s)}{s\bar{k}(s) + \bar{h}_0 c(s)} = \frac{s\bar{k}(s)}{(s+\gamma)\delta^{F^*}(s)} \approx \frac{s}{s+\gamma} \overset{\text{def}}{=} S_\gamma(s) \quad (9.5.26)$$

if $\bar{k}(s) \approx \delta^{F^*}(s)$, which will be the case if

$$\bar{h}_0 = \gamma \underbrace{\frac{\delta^{F^*}(0)}{c(0)}}_{\alpha} \qquad (9.5.27)$$

as defined by Eq. (9.5.22) is "small." A low loop gain (LLG) value for $\bar{h}_0$ can be obtained by choosing a "small" value for γ. Obviously, these magnitude measures, and the resulting sensitivity approximation $S_\gamma(s)$, are dependent on the numerical values associated with a particular example, and design trade-offs will often be required to determine an acceptable compensator.

The Waterbed Effect

It is of interest to note that when a system with unstable zeros is compensated by some $H(s)$, so that $|S(j\omega)| < 1$ over some frequency band, for example, when $L(s) = G(s)H(s)$ is type 1 or higher, then $|S(j\omega)| > 1$ over another frequency band, as depicted in Figure 9.33. In such cases, any closed-loop compensator is constrained by the so-called **waterbed effect**,[13] which states that it is impossible to obtain a $|S(j\omega)| < 1$ (the level of the "waterbed") over one frequency band without having $|S(j\omega)| > 1$ over another frequency band. The best that can be done is to "flatten out" $|S(j\omega)|$ over the band of frequencies where $|S(j\omega)| > 1$, as depicted by the dashed line in Figure 9.33, to prevent "peaking," as depicted by the solid line in Figure 9.33 at $\omega = \omega_p$, where $|S(j\omega)| > 2$, which usually would imply unacceptable stability margins. In view of Eq. (9.5.26), a LLG/IMP design represents one way of minimizing the $\mathbf{H}_\infty$-norm of the sensitivity function $S(s)$, thereby negating the potential peaking consequences associated with the waterbed effect.

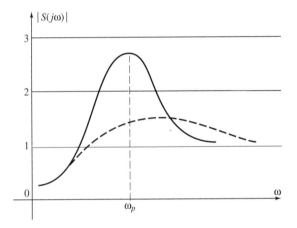

• **FIGURE 9.33**
The Waterbed Effect

EXAMPLE 9.5.28 To illustrate the waterbed effect, consider the design of an LLG/IMP control system for automotive fuel metering, which is motivated by the $\mathbf{H}_\infty$ design outlined by Kuraoka et al. [42]. At a temperature of 25°C, the nominal "plant" is defined by the stable, but *nonminimum* phase transfer function

$$G(s) = \frac{c(s)}{a(s)} = \frac{5.5s^2 + 401s - 444,400}{s^3 + 93.72s^2 + 9520s + 121,400} = \frac{5.5(s+323)(s-250)}{(s+14.5)(s+39.6 \pm j82.5)}$$

---

[13] A formal statement of the waterbed effect extends beyond the scope of this text and, therefore, will not be presented here. The interested reader is referred to References [24] and [19] for additional details.

To obtain a desirable closed-loop response characterized by a robust $e_{ss}(t) = 0$ for step changes in the reference input, we first determine an appropriate $\delta^{F^*}(s)$.

If $\rho = 1$, Eq. (9.4.13) implies that

$$a(s)a(-s) + c(s)c(-s) = (s \pm 46.2)(s \pm 43.5 \pm j89.9) = \delta^{F^*}(s)\delta^{F^*}(-s)$$

so that

$$\delta^{F^*}(s) = (s + 46.2)(s + 43.5 \pm j89.9) = s^3 + 133s^2 + 14{,}013s + 461{,}840$$

A resulting unit step response plot of

$$y(t) = \mathcal{L}^{-1}\left\{ y(s) = \frac{\alpha c(s)}{\delta^{F^*}(s)}\frac{1}{s} \right\}$$

as depicted in Figure 9.34, when

$$\alpha = \frac{\delta^{F^*}(0)}{c(0)} = \frac{461{,}840}{-444{,}400} = -1.039$$

displays very desirable response characteristics, so that this particular $\delta^{F^*}(s)$ will be selected.

If we now choose a $\gamma = 10$, which is "small" relative to the coefficients of $\delta^{F^*}(s)$, Eq. (9.5.22) implies that

$$s\underbrace{(s^3 + \bar{k}_2 s^2 + \bar{k}_1 s + \bar{k}_0)}_{\bar{k}(s)} + \bar{h}_0 \underbrace{(5.5s^2 + 401s - 444{,}400)}_{c(s)}$$

$$= s^4 + \bar{k}_2 s^3 + (\bar{k}_1 + 5.5\bar{h}_0)s^2 + (\bar{k}_0 + 401\bar{h}_0)s - 444{,}400\bar{h}_0$$

$$= (s + \gamma)\delta^{F^*}(s) = (s + 10)(s^3 + 133s^2 + 14{,}013s + 461{,}840)$$

$$= s^4 + 143s^3 + 15{,}345s^2 + 601{,}970s + 4{,}618{,}400$$

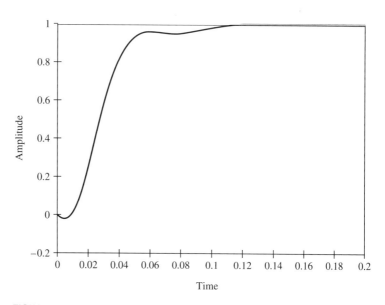

● **FIGURE 9.34**
The Unit Step Response of the $\rho = 1$ LLG/IMP Design

We therefore sequentially determine that $\bar{h}_0 = -10.39$, $\bar{k}_2 = 143$, $\bar{k}_1 = 15{,}402$, and $\bar{k}_0 = 606{,}140$, or that

$$k(s) = s\bar{k}(s) = s^4 + 143s^3 + 15{,}402s^2 + 606{,}140s$$

with $\bar{h}_0 = -10.39$ "small" relative to the other loop coefficients.

Bode diagrams of

$$G(s), \qquad H(s) = \frac{h(s)}{k(s)} = \frac{\bar{h}_0 a(s)}{s\bar{k}(s)}, \quad \text{and} \quad L(s) = G(s)H(s)$$

as depicted in Figure 9.35, display a $\Phi M \approx 80°$ at a gain crossover frequency $\omega_g \approx 7.5$ and a $GM \approx 1/0.125 = 8$ at a phase crossover frequency $\omega_\phi \approx 62$. These desirable stability margins are verified by a Bode plot of the magnitude of the actual sensitivity function

$$S(s) = \frac{s\bar{k}(s)}{(s+\gamma)\delta^{F^*}(s)} = \frac{s^4 + 143s^3 + 15{,}402s^2 + 606{,}140s}{(s+10)(s^3 + 133s^2 + 14{,}013s + 461{,}840)}$$

which is depicted by the solid line in Figure 9.36, along with a magnitude plot of the approximate sensitivity function

$$S_\gamma(s) = \frac{s}{s+10}$$

(the dashed line), as defined by Eq. (9.5.26). These magnitude plots do reflect a "flattened out" $|S(j\omega)| \approx |S_\gamma(j\omega)|$ in this LLG case, which tends to minimize the waterbed effect. Moreover, $\bar{S} = \max_\omega |S(j\omega)| = 1.18$, which is consistent with the stability margins noted above.

9.6 LOOP TRANSFER RECOVERY (LTR)

In Section 9.5 we showed that in the $\sigma \to 0$ limiting case of the spectral factorization defined by Eq. (9.4.16), a LLG design can produce desirable closed-loop performance for stable plants, especially those characterized by unstable zeros. In this section, we will show that **loop transfer recovery** (LTR) compensation [18] [58], which is a special case of LQR optimal control defined in the converse limiting case when $\sigma \to \infty$, can imply desirable closed-loop performance for plants whose zeros are stable, including those characterized by unstable poles.

As we will show, an LTR design "approximates" or *recovers* the *loop transfer function* associated with actual LQR optimal, linear state variable feedback (LSVF). To motivate LTR compensation, we will first derive an important property of the sensitivity function that characterizes an LQR optimal LSVF compensated system.

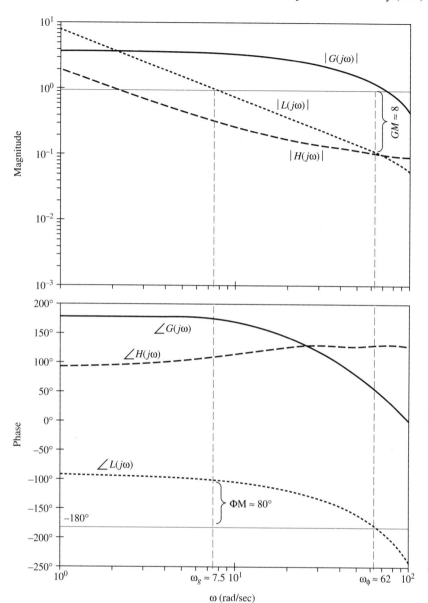

• FIGURE 9.35
Bode Diagrams of $G(j\omega)$, $H(j\omega)$, and $L(j\omega)$

The LQR Optimal LSVF Sensitivity Function

If a system is defined by the open-loop $G(s)$ of Eq. (9.4.2), then using Eq. (9.4.11), the Laplace transformed LQR optimal LSVF control law

$$u^*(s) = f^*(s)z(s) = [a(s) - \delta^{F^*}(s)]z(s) \qquad (9.6.1)$$

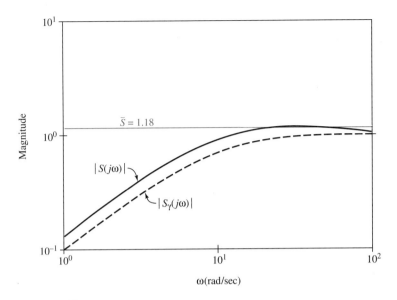

• **FIGURE 9.36**
Bode Plots of $|S(j\omega)|$ and $|S_\gamma(j\omega)|$

minimizes the LQR performance index defined by Eq. (9.4.1). The employment of the actual LQR optimal LSVF control law Eq. (9.6.1) implies the closed-loop system depicted in Figure 9.37, which is characterized by an LQR optimal LSVF loop gain

$$L(s) = -\frac{f^*(s)}{a(s)} \qquad (9.6.2)$$

and a corresponding LQR optimal LSVF sensitivity function

$$S(s) = \frac{1}{1 + L(s)} = \frac{a(s)}{a(s) - f^*(s)} = \frac{a(s)}{\delta^{F^*}(s)} \qquad (9.6.3)$$

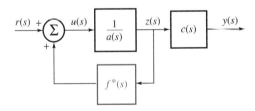

• **FIGURE 9.37**
Actual LQR Optimal LSVF Compensation

Referring to Eqs. (9.4.3) and (9.4.13), it now follows that

$$|S(j\omega)|^2 = \left| \frac{a(j\omega)}{\delta^{F^*}(j\omega)} \right|^2 = \frac{a(j\omega)a(-j\omega)}{\underbrace{a(j\omega)a(-j\omega) + \rho c(j\omega)c(-j\omega)}_{\delta^{F^*}(j\omega)\delta^{F^*}(-j\omega)}}$$

$$= \frac{|a(j\omega)|^2}{|a(j\omega)|^2 + \rho|c(j\omega)|^2} \leq 1 \quad \forall \quad \omega \tag{9.6.4}$$

which establishes the fact that the magnitude of an LQR optimal LSVF sensitivity function *never* exceeds unity. Moreover, since $\deg[f^*(s)] = n - 1 < n = \deg[a(s)]$ in Eq. (9.6.2), it follows that $|L(j\omega)| \to 0$ as $\omega \to \infty$. Therefore,

$$\|S\|_\infty = \max_\omega |S(j\omega)| = \bar{S} = 1 \tag{9.6.5}$$

if an LQR optimal LSVF control law is employed, which implies an infinite GM and a $\Phi M \geq 60°$, in light of Eq. (6.1.27) and Eq. (6.1.28), respectively.

As we noted earlier, however, an actual LSVF control law usually is impossible to implement in practice, an observation that now serves to motivate optimal observer based LTR compensation.

LTR Compensation

Consider a system defined by the minimal, strictly proper, rational transfer function

$$G(s) = \frac{y(s)}{u(s)} = \frac{c(s)}{a(s)} = \frac{c_m s^m + \cdots + c_1 s + c_0}{s^n + \cdots + a_1 s + a_0} \tag{9.6.6}$$

with all m zeros of $c(s)$ in the stable half-plane $Re(s) < 0$. As in the case of LLG compensation, assume that an LQR optimal $\delta^{F^*}(s)$ has been determined by means of Eq. (9.4.13). In light of Eq. (9.4.16), we then note that as $\sigma \to \infty$, m poles of an LQR optimal observer will move to the zeros of $c(s)$, while the remaining poles will approach ∞ in a Butterworth configuration.

We will therefore derive an LTR optimal observer of order $n - 1$ by setting the polynomial

$$\hat{q}(s) = c(s)\tilde{\delta}^{H^*}(s) \tag{9.6.7}$$

in Eq. (9.4.18), where $\tilde{\delta}^{H^*}(s)$ is a polynomial of degree $n - m - 1$ in *corner frequency factored form*, whose roots lie in a Butterworth configuration at an arbitrary distance from the origin in the stable half-plane $Re(s) < 0$. As a consequence,

$$a(s)k(s) + c(s)h(s) = \delta^{F^*}(s)c(s)\tilde{\delta}^{H^*}(s) \tag{9.6.8}$$

has a unique solution, as given by the compensator polynomials

$$h(s) = h_{n-1}s^{n-1} + \cdots + h_1 s + h_0 \tag{9.6.9}$$

and

$$k(s) = c(s) \underbrace{[\tilde{k}_{n-m-1}s^{n-m-1} + \cdots \tilde{k}_1 s + \tilde{k}_0]}_{\tilde{k}(s)} \tag{9.6.10}$$

Note that $c(s)$ must divide $k(s)$ since $a(s)$ and $c(s)$ are coprime. It therefore follows that the LTR observer poles defined by the roots of $c(s)$ represent closed-loop modes that are *both* uncontrollable and unobservable (see Problem 9-19).

If $c(s)$ is factored out of Eq. (9.6.8), $h(s)$ and $\tilde{k}(s)$ can be obtained by means of the relationship

$$a(s)\tilde{k}(s) + h(s) = \delta^{F^*}(s)\tilde{\delta}^{H^*}(s) = [a(s) - f^*(s)]\tilde{\delta}^{H^*}(s) \tag{9.6.11}$$

Therefore, if

$$q(s) = \alpha\hat{q}(s) = \alpha c(s)\tilde{\delta}^{H^*}(s) \tag{9.6.12}$$

an **LTR optimal observer**, as defined by the triple $\{h(s), k(s), q(s)\}$, will imply the closed-loop system depicted in Figure 9.38, which is characterized by the LQR optimal transfer function

$$T(s) = \frac{y(s)}{r(s)} = \frac{c(s)\alpha\hat{q}(s)}{a(s)c(s)\tilde{k}(s) + c(s)h(s)} = \frac{\alpha c(s)\tilde{\delta}^{H^*}(s)}{\delta^{F^*}(s)\tilde{\delta}^{H^*}(s)} = \frac{\alpha c(s)}{\delta^{F^*}(s)} \tag{9.6.13}$$

In this LTR case, the loop gain

$$L(s) = \frac{c(s)h(s)}{a(s)c(s)\tilde{k}(s)} = \frac{h(s)}{a(s)\tilde{k}(s)} \tag{9.6.14}$$

implies the sensitivity function

$$S(s) = \frac{1}{1 + L(s)} = \frac{a(s)\tilde{k}(s)}{a(s)\tilde{k}(s) + h(s)} = \frac{a(s)\tilde{k}(s)}{\delta^{F^*}(s)\tilde{\delta}^{H^*}(s)} \tag{9.6.15}$$

Referring to Eqs. (4.4.2) and (4.4.3), the farther into the left-half s-plane we position the roots of $\tilde{\delta}^{H^*}(s)$, the closer the corner frequency

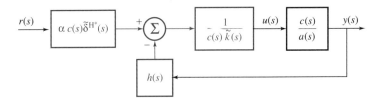

• **FIGURE 9.38**
An LTR Optimal Closed-Loop System

factored form $\tilde{\delta}^{H^*}(s) \to 1$. Therefore, Eq. (9.6.11) implies that in this limiting case,

$$a(s)\tilde{k}(s) + h(s) = \delta^{F^*}(s)\tilde{\delta}^{H^*}(s) \to \delta^{F^*}(s) = a(s) - f^*(s) \quad (9.6.16)$$

and since $a(s)$ and $\delta^{F^*}(s)$ are both monic polynomials of degree $n > \deg[h(s)] = n - 1$,

$$\tilde{k}(s) \to \tilde{\delta}^{H^*}(s) \to 1 \quad \text{and} \quad h(s) \to -f^*(s) \quad (9.6.17)$$

Note further that as $\tilde{\delta}^{H^*}(s) \to 1$, Eq. (9.6.15) implies that

$$S(s) \to \frac{a(s)}{\delta^{F^*}(s)}$$

the LQR optimal LSVF sensitivity function defined by Eq. (9.6.3), and

$$L(s) = \frac{h(s)}{a(s)\tilde{k}(s)} \to \frac{-f^*(s)}{a(s)} \quad (9.6.18)$$

the LQR optimal LSVF loop gain defined by Eq. (9.6.2).

....

EXAMPLE 9.6.19 Consider a third-order system defined by the controllable canonical form state-space representation

$$\underbrace{\begin{bmatrix} \dot{x}_1(t) \\ \dot{x}_2(t) \\ \dot{x}_3(t) \end{bmatrix}}_{\dot{\mathbf{x}}(t)} = \underbrace{\begin{bmatrix} 0 & 1 & 0 \\ 0 & 0 & 1 \\ -8 & 10 & -1 \end{bmatrix}}_{A} \underbrace{\begin{bmatrix} x_1(t) \\ x_2(t) \\ x_3(t) \end{bmatrix}}_{\mathbf{x}(t)} + \underbrace{\begin{bmatrix} 0 \\ 0 \\ 1 \end{bmatrix}}_{B} u(t); \qquad y(t) = \underbrace{[3, \ 1, \ 0]}_{C}\mathbf{x}(t)$$

Referring to Eq. (2.7.22), the system is equivalent to the differential operator system

$$\underbrace{(D^3 + D^2 - 10D + 8)}_{a(D)} z(t) = u(t); \qquad y(t) = \underbrace{(D+3)}_{c(D)} z(t)$$

which is characterized by the minimal transfer function

$$G(s) = \frac{y(s)}{u(s)} = C(sI - A)^{-1}B = \frac{c(s)}{a(s)} = \frac{s+3}{s^3 + s^2 - 10s + 8} = \frac{s+3}{(s-1)(s-2)(s+4)}$$

which has a stable zero at $s = -3$, a stable pole at $s = -4$, and two unstable poles at $s = +1$ and $+2$.

If $\rho = 10$ in Eq. (9.4.13),

$$a(s)a(-s) + 10c(s)c(-s) = (s \pm 3.95)(s \pm 1.71 \pm j.46) = \delta^{F^*}(s)\delta^{F^*}(-s)$$

so that

$$\delta^{F^*}(s) = (s+3.95)(s+1.71 \pm j.46) = s^3 + 7.37s^2 + 16.65s + 12.4$$

Equation Eq. (9.4.17) then implies an LQR optimal

$$T(s) = \frac{y(s)}{r(s)} = \frac{\alpha c(s)}{\delta^{F^*}(s)} = \frac{4.13(s+3)}{s^3 + 7.37s^2 + 16.65s + 12.4}$$

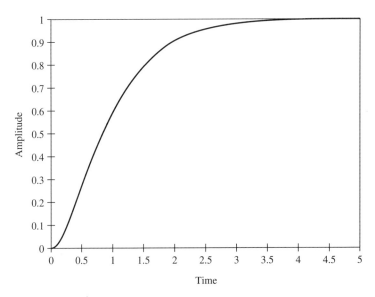

• **FIGURE 9.39**

The Unit Step Response of $T(s) = \dfrac{\alpha c(s)}{\delta^{F^*}(s)}$

with $\alpha = 4.13$, so that $\alpha c(0) = \delta^{F^*}(0) = 12.4$. The resulting unit step response, as depicted in Figure 9.39, displays a non-robust $y(t) \to 1 = r(t)$ as $t \to \infty$, with acceptable response times and no overshoot.

To obtain this $T(s)$, we will now determine an LTR observer of order $n - 1 = 2$. Eq. (9.6.7) implies that

$$\hat{q}(s) = \underbrace{(s+3)}_{c(s)} \underbrace{(\tau s + 1)}_{\tilde{\delta}^{H^*}(s)}$$

for some arbitrarily small positive scalar τ. The reader can now verify (see Problem 9-20) that if $\tau = 0.1$, the polynomial pair

$$\tilde{k}(s) = 0.1s + 1.64 \quad \text{and} \quad h(s) = 8.4s^2 + 33.46s - 0.67$$

solves Eq. (9.6.11), and if $\tau = 0.01$, the polynomial pair

$$\tilde{k}(s) = 0.01s + 1.06 \quad \text{and} \quad h(s) = 6.57s^2 + 27.33s + 3.89$$

solves Eq. (9.6.11).

Bode plots of the sensitivity function magnitudes in these two cases, as defined by Eq. (9.6.15), are depicted in Figure 9.40 by the dashed ($\tau = 0.1$) and dotted ($\tau = 0.01$) lines. They display a $\max_\omega |S(j\omega)| = \bar{S} = 1.6$ and 1.06, respectively, hence acceptable stability margins in both cases. The solid line depicts a magnitude plot of the LQR optimal LSVF sensitivity function, as defined by Eq. (9.6.3), for comparison purposes.

Bode plots of the loop gain magnitudes in these two cases, as defined by Eq. (9.6.14), are depicted in Figure 9.41 by the dashed ($\tau = 0.1$) and dotted ($\tau = 0.01$) lines. The solid line depicts

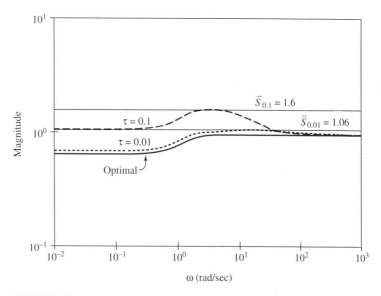

- **FIGURE 9.40**
Magnitude Plots of the Sensitivity Functions

a magnitude plot of the LQR optimal LSVF loop gain, as defined by Eq. (9.6.2), for comparison purposes. Note that the magnitude plots associated with the LTR optimal observer approach the actual LSVF optimal plots as $\tau \to 0$, that is, as the LTR observer pole at $s = -1/\tau$ is moved farther into the left-half s-plane.

Nyquist plots of the loop gain

$$L(j\omega) = \frac{h(j\omega)}{a(j\omega)\tilde{k}(j\omega)}$$

in these two cases are depicted in Figure 9.42 by the dashed ($\tau = 0.1$) and dotted ($\tau = 0.01$) lines. They display the first (for $\omega > 0$) of the two counterclockwise encirclements of the -1 point that are required to ensure closed-loop stability in this case via the Nyquist stability criterion, so that

$$Z = N + P = -2 + 2 = 0$$

The solid line represents the lower half of a unit circle centered at the -1 point.

We now recall the Section 6.1 observation that

> a polar plot of $L(j\omega)$ will just contact, but fail to penetrate, a circle of radius $\bar{S}^{-1}$, centered at -1 in the complex $L(j\omega)$-plane, as depicted in Figure 6.6.

This observation is illustrated here, in Figure 9.42, where the $\tau = 0.1$ loop gain trajectory remains outside a circle of radius $\bar{S}^{-1} = (1.6)^{-1} = 0.625$, while the $\tau = 0.01$ trajectory remains outside a circle of radius $\bar{S}^{-1} = (1.06)^{-1} = 0.943$. Moreover, both trajectories remain at a rather *uniform distance* away from the -1 point over the depicted frequency range, which implies a "flattened out" $|S(j\omega)|$ over this range of frequencies, as in Example 9.5.28.

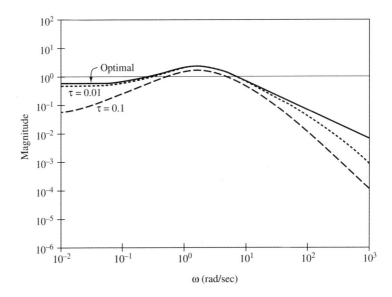

● **FIGURE 9.41**
Magnitude Plots of the Loop Gain

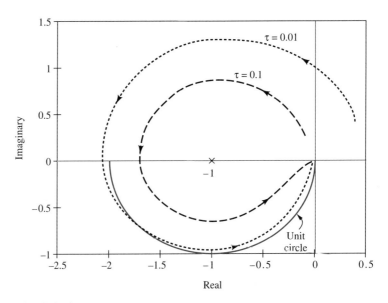

● **FIGURE 9.42**
Nyquist Plots of the Loop Gain

In general, larger values of τ will imply poorer stability margins and increased sensitivity to plant parameter variations but better noise attenuation and robust stability properties with respect to unmodeled high frequency dynamics. Conversely, smaller values of τ will improve the stability margins and the sensitivity properties of the closed-loop system, but there will be a resultant increase in the loop bandwidth, hence decreased noise attenuation and possible stability problems due to unmodeled high frequency dynamics. Design trade-offs via τ can be made among these conflicting performance goals to arrive at a suitable closed-loop design. The sensitivity function and loop gain magnitude plots associated with this example suggest how such trade-offs may be made.

LTR/IMP Robust Zero-Error Tracking

The preceding LTR compensation discussion now serves to motivate a LTR closed-loop design procedure for systems whose zeros are stable that invokes the internal model principle (IMP) to ensure both a robust $e_{ss}(t) = 0$ and the desirable sensitivity properties and stability margins associated with a LQR optimal LSVF design. As we will show, the resulting LTR/IMP design can be particularly effective in the case of unstable plants. The design is based on the robust $e_{ss}(t) = 0$ LQR tracking results presented in Section 9.4, when $\sigma \to \infty$, so that $\hat{q}(s) = c(s)\tilde{\delta}^{H^*}(s)$, as in Eq. (9.6.7).

To begin, let us assume that the reference input $r(t)$ is a known, nondiminishing signal, so that

$$r(s) = \mathcal{L}[r(t)] = \frac{m_r(s)}{p_r(s)}$$

as in Eq. (9.4.25), with all r roots of $p_r(s)$ on the $j\omega$-axis. We now can define an appropriate

$$\bar{\delta}^{F^*}(s) = \delta^{F^*}(s)\delta_r(s) \tag{9.6.20}$$

as in Eq. (9.4.29), with $\delta_r(s)$ an arbitrary stable polynomial of degree $\bar{r} \leq r$. Referring to Eq. (9.4.30), the polynomial equation

$$a(s)\underbrace{\bar{p}_r(s)c(s)\tilde{k}(s)}_{k(s)} + c(s)h(s) = \bar{\delta}^{F^*}(s)\underbrace{c(s)\tilde{\delta}^{H^*}(s)}_{\hat{q}(s)} \tag{9.6.21}$$

or

$$a(s)\bar{p}_r(s)\tilde{k}(s) + h(s) = \bar{\delta}^{F^*}(s)\tilde{\delta}^{H^*}(s) \tag{9.6.22}$$

has a unique solution, as given by the compensator polynomials

$$\tilde{k}(s) = \tilde{k}_{n-m-1}s^{n-m-1} + \cdots + \tilde{k}_1 s + \tilde{k}_0 \tag{9.6.23}$$

and

$$h(s) = h_{n+\bar{r}-1}s^{n+\bar{r}-1} + \cdots + h_1 s + h_0 \tag{9.6.24}$$

Once $h(s)$ has been determined, the polynomial equation

$$\underbrace{c(s)\tilde{\delta}^{H^*}(s)\,\bar{q}(s)}_{\hat{q}(s)} + p_r(s)\hat{h}(s) = h(s) \tag{9.6.25}$$

can be solved for the nonunique polynomials $\bar{q}(s)$ and $\hat{h}(s)$, of degrees $\bar{r}$ and $\bar{n} - r - 1$, respectively, as in Eq. (9.4.33).

The polynomials

$$h(s) = h(s),\ \ k(s) = \bar{p}_r(s)c(s)\tilde{k}(s),\ \ \text{and}\ \ q(s) = \bar{q}(s)c(s)\tilde{\delta}^{H^*}(s) \tag{9.6.26}$$

will then imply a proper 2 DOF compensator characterized by a robust $e_{ss}(t) = 0$, since

$$a(s)k(s) = a(s)\bar{p}_r(s)c(s)\tilde{k}(s) = \tilde{a}(s)c(s)\tilde{k}(s)p_r(s) \tag{9.6.27}$$

will imply Eq. (7.3.16), and Eq. (9.6.25) will imply Eq. (7.3.17). The resulting closed-loop system, as depicted in Figure 9.43, will be characterized by the output response transfer function

$$T(s) = \frac{c(s)q(s)}{a(s)k(s) + c(s)h(s)} = \frac{c(s)\bar{q}(s)\tilde{\delta}^{H^*}(s)}{\bar{\delta}^{F^*}(s)\tilde{\delta}^{H^*}(s)} = \frac{c(s)\bar{q}(s)}{\delta^{F^*}(s)\delta_r(s)} \tag{9.6.28}$$

which is similar to the LQR optimal $T(s)$ defined by Eq. (9.4.17).

Since the loop gain

$$L(s) = \frac{c(s)h(s)}{a(s)k(s)} = \frac{h(s)}{a(s)\bar{p}_r(s)\tilde{k}(s)} \tag{9.6.29}$$

with $a(s)\bar{p}_r(s) = \tilde{a}(s)p_r(s)$ via Eq. (9.4.26), the magnitude of the sensitivity function

$$S(s) = \frac{1}{1 + L(s)} = \frac{a(s)\bar{p}_r(s)\tilde{k}(s)}{a(s)\bar{p}_r(s)\tilde{k}(s) + h(s)} = \frac{\tilde{a}(s)p_r(s)\tilde{k}(s)}{\delta^{F^*}(s)\delta_r(s)\tilde{\delta}^{H^*}(s)} \tag{9.6.30}$$

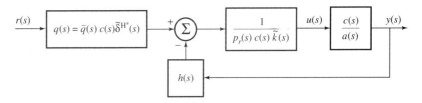

● **FIGURE 9.43**
2 DOF LTR/IMP Compensation

will be zero at those frequencies ω where $p_r(j\omega) = 0$.[14] Moreover, if the roots of $\delta_r(s)$ are positioned in the stable half-plane "close to" the roots of $\bar{p}_r(s)$, so that $\delta_r(s) \approx \bar{p}_r(s)$, it follows that

$$S(s) = \frac{a(s)\bar{p}_r(s)\tilde{k}(s)}{\delta^{F^*}(s)\delta_r(s)\tilde{\delta}^{H^*}(s)} \approx \frac{a(s)\tilde{k}(s)}{\delta^{F^*}(s)\tilde{\delta}^{H^*}(s)}$$

Therefore, the farther into the left-half s-plane the roots of $\tilde{\delta}^{H^*}(s)$ are positioned, the more closely $\tilde{k}(s) \to \tilde{\delta}^{H^*}(s) \to 1$ as in Eq. (9.6.17), with a resultant

$$S(s) \approx \frac{a(s)\tilde{k}(s)}{\delta^{F^*}(s)\tilde{\delta}^{H^*}(s)} \to \frac{a(s)}{\delta^{F^*}(s)} \qquad (9.6.31)$$

the LQR optimal LSVF sensitivity function defined by Eq. (9.6.3).

. . . .

EXAMPLE 9.6.32 Consider the system defined in Example 9.6.19, which is characterized by the minimal transfer function

$$G(s) = \frac{c(s)}{a(s)} = \frac{s+3}{s^3 + s^2 - 10s + 8} = \frac{s+3}{(s-1)(s-2)(s+4)}$$

Suppose that we wish to track sinusoidal reference input signals of frequency $\omega = 3$ with a robust $e_{ss}(t) = 0$, or alternatively, suppose we wish to completely reject output disturbance signals of frequency $\omega = 3$. In either case,

$$p_r(s) = \bar{p}_r(s) = s^2 + 9 = s \pm j3$$

Therefore, we will define $\bar{\delta}^{F^*}(s) = \delta^{F^*}(s)\delta_r(s)$, as in Eq. (9.6.20), with

$$\delta^{F^*}(s) = s^3 + 7.37s^2 + 16.65s + 12.4 = (s + 3.95)(s + 1.71 \pm j.46)$$

as in Example 9.6.19, and set

$$\delta_r(s) = s^2 + 2s + 10 = s + 1 \pm j3$$

so that its roots are "close to" those of $p_r(s)$. A choice of $\tilde{\delta}^{H^*}(s) = \tau s + 1 = 0.05s + 1$ will then imply that the polynomial Eq. (9.6.22)

$$\underbrace{(s^3 + s^2 - 10s + 8)}_{a(s)} \underbrace{(s^2 + 9)}_{\bar{p}_r(s)} \underbrace{(\tilde{k}_1 s + \tilde{k}_0)}_{\tilde{k}(s)} + \underbrace{h_4 s^4 + h_3 s^3 + h_2 s^2 + h_1 s + h_0}_{h(s)}$$

$$= \bar{\delta}^{F^*}(s)\tilde{\delta}^{H^*}(s) = \underbrace{(s^3 + 7.37s^2 + 16.65s + 12.4)}_{\delta^{F^*}(s)} \underbrace{(s^2 + 2s + 10)}_{\delta_r(s)} \underbrace{(0.05s + 1)}_{\tilde{\delta}^{H^*}(s)}$$

---

[14]Therefore, if certain of the roots of $p_r(s)$ correspond to output disturbance modes, this LTG/IMP design will imply a *complete disturbance rejection* of these modes, as in the LLG/IMP case.

or

$$(s^5 + s^4 - s^3 + 17s^2 - 90s + 72)(\tilde{k}_1 s + \tilde{k}_0) + h_4 s^4 + h_3 s^3 + h_2 s^2 + h_1 s + h_0$$

$$= 0.05s^6 + 1.47s^5 + 11.44s^4 + 47.36s^3 + 128.97s^2 + 197.5s + 124$$

can be solved uniquely for $\tilde{k}(s)$ and $h(s)$. In this particular case,

$$\tilde{k}(s) = 0.05s + 1.42 \quad \text{and} \quad h(s) = 10.07s^4 + 47.93s^3 + 109.36s^2 + 321.52s + 21.9$$

so that

$$k(s) = \underbrace{(s^2 + 9)}_{\bar{p}_r(s)} \underbrace{(s + 3)}_{c(s)} \underbrace{(0.05s + 1.42)}_{\tilde{k}(s)} = 0.05s^4 + 1.57s^3 + 4.71s^2 + 14.13s + 38.34$$

Figure 9.44 compares a Bode plot of the magnitude of the LTR/IMP sensitivity function (the dashed line) defined by Eq. (9.6.30) with the LQR optimal LSVF sensitivity function (the solid line) defined by Eq. (9.6.3). Note that the LTR/IMP $\|S\|_\infty = \max_\omega |S(j\omega)| = \bar{S} = 1.34$, thus implying excellent stability margins, with $|S(j\omega)| \to 0$ as $\omega \to 3$, thus implying the complete rejection of any sinusoidal output disturbances of frequency $\omega = 3$. If this were the only reason for the IMP $p_r(s) = s^2 + 9$, we could set $\bar{q}(s) = \alpha \delta_r(s)$ which, in light of Eq. (9.6.28), would imply the LQR optimal $T(s)$ defined by Eq. (9.4.17).

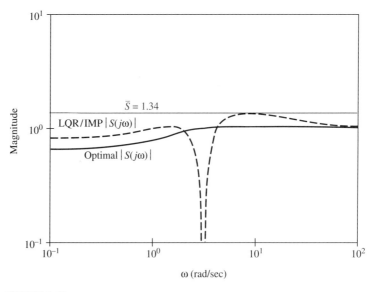

• **FIGURE 9.44**
Magnitude Plots of the Sensitivity Functions

However, to ensure a robust $e_{ss}(t) = 0$ for sinusoidal reference inputs of frequency 3 as well, $\bar{q}(s)$ and $\hat{h}(s)$ should be chosen to satisfy Eq. (9.6.25); in this case,

$$\underbrace{(\bar{q}_2 s^2 + \bar{q}_1 s + \bar{q}_0)}_{\bar{q}(s)} \underbrace{(0.05s^2 + 1.15s + 3)}_{c(s)\tilde{\delta}^{H^*}(s)} + \underbrace{(\hat{h}_2 s^2 + \hat{h}_1 s + \hat{h}_0)}_{\hat{h}(s)} \underbrace{(s^2 + 9)}_{p_r(s)}$$

$$= h(s) = 10.07s^4 + 47.93s^3 + 109.36s^2 + 321.52s + 21.9$$

for some nonunique $\bar{q}(s)$ and $\hat{h}(s)$. Since $\hat{h}(s)$ does not affect either the loop or the response performance, $\bar{q}(s)$ will be varied; for example, if we set $\bar{q}_2 = 0$ to minimize reference input "differentiation,"

$$\bar{q}(s) = -6.055s - 82.095 \quad \text{and} \quad \hat{h}(s) = 10.07s^2 + 48.233s + 29.798$$

will satisfy Eq. (9.6.25). A resultant

$$q(s) = \underbrace{c(s)\tilde{\delta}^{H^*}(s)}_{\hat{q}(s)} \bar{q}(s) = -0.3028s^3 - 11.068s^2 - 112.574s - 246.285$$

will ensure a robust $e_{ss}(t) = \lim_{t \to \infty}[r(t) - y(t)] = 0$, as depicted by the Figure 9.45 plot of the output $y(t)$ (the solid line) that results when a sinusoidal reference input $r(t) = \sin 3t$ (the dashed line) is applied.

A Bode plot of the loop gain magnitude defined by Eq. (9.6.29) is depicted in Figure 9.46 by the dashed line. The solid line depicts the LQR optimal loop gain magnitude for comparison purposes. A Nyquist plot of the loop gain, as depicted in Figure 9.47 by the dashed line, displays the first (for $\omega > 0$) of the two counterclockwise encirclements of the -1 point that are required to ensure closed-loop stability in this case. The solid line represents the lower half of a unit circle centered at -1.

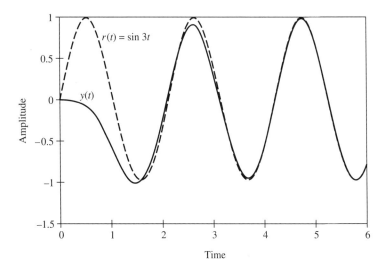

• **FIGURE 9.45**
  A Robust $e_{ss}(t) = 0$ for $r(t) = \sin 3t$

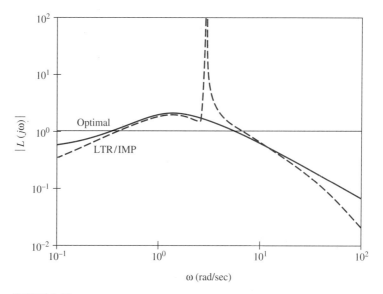

• **FIGURE 9.46**
Magnitude Plots of the Loop Gain

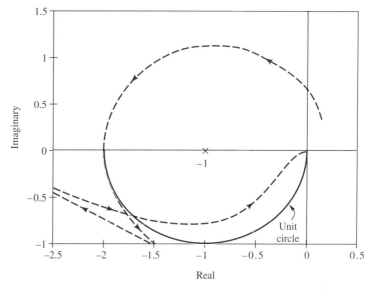

• **FIGURE 9.47**
A Nyquist Plot of the Loop Gain

It is of interest to note that the Nyquist plot remains outside a circle of radius $\bar{S}^{-1} = (1.34)^{-1} = 0.75$ at a rather uniform distance away from the -1 point over the depicted frequency range, which implies a "flattened out" $|S(j\omega)|$ over this range of frequencies. This illustrates that an LTR/IMP design represents another way of minimizing the $\mathbf{H}_\infty$-norm of the sensitivity function $S(s)$ to obtain acceptable stability margins.

LTR/IMP Robust Zero-Error Step Tracking

In the case of step changes in reference input $r(t)$, it is possible to obtain the exact LQR optimal closed-loop transfer function defined by Eq. (9.4.17), as well as a robust $e_{ss}(t) = 0$. If the plant $G(s)$ is type 0, it follows that $p_r(s) = \bar{p}_r(s) = s$ in Eq. (9.6.22), which then can be solved for the unique polynomials

$$\tilde{k}(s) = \tilde{k}_{n-m-1}s^{n-m-1} + \cdots + \tilde{k}_1 s + \tilde{k}_0 \quad \text{and} \quad h(s) = h_n s^n + \cdots + h_1 s + h_0 \tag{9.6.33}$$

Once $h(s)$ has been determined, a choice of

$$\bar{q}(s) = \alpha(s + \gamma), \quad \text{with} \quad \alpha = \frac{h_0}{\gamma c(0)}$$

will imply that

$$q(s) = \bar{q}(s)\hat{q}(s) = \frac{h_0}{\gamma c(0)}(s + \gamma) \underbrace{c(s)\tilde{\delta}^{H^*}(s)}_{\hat{q}(s)} \tag{9.6.34}$$

in light of Eq. (9.6.26). Therefore,

$$q(0) = \frac{h_0 \gamma c(0)}{\gamma c(0)} \underbrace{\tilde{\delta}^{H^*}(0)}_{1} = h_0 = h(0)$$

which will ensure a robust $e_{ss}(t) = 0$ for step inputs, in light of Eq. (7.3.18). The resulting 2 DOF LTR closed-loop system, as depicted in Figure 9.48, will be characterized by an output response transfer function identical to

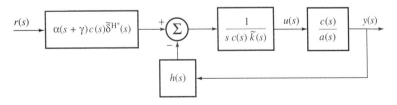

- **FIGURE 9.48**
  An LTR/IMP Robust $e_{ss}(t) = 0$ Step Response System

the LQR optimal $T(s)$ defined by Eq. (9.4.17), namely

$$T(s) = \frac{y(s)}{r(s)} = \frac{c(s)q(s)}{a(s)k(s) + c(s)h(s)} = \frac{\bar{q}(s)c(s)\tilde{\delta}^{H^*}(s)}{a(s)s\tilde{k}(s) + h(s)}$$

$$= \frac{\alpha(s + \gamma)c(s)\tilde{\delta}^{H^*}(s)}{\delta^{F^*}(s)(s + \gamma)\tilde{\delta}^{H^*}(s)} = \frac{\alpha c(s)}{\delta^{F^*}(s)} \qquad (9.6.35)$$

We finally note that it is not always possible to satisfy all of the desired performance goals, because an appropriate compensator sometimes does not exist. For example, one case in which it often is impossible to achieve acceptable stability margins is when open-loop plant poles that are "close to" open-loop zeros are moved to closed-loop locations "some distance away" from the zeros. In such cases, the compensator polynomials $h(s)$ and $k(s)$ could be characterized by large coefficients, thus implying excessive sensitivity function magnitudes and poor stability margins.

This usually is the case when $G(s)$ has unstable poles "close to" unstable zeros, because the unstable poles must be moved into the stable half-plane some distance away from the unstable zeros, as we will illustrate in the next section with Example 9.7.10. However, this situation also can occur when all of the poles and zeros of $G(s)$ are stable (see Problem 9-22).

9.7 IMPLEMENTABLE TRANSFER FUNCTIONS

In the previous sections, a variety of design techniques were presented that, after implementation, imply a final closed-loop transfer function of the general form:

$$T(s) = \frac{y(s)}{r(s)} = \frac{\alpha c(s)}{\delta^F(s)}$$

In certain situations, however, a desired or "model" closed-loop transfer function may be specified explicitly at the outset. For example, in aircraft controller design, the desired handling qualities of an aircraft may be specified by a complete *model transfer function*. In such cases, it obviously is of interest to determine if **exact model matching** [61] is possible, that is, whether or not the model transfer function can be obtained or "implemented" by an appropriate compensator.[15] Chen [14] has termed this an *inward approach* to control system design, in contrast to a more conventional *outward approach*, that "starts from internal compensators and works toward external overall transfer functions."

---

[15]Truxal [60] was one of the first to investigate such synthesis questions.

Motivated by model matching, we now will determine the set of all stable, minimal transfer functions

$$T(s) = \frac{y(s)}{r(s)} = \frac{m(s)}{p(s)} \qquad (9.7.1)$$

that can be obtained by compensating a given system that is defined by a minimal, strictly proper transfer function

$$G(s) = \frac{y(s)}{u(s)} = \frac{c(s)}{a(s)} = \frac{c_m s^m + \cdots + c_1 s + c_0}{s^n + \cdots + a_1 s + a_0} \qquad (9.7.2)$$

by a proper, 2 DOF linear compensator, as defined by the triple $\{h(s), k(s), q(s)\}$ and depicted in Figure 9.49. The resulting 2 DOF implementable system will be defined by the strictly proper transfer function

$$T(s) = \frac{y(s)}{r(s)} = \frac{c(s)q(s)}{\underbrace{a(s)k(s) + c(s)h(s)}_{\delta(s)}} = \frac{m(s)}{p(s)} \qquad (9.7.3)$$

with $\delta(s)$ stable. For convenience, and without loss of generality, $\delta(s)$, $p(s)$, and $k(s)$ will be assumed to be monic.

In light of Eq. (9.7.3) and the stability of $\delta(s)$, it now follows that

i. *Any unstable zeros of $G(s)$ must be (unstable) zeros of $T(s)$ as well.* Therefore, if we express

$$\frac{m(s)}{c(s)} = \frac{\bar{m}(s)}{\bar{c}(s)} \qquad (9.7.4)$$

with $\bar{m}(s)$ and $\bar{c}(s)$ coprime, and $\bar{c}(s)$ monic (by choice), then $\bar{c}(s)$ *must be stable.* Moreover, since $\deg[q(s)] \le \deg[k(s)]$, condition (ii) follows.

ii. *The relative degree of $T(s)$*

$$\deg[\delta(s)] - \deg[c(s)q(s)] =$$

$$\underbrace{\deg[a(s)] - \deg[c(s)]}_{n-m} + \underbrace{\deg[k(s)] - \deg[q(s)]}_{\ge 0} \qquad (9.7.5)$$

can be no less than $n - m$, the relative degree of $G(s)$.

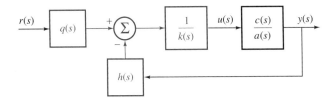

● **FIGURE 9.49**
2 DOF Proper Linear Compensation

Note that these two conditions are *necessary* for $T(s)$ to be 2 DOF implementable from a given $G(s)$. We will now prove that these two conditions are *sufficient* as well, by showing how $T(s)$ can be implemented from $G(s)$ when both conditions hold.

In particular, condition (i) and Eq. (9.7.4) imply that

$$T(s) = \frac{m(s)}{p(s)} = \frac{\bar{m}(s)c(s)}{\bar{c}(s)p(s)} \tag{9.7.6}$$

with $\bar{c}(s)p(s)$ both monic and stable. We now choose $\tilde{q}(s)$ to be any stable, monic polynomial of sufficient degree to ensure that

$$\upsilon \stackrel{\text{def}}{=} \deg[\bar{c}(s)p(s)\tilde{q}(s)] \geq 2n - 1 \tag{9.7.7}$$

so that $\deg[\tilde{q}(s)] \geq 2n - 1 - \deg[p(s)] - \deg[\bar{c}(s)]$. For example, if $\deg[\bar{c}(s)p(s)] \geq 2n - 1$, then a $\tilde{q}(s) = 1$ would be an appropriate choice.

Once $\tilde{q}(s)$ has been selected, the polynomial Diophantine equation

$$a(s)k(s) + c(s)h(s) = \delta(s) = p(s)\bar{c}(s)\tilde{q}(s) \tag{9.7.8}$$

can be solved uniquely for a monic $k(s)$ of degree $\upsilon - n \geq n - 1$, and a corresponding $h(s)$ of degree $n - 1$. Note that since $\bar{c}(s)$ divides $c(s)$ but not $a(s)$, $\bar{c}(s)$ must also divide $k(s)$, so that $k(s) = \bar{c}(s)\bar{k}(s)$, for some monic polynomial $\bar{k}(s)$.

Since $c(s)\bar{m}(s) = \bar{c}(s)m(s)$, using Eq. (9.7.4), a choice of $q(s) = \bar{m}(s)\tilde{q}(s)$ will then imply the desired, closed-loop transfer function. In particular,

$$T(s) = \frac{c(s)q(s)}{a(s)k(s) + c(s)h(s) = \delta(s)} = \frac{c(s)\bar{m}(s)\tilde{q}(s)}{p(s)\bar{c}(s)\tilde{q}(s)} = \frac{m(s)}{p(s)} \tag{9.7.9}$$

with $\deg[q(s)] \leq \deg[k(s)]$, in light of Eq. (9.7.5), which establishes the sufficiency of conditions (i) and (ii).

EXAMPLE 9.7.10 Consider the third-order ($n = 3$) state-space system:

$$\dot{\mathbf{x}}(t) = \underbrace{\begin{bmatrix} 2 & -1 & -2 \\ 0 & -3 & 0 \\ 1 & 0 & -1 \end{bmatrix}}_{A} \mathbf{x}(t) + \underbrace{\begin{bmatrix} 0 \\ -1 \\ 0 \end{bmatrix}}_{B} u(t); \qquad y(t) = \underbrace{[0, \quad -1, \quad -2]}_{C} \mathbf{x}(t)$$

which is characterized by the minimal transfer function

$$G(s) = \frac{y(s)}{u(s)} = C(sI - A)^{-1}B = \frac{c(s)}{a(s)} = \frac{s^2 - s - 2}{s^3 + 2s^2 - 3s} = \frac{(s + 1)(s - 2)}{s(s + 3)(s - 1)}$$

It now follows that any stable, minimal, strictly proper $T(s) = \frac{m(s)}{p(s)}$ is 2 DOF implementable from $G(s)$ if (i) $s - 2$ divides $m(s)$, and (ii) the relative degree of $T(s)$ is no less than 1, the relative degree of $G(s)$; that is, any arbitrary $T(s)$ of the general form

$$T(s) = \frac{(s - 2)\tilde{m}(s)}{\delta(s)}$$

with a stable $\delta(s)$ of degree $> \deg[(s-2)\tilde{m}(s)]$ would be 2 DOF implementable from $G(s)$. For example, a

$$T(s) = \frac{m(s)}{p(s)} = \frac{-2s+4}{s^3+4s^2+6s+4} = \frac{-2(s-2)}{(s+1\pm j)(s+2)} = \frac{(s-2)\tilde{m}(s)}{\delta(s)}$$

is implementable from $G(s)$ because $s - 2$ divides $m(s)$ and the relative degree of $T(s)$ is $2 > 1$, the relative degree of $G(s)$.

To explicitly determine a 2 DOF compensator $\{h(s), k(s), q(s)\}$ that yields this particular $T(s)$, we first note that

$$\frac{m(s)}{c(s)} = \frac{-2(s-2)}{(s+1)(s-2)} = \frac{-2}{s+1} = \frac{\bar{m}(s)}{\bar{c}(s)}$$

using Eq. (9.7.4). We now choose a stable $\tilde{q}(s)$ of degree $= 2n - 1 - \deg[p(s)] - \deg[\bar{c}(s)] = 6 - 1 - 3 - 1 = 1$ to ensure Eq. (9.7.7); for example, a $\tilde{q}(s) = s + 1$ of degree 1 implies an appropriate

$$\delta(s) = p(s)\bar{c}(s)\tilde{q}(s) = s^5 + 6s^4 + 15s^3 + 20s^2 + 14s + 4$$

of degree $\upsilon = 2n - 1 = 5$.

We next solve Eq. (9.7.8) for a monic $k(s) = \bar{c}(s)\bar{k}(s) = (s+1)(s+\bar{k}_0)$ of degree $\upsilon - n = 5 - 3 = 2$, and a corresponding $h(s)$ of degree $n - 1 = 2$; that is, in this case,

$$\delta(s) = \underbrace{(s^3+2s^2-3s)}_{a(s)}\underbrace{(s+1)(s+\bar{k}_0)}_{k(s)} + \underbrace{(s+1)(s-2)}_{c(s)}\underbrace{(h_2s^2+h_1s+h_0)}_{h(s)}$$

$$= (s+1)\left[s^4 + (\bar{k}_0+h_2+2)s^3 + (2\bar{k}_0-2h_2+h_1-3)s^2 + (h_0-3\bar{k}_0-2h_1)s - 2h_0\right]$$

$$= p(s)\bar{c}(s)\tilde{q}(s) = (s+1)(s^4+5s^3+10s^2+10s+4)$$

so that

$$h(s) = -7s^2 - 21s - 2 \quad \text{and} \quad k(s) = \underbrace{(s+1)}_{\bar{c}(s)}\underbrace{(s+10)}_{\bar{k}(s)} = s^2 + 11s + 10$$

(see Problem 9-24).

A corresponding

$$q(s) = \bar{m}(s)\tilde{q}(s) = -2(s+1) = -2s - 2$$

then implies the desired, closed-loop transfer function

$$T(s) = \frac{c(s)q(s)}{\delta(s)} = \frac{-2(s+1)(s-2)(s+1)}{\underbrace{(s+1)}_{\bar{c}(s)}\underbrace{(s+1\pm j)(s+2)}_{p(s)}\underbrace{(s+1)}_{\tilde{q}(s)}} = \frac{-2s+4}{s^3+4s^2+6s+4} = \frac{m(s)}{p(s)}$$

Since $G(s)$ is type 1 and $q(0) = -2 = h(0)$, Eq. (7.3.18) and the IMP imply that the closed-loop system will be characterized by a robust $e_{ss}(t) = 0$ for step changes in the reference input. However, the magnitude of the sensitivity function associated with *any* stabilizing linear compensator will be no less than 3 in this case (see Problem 9-27), because of the need to move the unstable plant pole

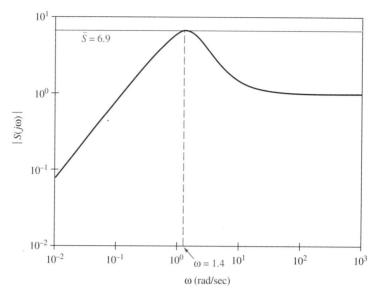

● **FIGURE 9.50**
$|S(j\omega)|$ for a System with Unstable Poles and Unstable Zeros

at $s = +1$ into the stable half-plane "some distance away" from the unstable plant zero at $s = +2$. A Bode plot of $|S(j\omega)|$ for the particular

$$H(s) = \frac{h(s)}{k(s)} = \frac{-7s^2 - 21s - 2}{s^2 + 11s + 10}$$

employed here, as depicted in Figure 9.50, displays a $\max_\omega |S(j\omega)| = \bar{S} = 6.9$ at $\omega = 1.4$, which implies that our 2 DOF compensator is characterized by a high sensitivity to plant parameter variations, hence poor stability margins.

1 DOF Implementations

If a particular $T(s)$ is 2 DOF implementable from a given $G(s)$, it is of interest to determine when it is 1 DOF implementable from $G(s)$ as well,[16] using a proper $H(s)$ closed under unity feedback, as depicted in Figure 9.51. We will resolve this question by first presenting an alternative way of characterizing the set of all $T(s)$ that are 2 DOF implementable from a given $G(s)$.

---

[16]Note that a 1 DOF implementable $T(s)$ is always 2 DOF implementable.

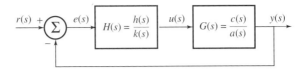

● **FIGURE 9.51**
1 DOF Proper Linear Compensation

In particular, note that condition (i) holds if and only if

$$\frac{T(s)}{G(s)} = \frac{a(s)}{c(s)} \frac{m(s)}{p(s)} = \frac{a(s)\bar{m}(s)}{p(s)\bar{c}(s)} \stackrel{\text{def}}{=} Q(s) \qquad (9.7.11)$$

is stable, and condition (ii) holds if and only if

$$\deg[p(s)] - \deg[m(s)] \geq \underbrace{\deg[a(s)] - \deg[c(s)]}_{n - m}$$

or if and only if the relative degree of $Q(s)$, namely,

$$deg[c(s)p(s)] - deg[a(s)m(s)] \geq 0 \qquad (9.7.12)$$

Stated as a solution to the exact model-matching question, we thus conclude that *a stable, strictly proper, desired or "model" T(s) is 2 DOF implementable from a given G(s) if and only if a minimal representation of*

$$Q(s) = \frac{T(s)}{G(s)} \qquad (9.7.13)$$

is both stable and proper.

If the $Q(s)$ defined by Eq. (9.7.13) is both stable and proper, so that $T(s)$ is 2 DOF implementable, then $T(s)$ will also be 1 DOF implementable by means of the $H(s)$ of Figure 9.51 if

$$\frac{y(s)}{r(s)} = \frac{G(s)H(s)}{1 + G(s)H(s)} = \frac{c(s)h(s)}{\underbrace{a(s)k(s) + c(s)h(s)}_{\delta(s)}} = T(s) = G(s)Q(s)$$

$$(9.7.14)$$

with $\delta(s)$ stable. We now will determine when such an $H(s)$ exists.

If an appropriate $H(s)$ exists, Eqs. (9.7.11) and (9.7.14) will imply that

$$H(s) = \frac{Q(s)}{1 - G(s)Q(s)} = \frac{\dfrac{\bar{m}(s)a(s)}{p(s)\bar{c}(s)}}{1 - \dfrac{c(s)}{a(s)} \dfrac{\bar{m}(s)a(s)}{p(s)\bar{c}(s)}} = \frac{\bar{m}(s)a(s)}{p(s)\bar{c}(s) - c(s)\bar{m}(s)}$$

$$= \frac{\bar{m}(s)a(s)}{\bar{c}(s)[p(s) - m(s)]} = \frac{\bar{m}(s)\bar{a}(s)}{\bar{c}(s)[\overline{p(s) - m(s)}]} \qquad (9.7.15)$$

where

$$\frac{a(s)}{p(s) - m(s)} = \frac{\bar{a}(s)}{\overline{p(s) - m(s)}} \qquad (9.7.16)$$

with $\bar{a}(s)$ and a monic (by choice) $\overline{p(s) - m(s)}$ coprime. Since $\bar{c}(s)$ and $\bar{m}(s)\bar{a}(s)$ also are coprime, it follows that

$$\frac{\bar{m}(s)\bar{a}(s)}{\bar{c}(s)[\overline{p(s) - m(s)}]} = \frac{h(s)}{k(s)} = H(s) \qquad (9.7.17)$$

will represent a unique, minimal representation of the 1 DOF loop compensator.

Referring to Eqs. (9.7.11) and (9.7.15), we next note that the relative degree of $H(s)$ is equal to the relative degree of $Q(s)$, namely, $\deg[\bar{c}(s)p(s)] - \deg[\bar{m}(s)a(s)]$, so that $H(s)$ is proper. Moreover, since

$$a(s)[\overline{p(s) - m(s)}] = \bar{a}(s)[p(s) - m(s)]$$

using Eq. (9.7.16), and $\bar{c}(s)m(s) = c(s)\bar{m}(s)$, Eq. (9.7.17) implies that

$$\delta(s) = a(s)k(s) + c(s)h(s) = a(s)\bar{c}(s)[\overline{p(s) - m(s)}] + c(s)\bar{m}(s)\bar{a}(s)$$
$$= \bar{a}(s)\{\bar{c}(s)[p(s) - m(s)] + \bar{c}(s)m(s)\} = \bar{a}(s)\bar{c}(s)p(s) \qquad (9.7.18)$$

Since both $\bar{c}(s)$ and $p(s)$ are stable, we thus conclude that *if $T(s)$ is 2 DOF implementable from $G(s)$, then it is 1 DOF implementable as well via the unique, minimal $H(s)$ defined by Eq. (9.7.15), if and only if $\bar{a}(s)$, as defined by Eq. (9.7.16), is stable.*

Even when a desired $T(s)$ is 1 DOF implementable from $G(s)$, it often is advantageous to employ a 2 DOF implementation, because a 2 DOF design allows a variation in the loop performance of the system by means of different choices for $\tilde{q}(s)$ while maintaining the same response performance. Alternatively, the response performance associated with a 1 DOF implementation of $T(s)$ that achieves a desired loop performance often can be improved by a 2 DOF configuration.

EXAMPLE 9.7.19 To illustrate this latter observation, consider the plant transfer function

$$G(s) = \frac{c(s)}{a(s)} = \frac{s^2 - s - 2}{s^3 + 2s^2 - 3s} = \frac{(s+1)(s-2)}{s(s+3)(s-1)}$$

of Example 9.7.10. If

$$T(s) = \frac{-7s^3 - 7s^2 + 40s + 4}{s^4 + 5s^3 + 10s^2 + 10s + 4} = \frac{-7(s-2)(s+2.9)(s+0.1)}{(s+1 \pm j)(s+2)(s+1)} = \frac{m(s)}{p(s)}$$

then it is 2 DOF implementable, since the only unstable zero of $G(s)$ at $s = +2$ also is a zero of $T(s)$, and the relative degrees of $T(s)$ and $G(s)$ are both equal to 1.

Since

$$p(s) - m(s) = s^4 + 12s^3 + 17s^2 - 30s = s(s+10)(s+3)(s-1),$$

Eq. (9.7.16) implies that

$$\frac{a(s)}{p(s) - m(s)} = \frac{s(s+3)(s-1)}{s(s+10)(s+3)(s-1)} = \frac{1}{s+10} = \frac{\bar{a}(s)}{\overline{p(s) - m(s)}}$$

so that $\bar{a}(s) = 1$ is stable. Therefore, $T(s)$ is also 1 DOF implementable in this case.

To explicitly determine $H(s)$, we next note that

$$\frac{m(s)}{c(s)} = \frac{-7(s-2)(s+2.9)(s+0.1)}{(s+1)(s-2)} = \frac{-7(s+2.9)(s+0.1)}{s+1} = \frac{\bar{m}(s)}{\bar{c}(s)}$$

so that

$$H(s) = \frac{\bar{m}(s)\bar{a}(s)}{\bar{c}(s)[\overline{p(s) - m(s)}]} = \frac{-7(s+2.9)(s+0.1)}{(s+1)(s+10)} = \frac{-7s^2 - 21s - 2}{s^2 + 11s + 10} = \frac{h(s)}{k(s)}$$

Note that the compensator polynomials $h(s)$ and $k(s)$ that define this $H(s)$ are the same as those obtained in Example 9.7.10. Therefore, the loop performance of this 1 DOF unity feedback system will be exactly the same as that of the 2 DOF system of Example 9.7.10, where

$$T(s) = \frac{-2s + 4}{s^3 + 4s^2 + 6s + 4}$$

However, this latter $T(s)$ is not 1 DOF implementable from $G(s)$ because $\bar{a}(s) = (s+3)(s-1)$ is unstable (see Problem 9-26).

Since $G(s)$ is type 1, both closed-loop systems will be characterized by a robust $e_{ss}(t) = 0$ for step changes in the reference input, although their unit step responses, as depicted in Figure 9.52, are quite different. In particular, the 1 DOF unity feedback step response of this example (the solid line) exhibits "excessive" transient behavior, when compared with that of the 2 DOF design

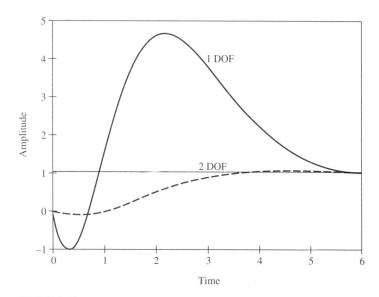

• **FIGURE 9.52**
1 DOF and 2 DOF Unit Step Response Comparisons

of Example 9.7.10 (the dashed line). Moreover, the 1 DOF response performance cannot be altered without changing the loop performance of the system, unlike the response performance in the 2 DOF case, which can be altered by changing $\tilde{q}(s)$.

···

A Parameterization of Stabilizing Compensators

Using Eqs. (9.7.15) and (9.7.16), we next note that if $G(s)$ is stable, so that the $\bar{a}(s)$ of Eq. (9.7.16) is stable, any 2 DOF implementable $T(s)$, as defined by the relationship

$$T(s) = G(s)Q(s) = \frac{m(s)}{p(s)}$$

will be 1 DOF implementable as well, by means of the proper loop compensator

$$H(s) = \frac{Q(s)}{1 - G(s)Q(s)} = \frac{\bar{m}(s)\bar{a}(s)}{\bar{c}(s)[\overline{p(s) - m(s)}]} = \frac{h(s)}{k(s)} \qquad (9.7.20)$$

Therefore, when $G(s) = \frac{c(s)}{a(s)}$ is stable, any proper compensator $H(s) = \frac{h(s)}{k(s)}$ that stabilizes the closed-loop system, so that

$$\frac{G(s)H(s)}{1 + G(s)H(s)} = \frac{c(s)h(s)}{a(s)k(s) + c(s)h(s) = \delta(s)}$$

is stable, can be expressed by Eq. (9.7.20) for some proper, stable

$$Q(s) = \frac{H(s)}{1 + G(s)H(s)} = \frac{a(s)h(s)}{\delta(s)} = \frac{a(s)\bar{m}(s)}{p(s)\bar{c}(s)} \qquad (9.7.21)$$

It may be noted that an arbitrary, stable, proper $Q(s)$ implies a parameterization of the set of *all* complementary sensitivity functions and sensitivity functions, that is,

$$C(s) = G(s)Q(s) \text{ and } S(s) = 1 - G(s)Q(s) \qquad (9.7.22)$$

respectively, that can be obtained by a proper loop compensator $H(s)$ when $G(s)$ is stable.[17] This parameterization has been employed in controller designs that minimize the $\mathbf{H}_\infty$-norm of "weighted" sensitivity or complementary sensitivity functions, such as those defined by Eqs. (6.3.6) and (6.2.14), respectively [50] [19].

Internal Model Control (IMC)

The loop compensator $H(s)$ defined by Eq. (9.7.20) can be implemented either as in Figure 9.53 (when $G(s)$ is stable; see Problem 9-28) or as in Figure 9.54. The particular control configuration depicted in Figure 9.53 has been termed **internal model control** (IMC),[18] because an explicit

---

[17]Such a parameterization is far more difficult to obtain when $G(s)$ is unstable [65].

[18]This should not be confused with the *internal model principle* of Section 7.3.

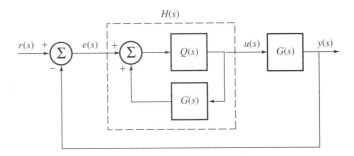

● **FIGURE 9.53**
An IMC Implementation of $H(s)$

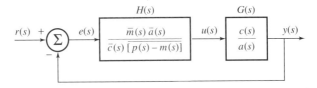

● **FIGURE 9.54**
A Minimal Implementation of $H(s)$

model of the nominal plant transfer function $G(s)$ is placed in the *internal* feedback control loop [50]. It may be noted that if the nominal $G(s)$ defines the dynamic behavior of the actual plant *exactly*, then the two feedback signals in Figure 9.53 will cancel one another. As a consequence,

$$\frac{y(s)}{r(s)} = T(s) = G(s)Q(s) \qquad (9.7.23)$$

which would be equivalent to the open-loop compensation of $G(s)$ by $Q(s)$, as depicted in Figure 9.55.

Although the IMC implementation of $H(s)$ in Figure 9.53 is conceptually appealing, the unity feedback, minimal implementation of $H(s)$ in Figure 9.54 is usually more practical, as we will now illustrate.

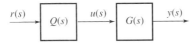

● **FIGURE 9.55**
Exact $G(s)$ IMC Compensation

. . . .

EXAMPLE 9.7.24 Consider a stable plant defined by the nominal transfer function

$$G(s) = \frac{c(s)}{a(s)} = \frac{s-2}{s^2 + s + 6} = \frac{s-2}{s + 0.5 \pm j2.398}$$

If the desired or model

$$T(s) = \frac{m(s)}{p(s)} = \frac{2s - 4}{s^2 + 3s + 2} = \frac{2(s-2)}{(s+1)(s+2)}$$

it follows that

$$\frac{m(s)}{c(s)} = \frac{2(s-2)}{s-2} = \frac{2}{1} = \frac{\bar{m}(s)}{\bar{c}(s)}$$

Therefore, the IMC implementation depicted in Figure 9.53 would imply an $H(s)$ of order 4, as defined by a

$$Q(s) = \frac{a(s)\bar{m}(s)}{p(s)\bar{c}(s)} = \frac{2(s^2 + s + 6)}{s^2 + 3s + 2} = \frac{2s^2 + 2s + 6}{s^2 + 3s + 2}$$

of order 2 in the forward path and a $G(s)$ of order 2 in the feedback path.

However, since

$$\frac{a(s)}{p(s) - m(s)} = \frac{s^2 + s + 6}{s^2 + s + 6} = \frac{1}{1} = \frac{\bar{a}(s)}{[p(s) - m(s)]}$$

the minimal implementation of Figure 9.54 would imply an

$$H(s) = \frac{\bar{m}(s)\bar{a}(s)}{\bar{c}(s)[\overline{p(s) - m(s)}]} = \frac{2}{1} = 2$$

of order 0; that is, a simple proportional compensator will implement the desired $T(s)$ in this case by means of a 1 DOF unity feedback configuration.

It may be noted that the magnitude of the resulting sensitivity function

$$S(s) = \frac{1}{1 + G(s)H(s)} = \frac{s^2 + s + 6}{s^2 + 3s + 2}$$

will be $3 = \bar{S}$ at $\omega = 0$, which could imply unacceptable stability margins. Improved stability margins can be obtained using a LLG implementation similar to that depicted in Figure 9.29, with $\alpha = 2$ and $\delta^{F^*}(s) = s^2 + 3s + 2$.

In particular, if a second-order compensator defined by the transfer function

$$H(s) = \frac{h_n(s^2 + s + 6)}{s^2 + 3s + 2 - h_n(s - 2)}$$

is placed in series with $G(s)$, as depicted in Figure 9.29, with $\hat{\alpha} = \frac{2}{h_n}$, then

$$\frac{y(s)}{r(s)} = T(s) = \frac{2(s-2)}{s^2 + 3s + 2}$$

as desired. Moreover,

$$L(s) = \frac{h_n(s-2)}{s^2 + 3s + 2 - h_n(s-2)} \implies S(s) = \frac{1}{1 + L(s)} = \frac{s^2 + 3s + 2 - h_n(s-2)}{s^2 + 3s + 2}$$

with

$$\max_{\omega} |S(j\omega)| = \bar{S} = S(j0) = 1 + h_n$$

which will be < 2 provided $h_n < 1$.

Smith Predictors

One situation in which the IMC configuration of Figure 9.53 does prove to be practical is when a given, *stable* system is characterized by transportation lag, as defined by Eq. (4.4.55), so that its transfer function

$$\bar{G}(s) = \underbrace{e^{-s\tau}}_{G_\tau(s)} \underbrace{\frac{c(s)}{a(s)}}_{G(s)} \qquad (9.7.25)$$

If $G(s)$ is stable, then for any stable and proper $Q(s)$, Eq. (9.7.20) implies that the transfer function

$$T(s) = \bar{G}(s)Q(s) = e^{-s\tau}G(s)Q(s) \qquad (9.7.26)$$

is 1 DOF implementable via an IMC

$$H(s) = \frac{Q(s)}{1 - \bar{G}(s)Q(s)} \qquad (9.7.27)$$

as depicted in Figure 9.56.

This procedure for compensating a stable system that is characterized by transportation lag $e^{-s\tau}$, independent of the magnitude τ of the time delay, is known as a **Smith predictor** [57]. Smith predictors have found rather widespread use in the process control industry, where they readily can be implemented on digital computers.

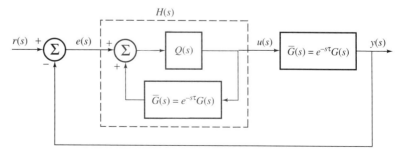

● **FIGURE 9.56**
A Smith Predictor

9.8   SUMMARY

This chapter has served to present some of the most recent and important new design techniques that can be employed to obtain both the response performance and the loop performance goals that were outlined in the earlier chapters. We began by introducing linear state variable feedback (LSVF) compensation and illustrating how a LSVF control law can be used to completely and arbitrarily position all of the controllable poles of a closed-loop system, without affecting its zeros. By employing the controllable canonical form, an equivalent LSVF design procedure was presented for differential operator systems.

A general inability to directly measure all n state variables then motivated the need for a state observer, a dynamic system whose state exponentially approaches that of the given system when it is "driven" by both the plant input and the plant output. Observer design was illustrated for both state-space and differential operator systems. The implementation of observer state feedback, in lieu of actual state feedback, was then shown to imply the same closed-loop transfer function $T(s)$ that would be obtained using actual LSVF. More importantly, differential operator observer feedback was shown to be equivalent to, hence to directly imply, the general 2 DOF design procedure of Section 7.2.

The question of LSVF closed-loop pole placement then was addressed in Section 9.3, where a variety of "error minimizing" performance indices were introduced. The integral of the time multiplied by the absolute error (ITAE) performance index was shown to be the most sensitive to changes in the damping ratio ζ of a system, and appropriate ITAE optimal closed-loop pole locations were given for zeroless systems of dynamic order $n = 1$ through 6.

A minimization of the very useful and commonly employed linear quadratic (output) regulator (LQR) performance index, namely,

$$J_{\text{LQR}} = \int_0^\infty \{\rho y^2(t) + u^2(t)\}\, dt$$

was then shown to imply not only an optimal LSVF feedback control law but also an appropriate positioning of the poles of an LQR optimal observer. A Laplace transformed differential operator approach, based on a spectral factorization and a corresponding root-square locus plot, was employed to develop a general 2 DOF LQR optimal design procedure. The "difficult to control" orbiting satellite of Chapter 2 served to illustrate LQR optimal compensation.

Conditions were then presented under which it is possible to obtain both an LQR optimal closed-loop transfer function and a robust $e_{ss}(t) = 0$, due to the application of a nondiminishing reference input signal. It was shown that a restrictive, 1 DOF unity feedback design need not be employed. Moreover, for step changes in the reference input, the 2 DOF design procedure that was presented was shown to imply the same LQR optimal closed-loop transfer function that would be obtained independent of a robust $e_{ss}(t) = 0$, namely,

$$T(s) = \frac{y(s)}{r(s)} = \frac{\alpha c(s)}{\delta^{F^*}(s)}$$

We then observed that a plant defined by a nominal transfer function $G(s) = \frac{c(s)}{a(s)}$, which is both stable and minimum phase, is relatively easy to control; that is, in light of the results presented in Section 9.7, a 1 DOF compensator $H(s)$ can be used to obtain any arbitrary output response transfer function $T(s)$ of relative degree $\geq \deg[a(s)] - \deg[c(s)] = n - m$. For example, by "cancelling" the open-loop plant poles and zeros by means of an

$$H(s) = \frac{a(s)}{c(s)[\tilde{\delta}^{H^*}(s) - 1]} \tag{9.8.1}$$

we can obtain a closed-loop

$$T(s) = \frac{y(s)}{r(s)} = \frac{G(s)H(s)}{1 + G(s)H(s)} = \frac{1}{\tilde{\delta}^{H^*}(s)} \tag{9.8.2}$$

where $\tilde{\delta}^{H^*}(s)$ is a Butterworth polynomial of degree $n - m$, whose roots can be positioned any arbitrary distance away from the $j\omega$-axis. The magnitude of the corresponding sensitivity function

$$S(s) = 1 - \underbrace{C(s)}_{T(s)} = \frac{\tilde{\delta}^{H^*}(s) - 1}{\tilde{\delta}^{H^*}(s)} \tag{9.8.3}$$

can then be made arbitrarily small over as wide a bandwidth as desired.[19]

Control becomes considerably more difficult when $G(s)$ has unstable poles or unstable zeros. In particular, when $G(s)$ is stable, but $G^{-1}(s)$ is not, the LLG/IMP procedure outlined in Section 9.5 can be employed to design a robust $e_{ss}(t) = 0$ tracking system with an $\bar{S} < 2$, despite the waterbed effect and without the explicit utilization of weighted sensitivity or complementary sensitivity functions. Conversely, when $G(s)$ is not stable, but $G^{-1}(s)$ is stable, the LTR/IMP procedure of Section 9.6 can be employed to obtain analogous results.

Generally speaking, the LQR optimal LLG/IMP and LTR/IMP procedures presented in Sections 9.5 and 9.6 of this chapter represent effective 2 DOF design techniques for ensuring both robust loop goal performance and desirable transient and steady-state response performance in the more

---

[19]Indeed, as noted in Reference [45], a meaningful minimization of the "nominal performance" criteria $\|SW_S\|_\infty$ requires that $G(s)$ have unstable zeros.

difficult to control situations, because they provide the control system designer with the trade-off flexibility of independently altering and evaluating either the loop performance or the response performance of a controlled system.

When both $G(s)$ and $G^{-1}(s)$ are unstable, it may be impossible to obtain desirable closed-loop performance regardless of the compensator employed because the unstable poles must be moved "some distance away" from the unstable zeros. As shown in References [24] and [49], if the nominal plant transfer function

$$G(s) = \frac{\prod_{i=1}^{l}(s - z_i)}{\prod_{j=1}^{k}(s - p_j)} G_{mp}(s) \qquad (9.8.4)$$

where $Re(z_i) > 0$ and $Re(p_j) > 0$, for $i = 1, 2, \ldots, l$, and $j = 1, 2, \ldots, k$, and the minimum phase $G_{mp}(s)$ has no poles or zeros in the half-plane $Re(s) \geq 0$, then the maximum magnitude of the sensitivity function, namely,

$$\|S\|_{\infty} = \max_{\omega} |S(j\omega)| = \bar{S} \geq \max_{i} \prod_{j=1}^{k} \frac{|z_i + p_j|}{|z_i - p_j|} \qquad (9.8.5)$$

In such cases, a $|S(j\omega)| \leq 2$ at all frequencies may be impossible to achieve and, as a consequence, it may be necessary to "redesign" the plant by adding or moving sensors or actuators to obtain satisfactory closed-loop performance (see Problem 9-27).

Finally, Section 9.7 illustrated how any desired closed-loop transfer function

$$T(s) = \frac{y(s)}{r(s)} = \frac{m(s)}{p(s)}$$

can be obtained or implemented by means of 2 DOF compensation, provided $T(s)$ has the same unstable zeros as the plant transfer function $G(s)$ and no lower relative degree. A 1 DOF implementation was then shown to be possible as well, provided any unstable poles of $G(s)$ are "canceled" by corresponding roots of $p(s) - m(s)$. Therefore, if $G(s)$ is stable, any 2 DOF implementable $T(s)$ will also be 1 DOF implementable by the unique loop compensator

$$H(s) = \frac{Q(s)}{1 - G(s)Q(s)}, \quad \text{where} \quad Q(s) = \frac{T(s)}{G(s)}$$

Moreover, when $G(s)$ is stable, it was shown that any stable, proper $Q(s)$ parameterizes the set of all sensitivity functions $S(s) = 1 - G(s)Q(s)$ and complementary sensitivity functions $C(s) = G(s)Q(s)$ that can be obtained using 1 DOF compensation.

PROBLEMS

9-1.* Illustrate the arbitrary pole placement formulas defined by Eqs. (9.1.16) and (9.1.17); that is, determine the LSVF gain vector F for the single input, controllable, state-space system defined by

$$A = \begin{bmatrix} 0 & 0 & -1 \\ 1 & 1 & 2 \\ 0 & 1 & 4 \end{bmatrix}$$

and $B = \begin{bmatrix} 1 \\ 0 \\ 1 \end{bmatrix} \implies \underbrace{[B, \ AB, \ \ldots, \ A^{n-1}B]}_{\mathcal{C}} = \begin{bmatrix} 1 & -1 & -4 \\ 0 & 3 & 10 \\ 1 & 4 & 19 \end{bmatrix}$

whose open-loop poles are given by the roots of

$$a(s) = (s + 0.285)(s - 4.507)(s - 0.778) = s^3 - 5s^2 + 2s + 1$$
$$= s^3 + a_2 s^2 + a_1 s + a_0$$

and whose desired LSVF closed-loop poles are defined by the roots of

$$\delta^F(s) = (s + 1)(s + 2 \pm j) = s^3 + 5s^2 + 9s + 5 = s^3 + \delta_2^F s^2 + \delta_1^F s + \delta_0^F$$

Now note that $\hat{\delta}^F Q_c$ in Eq. (9.1.17) can be written as

$$\hat{\delta}^F Q_c = q_c[\delta_0^F I + \delta_1^F A + \cdots + \delta_{n-1}^F A^{n-1}]$$

As a consequence, verify that the unique state feedback gain vector $F = [f_1 \ \ f_2 \ \ f_3]$ also can be determined by **Ackermann's formula** [1], namely,

$$F = -q_c[\delta_0^F I + \delta_1^F A + \cdots + \delta_{n-1}^F A^{n-1} + A^n]$$

9-2. Consider a controllable, single-input, state-space system $\dot{\mathbf{x}}(t) = A\mathbf{x}(t) + Bu(t)$. If an equivalent state

$$\underbrace{\begin{bmatrix} \hat{x}_1(t) \\ \hat{x}_2(t) \\ \vdots \\ \hat{x}_n(t) \end{bmatrix}}_{\hat{\mathbf{x}}(t)} = \underbrace{\begin{bmatrix} q_c \\ q_c A \\ \vdots \\ q_c A^{n-1} \end{bmatrix}}_{Q_c} \underbrace{\begin{bmatrix} x_1(t) \\ x_2(t) \\ \vdots \\ x_n(t) \end{bmatrix}}_{\mathbf{x}(t)}$$

with q_c defined by the relationship

$$q_c[B, \ AB, \ \ldots, \ A^{n-1}B] = [0 \ \ 0 \ldots 0 \ \ 1]$$

in light of Eqs. (9.1.6), (9.1.7), and (9.1.8), show that $\dot{\hat{x}}_i(t) = \hat{x}_{i+1}(t)$ for $i = 1, 2, \ldots, n - 1$, and that $\dot{\hat{x}}_n(t) = q_c A^n \mathbf{x}(t) + u(t)$. Now use the Cayley-Hamilton Theorem, as defined by Eq. (9.1.18), to verify that

$$\dot{\hat{x}}_n(t) = -a_0 \hat{x}_1(t) - a_1 \hat{x}_2(t) - \cdots - a_{n-1}\hat{x}_n(t) + u(t)$$

thus establishing Eq. (9.1.9).

9-3.* Verify that

$$Q_c = \begin{bmatrix} -0.093 & 0.0465 & -0.0233 \\ 0.0465 & -0.0233 & 0.5116 \\ -0.0233 & 0.5116 & -2.256 \end{bmatrix}$$

transforms the

$$A = \begin{bmatrix} 0 & 0 & -5 \\ 1 & 0 & -1 \\ 0 & 1 & -4 \end{bmatrix}; \qquad B = \begin{bmatrix} 1 \\ 2 \\ 0 \end{bmatrix}$$

pair of Example 9.1.19 to the controllable canonical form:

$$Q_c A Q_c^{-1} = \hat{A} = \begin{bmatrix} 0 & 1 & 0 \\ 0 & 0 & 1 \\ -5 & -1 & -4 \end{bmatrix} \quad \text{and} \quad Q_c B = \hat{B} = \begin{bmatrix} 0 \\ 0 \\ 1 \end{bmatrix}$$

9-4.* Verify that

$$Q_o = \begin{bmatrix} -9 & 4 & 1 \\ 4 & 9 & 2 \\ 1 & 2 & 0 \end{bmatrix}$$

transforms the

$$A = \begin{bmatrix} 0 & 1 & 0 \\ 0 & 0 & 1 \\ -5 & -1 & -4 \end{bmatrix}; \qquad C = [1, \ 2, \ 0]$$

pair of Example 9.2.12 to the observable canonical form:

$$Q_o A Q_o^{-1} = \hat{A} = \begin{bmatrix} 0 & 0 & -5 \\ 1 & 0 & -1 \\ 0 & 1 & -4 \end{bmatrix} \quad \text{and} \quad C Q_o^{-1} = \hat{C} = [0, \ 0, \ 1]$$

9-5. Using Eq. (9.2.19), verify that the output response of a minimal state-space system, compensated by observer state feedback, will be identical to that obtained if the system is compensated by actual LSVF, provided $\mathbf{x}(t_0) = \bar{\mathbf{x}}(t_0)$.

9-6. Show that the five unknown coefficients of $k(D)$ and $h(D)$ in Example 9.2.38 as defined by the relationship

$$(D^3 + 4D^2 + D + 5)(D^2 + k_1 D + k_0) + (2D + 1)(h_2 D^2 + h_1 D + h_0)$$

$$= D^5 + 9D^4 + 33D^3 + 63D^2 + 64D + 30$$

can be determined by solving the matrix/vector equation defined by Eq. (B7) in Appendix B, namely,

$$\underbrace{\begin{bmatrix} 5 & 0 & 0 & 1 & 0 & 0 \\ 1 & 5 & 0 & 2 & 1 & 0 \\ 4 & 1 & 5 & 0 & 2 & 1 \\ 1 & 4 & 1 & 0 & 0 & 2 \\ 0 & 1 & 4 & 0 & 0 & 0 \\ 0 & 0 & 1 & 0 & 0 & 0 \end{bmatrix}}_{S} \underbrace{\begin{bmatrix} k_0 \\ k_1 \\ 1 \\ h_0 \\ h_1 \\ h_2 \end{bmatrix}}_{W} = \underbrace{\begin{bmatrix} 30 \\ 64 \\ 63 \\ 33 \\ 9 \\ 1 \end{bmatrix}}_{\bar{\delta}}$$

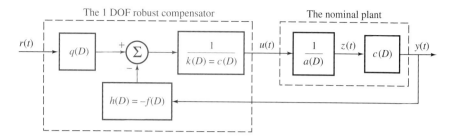

The 1 DOF robust compensator   The nominal plant

- **FIGURE 9.57**
2 DOF Robust Differential Operator Compensation

9-7. If $c(D)$ in Eq. (9.2.23) is a stable polynomial of degree $n - 1$, then a special *robust differential operator observer* [17] can be defined by setting $k(D) = \hat{q}(D) = c(D)$ and $h(D) = -f(D) = \delta^F(D) - a(D)$ in Eq. (9.2.29), so that $m(D) = k(D) - \hat{q}(D) = 0$ in Figure 9.4. Compare such a design with the design of an LTR optimal observer when $m = n-1$, as defined by Eqs. (9.6.7) through Eq. (9.6.12). In particular, show that the resulting, nominal, loop gain transfer function $L(s)$ will be identical to that associated with actual LSVF compensation.

In light of Figure 9.5, a corresponding 2 DOF *robust differential operator compensator*, as depicted in Figure 9.57, but with $q(D)$ arbitrary, would also be characterized by the same nominal loop gain transfer function as that associated with actual LSVF compensation.

(a) Show that the output response transfer function of the Figure 9.57 system is given by

$$T(s) = \frac{y(s)}{r(s)} = \frac{q(s)}{a(s) - f(s) = \delta^F(s)}$$

(b) Illustrate this 2 DOF robust compensator by determining appropriate $k(D)$, $h(D)$, and $q(D)$ polynomials for the differential operator system defined in Example 9.2.38, if $c(D) = D^2+4D+5$ (instead of $2D + 1$), which will yield the same output response transfer function obtained in that example, namely,

$$T(s) = \frac{y(s)}{r(s)} = \frac{2s + 1}{s^3 + 5s^2 + 8s + 6}$$

(c) Determine appropriate $k(D)$, $h(D)$, and $q(D)$ polynomials that will yield the alternative output response transfer function

$$T(s) = \frac{y(s)}{r(s)} = \frac{3}{s + 3}$$

and discuss the controllability and observability properties associated with such a design.

● **FIGURE 9.58**
The F-16 Aircraft in a Landing Approach Configuration.
©Robert P. Morrison/FPG International Corp.

9-8.* The longitudinal motion of the F-16 fighter aircraft in the landing
approach configuration depicted in Figure 9.58 can be defined by the
linearized state-space representation [26]:

$$
\begin{bmatrix} \dot{x}_1(t) \\ \dot{x}_2(t) \\ \dot{x}_3(t) \\ \dot{x}_4(t) \end{bmatrix} = \underbrace{\begin{bmatrix} -0.0507 & -3.861 & 0 & -32.17 \\ -0.00117 & -0.5164 & 1 & 0 \\ -0.00013 & 1.4168 & -0.4932 & 0 \\ 0 & 0 & 1 & 0 \end{bmatrix}}_{A} \begin{bmatrix} x_1(t) \\ x_2(t) \\ x_3(t) \\ x_4(t) \end{bmatrix}
$$

$$
+ \underbrace{\begin{bmatrix} 0 \\ -0.0717 \\ -0.1645 \\ 0 \end{bmatrix}}_{B} u(t)
$$

where $u(t) = \delta_e(t)$, the elevator deflection. If a rate gyro measures the
pitch rate $x_3(t) = y(t)$, verify that the open-loop transfer function

$$
G(s) = \frac{y(s)}{u(s)} = \frac{c(s)}{a(s)} = \frac{-0.1645s^3 - 0.1949s^2 - 0.0087s}{s^4 + 1.0603s^3 - 1.1154s^2 - 0.0658s - 0.0555}
$$

$$
= \frac{-0.1645s(s + 1.1379)(s + 0.0467)}{(s + 1.7036)(s + 0.0438 \pm j.2065)(s - 0.7309)}
$$

has an unstable pole at $s = +0.7309$, so that a stability augmentation
system is required to safely land this aerodynamically unstable aircraft.

    (a) Determine a third-order, differential operator observer with poles
at $s = -1.1379, -0.0467$ (the stable plant zeros), and at $s =$

−2, which will position the closed-loop poles of the system at the desired $s = -1.25 \pm j2.165$ "short-period" locations, and the desired $s = -0.01 \pm j.1$ "phugoid" locations suggested by Rynaski [56].

(b) Now assume that the pitch angle $x_4(t)$ is employed as a second feedback signal $y_2(t)$, in addition to the pitch rate $x_3(t) = y_1(t)$. Referring to Problem 9-7 and (9.2.41), design a 2 DOF robust compensator of order $r = 2$ $(n - p = 4 - 2)$ in this case, which will imply the same short period and phugoid poles obtained in Part (a).

(c) In this two output case, show that $q(s = D)$ can be chosen to cancel either the short-period poles or the phugoid poles, thus implying an output response transfer function of minimal order 2.

9-9. Show that a choice of $h_2(D) = h_3 D + h_2$ in Example 9.2.44, with $k(D) = D + k_0$ and $h_1(D) = h_1 D + h_0$, will imply an infinite number of solutions to the Diophantine equation:

$$\underbrace{(D^3+4D^2+D+5)}_{a(D)} \underbrace{(D+k_0)}_{k(D)} + \underbrace{(2D+1)}_{c_1(D)} \underbrace{(h_1 D+h_0)}_{h_1(D)} + \underbrace{(D^2-1)}_{c_2(D)} \underbrace{(h_3 D+h_2)}_{h_2(D)}$$

$$= D^4 + 7D^3 + 18D^2 + 22D + 12$$

9-10. Consider the closed-loop MO system depicted in Figure 9.6 and defined by Eq. (9.2.40). Verify that the error $e_1(s)$ between a nondiminishing reference input $r(s) = \dfrac{m_r(s)}{p_r(s)}$ and the first component $y_1(s) = c_1(s)z(s)$ of the output is defined by the relationship:

$$e_1(s) = \frac{\overbrace{\{a(s)k(s)+c_2(s)h_2(s)+\cdots+c_p(s)h_p(s)+c_1(s)[h_1(s)-q(s)]\}m_r(s)}^{\overset{\text{def}}{=}\, m_{e1}(s)}}{\delta(s)p_r(s)}$$

so that using Eq. (7.3.3), $p_r(s)$ must divide $m_{e1}(s)$ to ensure a robust $e_{ss}(t) = 0$.

Therefore, if $c_1(s) = 1$, $c_2(s) = s$, ..., $c_p(s) = s^{p-1}$, show that a robust $e_{ss}(t) = 0$ will be ensured for step changes in the reference input provided both the IMP is satisfied, that is, if

$$a(s)k(s) = \bar{m}_e(s) \underbrace{p_r(s)}_{s}$$

as in Eq. (7.3.16), and if $h_1(0) = q(0)$, as in Eq. (7.3.18).

9-11. In light of Problem 9-10 and Eq. (9.2.41), consider the $p = 2$ output X-Y plotter of Problem 8-5. Show that an observer of order $r = 1$ $(n - p = 3 - 2)$ can be employed to arbitrarily position all of the closed-loop poles of the system and to ensure a robust $e_{ss}(t) = 0$ for step changes in the reference input.

9-12. Show that the loop gain transfer function of the closed-loop MO system depicted in Figure 9.6 and defined by Eq. (9.2.40) is given by

$$L(s) = \frac{c_1(s)h_1(s) + c_2(s)h_2(s) + \cdots + c_p(s)h_p(s)}{a(s)k(s)}.$$

Using this relationship, derive a corresponding expression for the sensitivity function $S(s)$ and compare it to the SISO sensitivity function defined by Eq. (5.1.28).

9-13. Consider the ideal PID compensator defined by Eq. (8.1.19),

$$H_{PID}(s) = \frac{K_D s^2 + K_P s + K_I}{s}.$$

Show that the employment of such a compensator to control a zeroless, second-order system is equivalent to complete and arbitrary LSVF. In such cases, note that $H_{PID}(s)$ can be used to arbitrarily position all three closed-loop poles, as verified by Example 8.1.21.

9-14. Consider a SISO system in controllable canonical form, with a scalar $E \neq 0$,

$$\begin{bmatrix} \dot{x}_1(t) \\ \dot{x}_2(t) \\ \vdots \\ \dot{x}_n(t) \end{bmatrix} = \underbrace{\begin{bmatrix} 0 & 1 & 0 & \cdots & 0 \\ 0 & 0 & 1 & \cdots & 0 \\ \vdots & \vdots & & \ddots & \vdots \\ -a_0 & -a_1 & & \cdots & -a_{n-1} \end{bmatrix}}_{A} \begin{bmatrix} x_1(t) \\ x_2(t) \\ \vdots \\ x_n(t) \end{bmatrix} + \underbrace{\begin{bmatrix} 0 \\ \vdots \\ 0 \\ 1 \end{bmatrix}}_{B} u(t)$$

$$y(t) = \underbrace{[c_0 \; c_1 \; \ldots \; c_{n-1}]}_{C} \begin{bmatrix} x_1(t) \\ x_2(t) \\ \vdots \\ x_n(t) \end{bmatrix} + E u(t)$$

or, in view of Table 3.1, the equivalent differential operator system:

$$\underbrace{(D^n + a_{n-1}D^{n-1} + \ldots + a_1 D + a_0)}_{a(D)} z(t) = u(t);$$

$$y(t) = \underbrace{(c_{n-1}D^{n-1} + \cdots + c_1 D + c_0)}_{c(D)} z(t) + (e = E)u(t)$$

with open-loop transfer function

$$G(s) = \frac{y(s)}{u(s)} = \frac{c(s)}{a(s)} + e = \frac{ea(s) + c(s)}{a(s)}$$

$$= \frac{es^n + (ea_{n-1} + c_{n-1})s^{n-1} + \cdots + (ea_1 + c_1)s + (ea_0 + c_0)}{s^n + a_{n-1}s^{n-1} + \cdots + a_1 s + a_0}$$

If the system is compensated by LSVF, so that

$$u(t) = F\mathbf{x}(t) + r(t) = f(D)z(t) + r(t)$$

show that the resulting closed-loop transfer function

$$T(s) = \frac{y(s)}{r(s)} = \frac{ea(s) + c(s)}{a(s) - f(s) = \delta^F(s)}$$

Therefore, as in the case when $E = 0$, the zeros of an $E \neq 0$ system also are not affected by LSVF compensation.

9-15. The photograph that introduces Part III of the text depicts astronaut Bruce McCandless taking the first untethered "walk in space" from the shuttle *Challenger* on February 7, 1984. A simplified block diagram of his PD controlled, gas jet manned maneuvering unit is shown in Figure 9.59. Determine the values of K_P and K_D that will imply an ITAE optimal closed-loop system for step changes in the desired position, as well as a peak response time t_p of 2.5 seconds.

9-16.* Consider any minimal SISO state-space system

$$\dot{\mathbf{x}}(t) = A\mathbf{x}(t) + Bu(t); \qquad y(t) = C\mathbf{x}(t)$$

and the LQR performance index defined by Eq. (9.4.1),

$$J_{LQR} = \int_0^\infty \{\rho y^2(t) + u^2(t)\} \, dt$$

Kalman [39] was first to show that if P represents the unique positive definite solution to the so-called **algebraic Riccati equation**,

$$PA + A^T P - PBB^T P + \rho C^T C = 0$$

then the unique LSVF feedback control law

$$u^*(t) = -B^T P\mathbf{x}(t) \stackrel{\text{def}}{=} F^*\mathbf{x}(t)$$

minimizes J_{LQR} for any arbitrary set of nonzero initial conditions $\mathbf{x}(0)$.

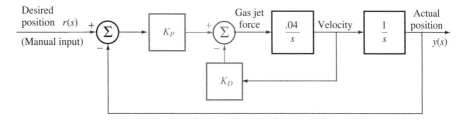

• **FIGURE 9.59**
The PD Controlled, Manned Maneuvering Unit of Problem 9-15.

Verify this result via the orbiting satellite of Example 9.4.24, where

$$A = \begin{bmatrix} 0 & 1 & 0 & 0 \\ 3 & 0 & 0 & 4 \\ 0 & 0 & 0 & 1 \\ 0 & -1 & 0 & 0 \end{bmatrix}, \qquad B = B_2 = \begin{bmatrix} 0 \\ 0 \\ 0 \\ 0.25 \end{bmatrix},$$

and $C = C_2 = [0 \ \ 0 \ \ 1 \ \ 0]$

In particular, show that when $\rho = 10$, the unique, positive definite

$$P = \begin{bmatrix} 81.08 & 33.97 & -28.97 & 57.11 \\ 33.97 & 16.58 & -15.05 & 20.65 \\ -28.97 & -15.05 & 22.80 & -12.65 \\ 57.11 & 20.65 & -12.65 & 47.31 \end{bmatrix}$$

solves the algebraic Riccati equation, so that

$$u^*(t) = -B^T P \mathbf{x}(t) = \underbrace{[-14.28, \ -5.16, \ 3.16, \ -11.83]}_{F^*} \mathbf{x}(t)$$

Now verify that

$$|sI - A - BF^*| = \delta^{F^*}(s) = s^4 + 2.958s^3 + 5.373s^2 + 5.409s + 2.374$$

as in Example 9.4.24. Also show that when $\rho = 1$, the resulting $\delta^{F^*}(s)$ will correspond to the $\delta^{H^*}(s)$ obtained in Example 9.4.24, namely,

$$\delta^{H^*}(s) = (s + 0.304 \pm j1.161)(s + 0.623 \pm j.365)$$

9-17. Recall the design of an LQR optimal controller for the orbiting satellite in Example 9.4.24, where a SISO observer of order 3, with poles given by the roots of

$$\hat{q}(s) = (s + 0.304)(s + 0.623 \pm j.365) = s^3 + 1.55s^2 + 0.9s + 0.1585$$

was employed to obtain the optimal closed-loop transfer functions

$$T_2^*(s) = \frac{y_2(s)}{r(s)} = \frac{\alpha c_2(s)}{\delta^{F^*}(s)} = \frac{-0.79(s + 1.732)(s - 1.732)}{s^4 + 2.958s^3 + 5.373s^2 + 5.409s + 2.374}$$

and

$$T_1^*(s) = \frac{y_1(s)}{r(s)} = \frac{\alpha c_1(s)}{\delta^{F^*}(s)} = \frac{-3.161s}{s^4 + 2.958s^3 + 5.373s^2 + 5.409s + 2.374}$$

Show that if $y_1(s) = c_1(s)z(s) = sz(s)$ is used for feedback control, in addition to $y_2(s) = c_2(s)z(s) = 0.25(s^2 - 3)z(s)$, then in view of Eq. (9.2.41), the same optimal closed-loop transfer functions can be obtained by an observer of order $r = 1 \left(\dfrac{n-p}{p} = \dfrac{4-2}{2} \right)$ in this case. In particular, if $\hat{q}(s) = s + 0.304$, and $\delta^{F^*}(s)$ is as defined above, determine a monic $k(s)$ of degree 1, and corresponding first-degree polynomials $h_1(s)$ and $h_2(s)$, such that the 2 DOF MO observer design of Figure 9.6 will imply the optimal $T_1^*(s)$ and $T_2^*(s)$ transfer functions noted above.

9-18. Consider the system defined in Example 4.4.45 by the minimum phase transfer function

$$G(s) = \frac{2.4s^3 + 15.36s^2 + 19.2s + 12}{s^4 + 30.08s^3 + 2.44s^2 + 1.2s}$$

which is characterized by three zeros at $s = -0.7 \pm j.714$ and -5, and four poles at $s = 0, -0.04 \pm j.196$, and -30. Determine a 2 DOF robust compensator for this system, as depicted in Figure 9.57, which will ensure a robust $e_{ss}(t) = 0$ for step changes in the reference input $r(t)$, as well as the nominal closed-loop transfer function

$$T(s) = \frac{y(s)}{r(s)} = \frac{4(s^3 + 6s^2 + 11s + 6)}{(s+4)(s^3 + 6s^2 + 11s + 6)} = \frac{4}{s+4}$$

9-19. Show that the LTR optimal observer poles defined by the roots of $c(s)$ in Eq. (9.6.8), with $k(s) = c(s)\tilde{k}(s)$ in light of Eq. (9.6.10), represent closed-loop modes that are both uncontrollable and unobservable.

9-20.* Assume that $a(s) = s^3 + s^2 - 10s + 8$, $\delta^{F^*}(s) = s^3 + 7.37s^2 + 16.65s + 12.4$, and $\tilde{\delta}^{H^*}(s) = \tau s + 1$, as in Example 9.6.19. Show that when $\tau = 0.1$, the polynomial pair

$$\tilde{k}(s) = 0.1s + 1.64 \quad \text{and} \quad h(s) = 8.4s^2 + 33.46s - 0.67$$

solves Eq. (9.6.11).

9-21. Show that if $G(s)$ is a type 1 (or higher) transfer function, then the LTR/IMP robust $e_{ss}(t) = 0$ compensator design for step inputs, as defined by Eqs. (9.6.33) and (9.6.34), can be simplified somewhat. In particular, derive appropriate expressions for the LTR/IMP compensator polynomials $k(s)$, $h(s)$, and $q(s)$ for step inputs when $G(s)$ is type 1.

9-22.* Motivated by an example in Reference [17], consider a dynamic system defined by the state-space representation: $\dot{\mathbf{x}}(t) = A\mathbf{x}(t) + B\mathbf{u}(t)$; $\mathbf{y}(t) = C\mathbf{x}(t)$, with

$$A = \begin{bmatrix} 0 & 1 \\ -3 & -4 \end{bmatrix}, \ B = \begin{bmatrix} 0 \\ 1 \end{bmatrix} \text{ and } C = \begin{bmatrix} C_1 \\ C_2 \end{bmatrix} = \begin{bmatrix} 2 & 1 \\ 52.913 & 8.943 \end{bmatrix}$$

(a) Verify that the open-loop transfer function

$$G(s) = C(sI - A)^{-1}B = \frac{\begin{bmatrix} y_1(s) \\ y_2(s) \end{bmatrix}}{u(s)} = \frac{\begin{bmatrix} c_1(s) \\ c_2(s) \end{bmatrix}}{a(s)}$$

$$= \frac{\begin{bmatrix} s+2 \\ 8.944(s + 5.916) \end{bmatrix}}{s^2 + 4s + 3}$$

so that the system has two stable poles at $s = -1$ and -3, a $y_1(s)$ output zero at $s = -2$, and a $y_2(s)$ output zero at $s = -5.916$.

(b) Show that a minimization of the LQR performance index defined by the *second output* $y_2(t)$, namely,

$$J_{LQR} = \int_0^\infty \{\rho y_2^2(t) + u^2(t)\}\, dt$$

when $\rho = 1$, implies an LQR optimal $\delta^{F^*}(s) = s^2 + 14s + 53 = (s+7\pm j2)$, as defined by the spectral factorization Eq. (9.4.13).

(c) Determine a first-order 2 DOF compensator (observer), with a pole at $s = -5$, that is driven by the plant input $u(s)$ and the *first output* $y_1(s)$, as depicted in Figure 9.60, that will imply the LQR optimal, closed-loop transfer function:

$$T(s) = \frac{\begin{bmatrix} y_1(s) \\ y_2(s) \end{bmatrix}}{r(s)} = \frac{\begin{bmatrix} c_1(s)q(s) \\ c_2(s)q(s) \end{bmatrix}}{\underbrace{a(s)k(s) + c_1(s)h(s)}_{\delta^{F^*}(s)q(s)}} = \frac{\begin{bmatrix} s+2 \\ 8.944(s+5.916) \end{bmatrix}}{s^2 + 14s + 53}$$

(d) Verify that the resulting loop sensitivity function

$$S(s) = \frac{a(s)k(s)}{\delta^{F^*}(s)q(s)}$$

has a maximum value of 5.54 at $\omega = 8.5$, that is, that $\|S\|_\infty = |S(j8.5)| = \bar{S} = 5.54$ in this case, which implies very poor stability margins.

(e) Explain why such a large $\bar{S}$ is associated with this LQR optimal design. (*Hint:* Plot a root locus of the loop gain transfer function, noting that although J_{LQR} is defined by $y_2(t)$, the resulting optimal design is implemented by $y_1(t)$.)

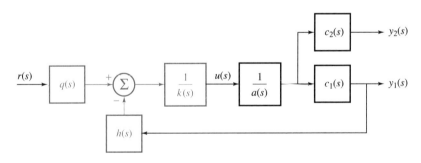

The LQR Optimal Compensator of Problem 9-22

9-23. Consider the 2 DOF compensator polynomials $k(s) = \bar{c}(s)\bar{k}(s)$ and $h(s)$, which satisfy Eq. (9.7.8), together with a $q(s) = \bar{m}(s)\tilde{q}(s)$, which define the closed-loop $T(s)$ of Eq. (9.7.9), as depicted in Figure 9.49. Prove that the zeros of $\tilde{q}(s)$ represent uncontrollable observer modes and that the zeros of $\bar{c}(s)$ represent unobservable compensator modes.

9-24.* Solve the polynomial equation of Example 9.7.10, namely,

$$(s+1)\left[s^4 + (\bar{k}_0 + h_2 + 2)s^3 + (2\bar{k}_0 - 2h_2 + h_1 - 3)s^2 \right.$$

$$\left. + (h_0 - 3\bar{k}_0 - 2h_1)s - 2h_0\right]$$

$$= p(s)\bar{c}(s)\tilde{q}(s) = (s+1)(s^4 + 5s^3 + 10s^2 + 10s + 4)$$

for $h(s) = h_2 s^2 + h_1 s + h_0$ and $k(s) = (s+1)(s+\bar{k}_0)$.

9-25. Show that if

$$G(s) = \frac{3}{s^2 - s}$$

as in Example 7.2.13, then the 2 DOF implementable

$$T(s) = \frac{\omega_n^2}{s^2 + 2\zeta\omega_n s + \omega_n^2} = \frac{2.44}{s^2 + 2.4s + 2.44}$$

of that example is *not* 1 DOF implementable.

9-26. Show that the 2 DOF implementable $T(s)$ of Example 9.7.19, namely,

$$T(s) = \frac{-2s + 4}{s^3 + 4s^2 + 6s + 4}$$

is not 1 DOF implementable from $G(s)$, because $\bar{a}(s) = (s+3)(s-1)$ is unstable.

9-27. Consider the third-order state-space system of Example 9.7.10, which is characterized by the minimal transfer function

$$G(s) = \frac{(s+1)(s-2)}{s(s+3)(s-1)}$$

Using Eq. (9.8.5), show that $\bar{S} \geq 3$, for any proper, rational compensator $H(s)$.

9-28. Show that if the IMC $G(s)$ depicted in Figure 9.53 matches the plant $G(s)$ exactly, then the closed-loop poles of the compensated system will correspond to those of the plant and those of $Q(s)$, so that *an IMC design implies robust closed-loop stability only for stable plants*.

MATRIX/VECTOR ALGEBRA

A **matrix** A, is a two-dimensional, rectangular ($n \times m$) array of nm **components** or **elements**, a_{ij}, which can represent almost any mathematical quantity, such as a number, symbol, function, and so forth. More specifically,

$$A = [a_{ij}] = \begin{bmatrix} a_{11} & a_{12} & \cdots & a_{1m} \\ a_{21} & a_{22} & \cdots & a_{2m} \\ \vdots & \vdots & \ddots & \vdots \\ a_{n1} & a_{n2} & \cdots & a_{nm} \end{bmatrix} \tag{A1}$$

where the a_{ij} can be real, complex, or binary numbers, functions of one or more variables, partial derivatives of n-defined functions of m variables, and so on. Matrix multiplication by a scalar and matrix addition are both defined component-wise, so that $kA = [ka_{ij}]$ for any scalar k. Furthermore, if $B = [b_{ij}]$ has the same dimensions ($n \times m$) as A, then $A + B = B + A = [a_{ij} + b_{ij}]$.

Matrices are employed in virtually all fields of engineering. Perhaps one of their most common uses is in the representation and solution of sets of simultaneous linear equations.[1] Suppose we have n linear equations that relate n variables, namely, $x_1, x_2, \ldots, x_n$, to m variables, namely, $y_1, y_2, \ldots, y_m$.

$$x_1 = a_{11}y_1 + a_{12}y_2 + \cdots + a_{1m}y_m$$

---

[1]The definition of matrices, and their use, is due to the English mathematician James J. Sylvester, who first employed them in 1850 to define the coefficient arrays associated with sets of simultaneous linear equations.

$$x_2 = a_{21}y_1 + a_{22}y_2 + \cdots + a_{2m}y_m$$

$$\vdots$$

$$x_n = a_{n1}y_1 + a_{n2}y_2 + \cdots + a_{nm}y_m \qquad \text{(A2)}$$

This set of simultaneous linear equations can be written in the form

$$
\begin{bmatrix} x_1 \\ x_2 \\ \vdots \\ x_n \end{bmatrix}
=
\begin{bmatrix}
a_{11} & a_{12} & \cdots & a_{1m} \\
a_{21} & a_{22} & \cdots & a_{2m} \\
\vdots & \vdots & \ddots & \vdots \\
a_{n1} & a_{n2} & \cdots & a_{nm}
\end{bmatrix}
\begin{bmatrix} y_1 \\ y_2 \\ \vdots \\ y_m \end{bmatrix}
\qquad \text{(A3)}
$$

if multiplication of the matrix A "times" the $(m \times 1)$ vector[2]

$$
y = \begin{bmatrix} y_1 \\ y_2 \\ \vdots \\ y_m \end{bmatrix}
$$

is defined as in Eq. (A2); that is, if for each ith row of A, the corresponding jth column element, a_{ij}, multiplies y_j, and the m resulting terms are then summed.

Matrix multiplication defined in this manner extends to those cases where the second matrix (associated with the multiplication) has more than one column. In particular, if A, B, and C are $(n \times m)$, $(m \times p)$, and $(n \times p)$ matrices, respectively, then matrix multiplication is defined by the relationship

$$
\underbrace{\begin{bmatrix}
c_{11} & c_{12} & \cdots & c_{1p} \\
c_{21} & c_{22} & \cdots & c_{2p} \\
\vdots & \vdots & \ddots & \vdots \\
c_{n1} & c_{n2} & \cdots & c_{np}
\end{bmatrix}}_{C}
=
\underbrace{\begin{bmatrix}
a_{11} & a_{12} & \cdots & a_{1m} \\
a_{21} & a_{22} & \cdots & a_{2m} \\
\vdots & \vdots & \ddots & \vdots \\
a_{n1} & a_{n2} & \cdots & a_{nm}
\end{bmatrix}}_{A}
\underbrace{\begin{bmatrix}
b_{11} & b_{12} & \cdots & b_{1p} \\
b_{21} & b_{22} & \cdots & b_{2p} \\
\vdots & \vdots & \ddots & \vdots \\
b_{m1} & b_{m2} & \cdots & b_{mp}
\end{bmatrix}}_{B}
$$

or as

$$C = AB \qquad \text{(A4)}$$

with each component of C given by the relationship:

$$c_{ij} = a_{i1}b_{1j} + a_{i2}b_{2j} + \cdots + a_{im}b_{mj} = \sum_{k=1}^{m} a_{ik}b_{kj} \qquad \text{(A5)}$$

---

[2] Any $n \times m$ matrix with either $n = 1$ or $m = 1$ is called a **vector**, and if both $n = 1$ and $m = 1$, then the vector is called a **scalar**.

The reader should note the relative simplicity and "compactness" associated with the matrix representation for multiplication, as given by Eq. (A4), which implies the more detailed relations given by Eq. (A5).

To define the matrix product AB, the number of columns (m) of A must equal the number of rows (m) of B. This compatibility condition must be satisfied by any matrix product. Note further, that if $p \neq n$, the matrix "product" BA is undefined. Indeed, even if $p = n$, BA would be an ($m \times m$) matrix and $C = AB$ would be ($n \times n$) = ($p \times p$), so assuming that $m \neq n = p$,

$$AB \neq BA \tag{A6}$$

It may be noted that Eq. (A6) is generally true even when $m = p = n$; that is, *matrix multiplication is noncommutative.*

A particularly useful concept in matrix/vector algebra is that of "matrix transposition." In particular, looking at Eq. (A1), the **transpose** of a matrix A, which we will denote as A^T, is defined as the ($m \times n$) matrix that results from an ordered interchange of the rows and columns of A. More specifically, using Eq. (A1),

$$A^T \overset{\text{def}}{=} \begin{bmatrix} a_{11} & a_{21} & \cdots & a_{n1} \\ a_{12} & a_{22} & \cdots & a_{n2} \\ \vdots & \vdots & \ddots & \vdots \\ a_{1m} & a_{2m} & \cdots & a_{nm} \end{bmatrix}$$

In view of this definition, both AA^T and A^TA always exist, although they are generally not equal to one another.

The following specific properties associated with matrix transposition also hold:

$$\left[A^T\right]^T = A$$

$$[A + B]^T = A^T + B^T$$

$$[kA]^T = kA^T, \quad \text{for any scalar} \quad k$$

and

$$[AB]^T = B^T A^T$$

The ($n \times n$) **identity matrix** I, or I_n, is defined as a square matrix whose diagonal elements are 1 and whose off-diagonal elements are zero,

$$I \quad (\text{or} \quad I_n) \overset{\text{def}}{=} \begin{bmatrix} 1 & 0 & 0 & \cdots \\ 0 & 1 & 0 & \cdots \\ \vdots & & \ddots & \\ 0 & 0 & \cdots & 1 \end{bmatrix}$$

with n defined as the number of both rows and columns of I. In light of this definition and Eq. (A1), it follows that

$$I_n A = A I_n = A$$

We can associate with any square ($n \times n$) matrix A a scalar quantity, called the **determinant** of A, that is denoted by either det(A), or by parallel bars around A; that is, both det(A) and $|A|$ will be used to denote the determinant of A. The evaluation of $|A|$ is relatively straightforward when n is either 2 or 3. When $n = 2$,

$$|A| = a_{11}a_{22} - a_{12}a_{21} \tag{A7}$$

and when $n = 3$,

$$|A| = a_{11}a_{22}a_{33} + a_{13}a_{21}a_{32} + a_{12}a_{23}a_{31} - a_{13}a_{22}a_{31} - a_{11}a_{23}a_{32} - a_{12}a_{21}a_{33} \tag{A8}$$

For $n > 3$, it is convenient to introduce the notion of "cofactors" to explicitly evaluate $|A|$. For any square ($n \times n$) matrix $A = [a_{ij}]$, the **cofactor** of element a_{ij}, which will be denoted as a_{ij}^*, is defined as $(-1)^{i+j}$ times the determinant of the ($n-1 \times n-1$) matrix obtained by deleting the ith row and the jth column of A. In light of this definition, the determinant of any ($n \times n$) matrix A, when $n > 3$, can be evaluated using Eq. (A8) and any one of the following $2n$ relationships

$$|A| = \sum_{i=1}^{n} a_{ij}a_{ij}^* = \sum_{j=1}^{n} a_{ij}a_{ij}^* \tag{A9}$$

for any i or j between 1 and n. It may be noted that either summation defined by Eq. (A9), together with Eq. (A8), can be used to determine $|A|$, and that both summations can be evaluated in n different ways, as we will now show.

Suppose we wish to determine $|A|$ when

$$A = \begin{bmatrix} 2 & 0 & 3 & 2 \\ -2 & 1 & 0 & 0 \\ 0 & 4 & 0 & 1 \\ -1 & -2 & 1 & 0 \end{bmatrix} \tag{A10}$$

In light of the second summation defined by Eq. (A9), with $i = 3$,

$$|A| = \sum_{j=1}^{4} a_{3j}a_{3j}^* = a_{31}a_{31}^* + a_{32}a_{32}^* + a_{33}a_{33}^* + a_{34}a_{34}^* = 4a_{32}^* + a_{34}^* \tag{A11}$$

since $a_{31} = a_{33} = 0$, $a_{32} = 4$, and $a_{34} = 1$. We next note that by definition

$$a_{32}^* = (-1)^5 \det \begin{bmatrix} 2 & 3 & 2 \\ -2 & 0 & 0 \\ -1 & 1 & 0 \end{bmatrix} = -1(-4) = 4 \tag{A12}$$

in view of Eq. (A8). Furthermore,

$$a_{34}^* = (-1)^7 \det \begin{bmatrix} 2 & 0 & 3 \\ -2 & 1 & 0 \\ -1 & -2 & 1 \end{bmatrix} = -1(2 + 12 + 3) = -17 \tag{A13}$$

so that

$$|A| = 4(4) - 17 = -1 \qquad (A14)$$

We can also determine $|A|$ in $(7 = 2n-1)$ "other ways." For example, if we chose $j = 3$ in the first summation defined by Eq. (A9), it follows that

$$|A| = \sum_{i=1}^{4} a_{i3}a_{i3}^* = 3a_{13}^* + a_{43}^*$$

since $a_{23} = a_{33} = 0$, $a_{13} = 3$, and $a_{43} = 1$. We next employ Eq. (A8) to determine that

$$a_{13}^* = (-1)^4 \det \begin{bmatrix} -2 & 1 & 0 \\ 0 & 4 & 1 \\ -1 & -2 & 0 \end{bmatrix} = -1 - 4 = -5 \qquad (A15)$$

and that

$$a_{43}^* = (-1)^7 \det \begin{bmatrix} 2 & 0 & 2 \\ -2 & 1 & 0 \\ 0 & 4 & 1 \end{bmatrix} = -1(2 - 16) = 14 \qquad (A16)$$

so that

$$|A| = 3(-5) + 14 = -1$$

which verifies Eq. (A14); that is, *a matrix has only one unique determinant.*

Perhaps the most important use of determinants and cofactors is that associated with the evaluation of "inverse" of A. If (and only if) $|A| \neq 0$, then A has a unique **inverse**, denoted as A^{-1}, that satisfies the relationship

$$AA^{-1} = A^{-1}A = I$$

so that

$$|AA^{-1}| = |A||A^{-1}| = |I| = 1 \qquad \Longrightarrow \qquad |A^{-1}| = \frac{1}{|A|}$$

To determine[3] A^{-1}, we first define A^+ as the transpose of the matrix of cofactors of A,

$$A^+ \overset{\text{def}}{=} [A^*]^T = [a_{ij}^*]^T = \begin{bmatrix} a_{11}^* & a_{21}^* & \cdots & a_{n1}^* \\ a_{12}^* & a_{22}^* & \cdots & a_{n2}^* \\ \vdots & \vdots & \ddots & \vdots \\ a_{1n}^* & a_{2n}^* & \cdots & a_{nn}^* \end{bmatrix}$$

It then follows that

$$A^{-1} = \frac{A^+}{|A|} \qquad (A17)$$

---

[3]There are several computer-aided design packages, such as MATLAB or MATRIX$_x$, that can be used to directly determine a variety of matrix functions, including the inverse.

A square matrix for which $|A| \neq 0$ is said to be **invertible** or **nonsingular**.

If $|A| = 0$, then A is said to be **singular**, and A^{-1} does not exist. In such cases, the **rank** of A (or the rank of any matrix with any number of rows and columns) is defined as the dimension of the largest *submatrix* of the given matrix with a nonzero determinant. A submatrix of A is any matrix formed from A by eliminating certain rows and columns of A. If A is square ($n \times n$) and nonsingular, then A also is said to be of **full rank** (n).

If $|A| \neq 0$, so that A^{-1} does exist, simultaneous linear equations, such as Eq. (A3), with $m = n$, can be solved for $y = \begin{bmatrix} y_1 \\ y_2 \\ \vdots \\ y_n \end{bmatrix}$ in terms of

$x \stackrel{\text{def}}{=} \begin{bmatrix} x_1 \\ x_2 \\ \vdots \\ x_n \end{bmatrix}$ directly, by simply premultiplying both sides of the equation

by A^{-1}. More specifically, if Eq. (A3) is written in vector/matrix form as

$$x = Ay$$

then a premultiplication by A^{-1} implies that $A^{-1}x = A^{-1}Ay = y$ or that

$$y = A^{-1}x.$$

To illustrate the notion of a matrix inverse, we now note that for the particular A defined by Eq. (A10),

$$A^+ = [A^*]^T = \begin{bmatrix} -1 & -2 & 2 & 3 \\ -2 & -5 & 4 & 6 \\ -5 & -12 & 10 & 14 \\ 8 & 20 & -17 & -24 \end{bmatrix}$$

using Eqs. (A12), (A13), (A15), (A16), and so on. Since $|A| = -1$, in light of Eq. (A14), Eq. (A17) now implies that

$$A^{-1} = \frac{A^+}{|A|} = \begin{bmatrix} 1 & 2 & -2 & -3 \\ 2 & 5 & -4 & -6 \\ 5 & 12 & -10 & -14 \\ -8 & -20 & 17 & 24 \end{bmatrix}$$

An **eigenvalue** of a square ($n \times n$) matrix A is any scalar λ such that

$$Av = \lambda v \tag{A18}$$

for any ($n \times 1$) **eigenvector** v, of A. Note that Eq. (A18) can be written as

$$(\lambda I - A)v = 0$$

Therefore, v is a nonzero vector if and only if the determinant of $(\lambda I - A)$ is zero, which defines the **characteristic equation of** A, namely,

$$|\lambda I - A| = \lambda^n + a_{n-1}\lambda^{n-1} + \cdots + a_1\lambda + a_0 = 0$$

with

$$a(\lambda) \overset{\text{def}}{=} \lambda^n + a_{n-1}\lambda^{n-1} + \cdots + a_1\lambda + a_0$$

the **characteristic polynomial** of A. Note that $a(\lambda)$ has exactly n roots, λ_1, λ_2, ... , λ_n, which correspond to the n eigenvalues of A, some of which may be repeated. Moreover, each eigenvalue λ_i of A defines a corresponding eigenvector v_i. When there are repeated eigenvalues, these eigenvectors also may be repeated.

We next establish the **Cayley-Hamilton Theorem**, which states that *any square matrix A satisfies its own characteristic equation*, or that

$$a(A) = A^n + a_{n-1}A^{n-1} + \ldots + a_1A + a_0I = 0$$

To establish this fact, we first note that

$$(\lambda I - A)^{-1} = \frac{(\lambda I - A)^+}{a(\lambda)} \tag{A19}$$

where $(\lambda I - A)^+$ can be expressed as

$$(\lambda I - A)^+ = B_0 + B_1\lambda + \cdots + B_{n-1}\lambda^{n-1} \tag{A20}$$

If we now multiply Eq. (A19) by $a(\lambda)(\lambda I - A)$, and substitute the right side of Eq. (A20) for $(\lambda I - A)^+$, we obtain the relationship

$$a_0I + a_1I\lambda + \cdots + a_{n-1}I\lambda^{n-1} + I\lambda^n$$
$$= (\lambda I - A)(B_0 + B_1\lambda + \cdots + B_{n-1}\lambda^{n-1})$$
$$= -AB_0 + (B_0 - AB_1)\lambda + \cdots + (B_{n-2} - AB_{n-1})\lambda^{n-1} + B_{n-1}\lambda^n$$

By equating coefficients of identical λ powers, we obtain the following relationships:

$$AB_0 = -a_0I$$

$$B_0 = AB_1 + a_1I$$

$$B_1 = AB_2 + a_2I$$

$$\vdots$$

$$B_{n-1} = I \tag{A21}$$

Recursively substituting $AB_1 + a_1I$ for B_0 in the relationship $AB_0 = -a_0I$, then $AB_2 + a_2I$ for B_1, and so on, up to and including I for B_{n-1}, as given by relationship Eq. ($A21$), we find that

$$A^n + a_{n-1}A^{n-1} + \cdots + a_1A + a_0I = 0$$

which establishes the Cayley-Hamilton theorem.

If all n eigenvalues of A are distinct, they define n corresponding distinct eigenvectors, $v_1, v_2, \ldots, v_n$. In such cases, Eq. (A18) implies that

$$A [\, v_1 \quad v_2 \quad \cdots \quad v_n \,] = [\, v_1 \quad v_2 \quad \cdots \quad v_n \,] \begin{bmatrix} \lambda_1 & 0 & 0 & \cdots \\ 0 & \lambda_2 & 0 & \cdots \\ \vdots & & & \ddots \\ 0 & 0 & \cdots & \lambda_n \end{bmatrix}$$

where

$$[\, v_1 \quad v_2 \quad \cdots \quad v_n \,] \overset{\text{def}}{=} V$$

is a nonsingular $(n \times n)$ **similarity transformation** matrix composed of n distinct eigenvectors of A that transforms A to a diagonal, or a **Jordan canonical form**,

$$V^{-1}AV = \begin{bmatrix} \lambda_1 & 0 & 0 & \cdots \\ 0 & \lambda_2 & 0 & \cdots \\ \vdots & & \ddots & \\ 0 & 0 & \cdots & \lambda_n \end{bmatrix} \overset{\text{def}}{=} \Lambda$$

Any two square matrices A and $\hat{A}$ that are **similar**, in the sense that $QAQ^{-1} = \hat{A}$ for some nonsingular (similarity transformation) matrix Q, have the same characteristic equation, hence the same eigenvalues, since

$$|\lambda I - \hat{A}| = |\lambda I - QAQ^{-1}| = |Q||\lambda I - A||Q^{-1}| = |\lambda I - A|$$

A square $(n \times n)$ matrix A is in **bottom-row companion form** if the upper right $(n-1 \times n-1)$ submatrix of A is I_{n-1} and the bottom nth row contains the only nontrivial elements, that is, if

$$A = \begin{bmatrix} 0 & 1 & 0 & \cdots & 0 \\ 0 & 0 & 1 & \cdots & 0 \\ \vdots & \vdots & & \ddots & \vdots \\ -a_0 & -a_1 & & \cdots & -a_{n-1} \end{bmatrix}$$

If such is the case,

$$|\lambda I - A| = \begin{vmatrix} \lambda & -1 & 0 & \cdots & 0 \\ 0 & \lambda & -1 & \cdots & 0 \\ \vdots & \vdots & & \ddots & \vdots \\ a_0 & a_1 & & \cdots & \lambda + a_{n-1} \end{vmatrix}$$

If we now evaluate $|\lambda I - A|$ via its nth row **minors**, that is, the second summation defined by Eq. (A9), with $i = n$, it follows that the characteristic polynomial of a companion form matrix A is immediately apparent by inspection, since

$$|\lambda I - A| = \sum_{j=1}^{n} a_{nj} a_{nj}^{*} = a_0 + a_1 \lambda + \cdots + a_{n-1} \lambda^{n-1} + \lambda^n = a(\lambda)$$

LAPLACE TRANSFORMS AND POLYNOMIAL ALGEBRA

The **Laplace transform**[1] is a very useful mathematical tool that is employed to a great extent in engineering and applied mathematics. If a function of time $f(t)$ satisfies the condition

$$\int_0^\infty |f(t)e^{-\sigma t}|\, dt < \infty \tag{B1}$$

for some positive scalar σ, then the Laplace transform of $f(t)$, which will be denoted as[2] $f(s) = \mathcal{L}[f(t)]$, is formally defined by means of the integral formula

$$f(s) = \mathcal{L}[f(t)] \stackrel{\text{def}}{=} \int_0^\infty f(t)e^{-st}\, dt \tag{B2}$$

where $s = \sigma + j\omega$ is defined to be the complex **Laplace variable**. For virtually all of the time functions of interest to us in this text, σ can and will be chosen large enough to insure that (B1) holds, as we will show.

---

[1] So named for the Marquis Pierre-Simon de Laplace (1749–1827), a famous French mathematician who contributed much to the field of applied mathematics.

[2] Note that the same letter designation (in this case f) will be used to denote both $f(t)$ and $f(s)$.

The Laplace transform defined by Eq. (B2) is sometimes termed "one-sided" because all information regarding $f(t)$ prior to $t = 0$ is ignored, or assumed to be zero, since the analysis of physical systems usually assumes that the dynamic behavior begins at some initial time t_0 that, for convenience, is often chosen to be 0.

If $f(t) = t^n$, for integer values of n (which would include the step, ramp, and parabolic input functions defined in Section 4.2) it follows that

$$\mathcal{L}\left[t^n\right] = \int_0^\infty t^n e^{-st}\, dt$$

A change of variables, with

$$x \stackrel{\text{def}}{=} st \quad\Longrightarrow\quad dt = \frac{dx}{s} \quad\text{and}\quad t^n = \frac{x^n}{s^n}$$

then implies that

$$\mathcal{L}\left[t^n\right] = \frac{1}{s^{n+1}} \underbrace{\int_0^\infty x^n e^{-x}\, dx}_{\stackrel{\text{def}}{=}\ \Gamma(n+1)} \tag{B3}$$

An integration by parts then implies that

$$\underbrace{\int_0^\infty x^{n-1} e^{-x}\, dx}_{\Gamma(n)} = \underbrace{\frac{1}{n} e^{-x} x^n \Big|_0^\infty}_{0} + \frac{1}{n} \underbrace{\int_0^\infty x^n e^{-x}\, dx}_{\Gamma(n+1)}$$

or that

$$\Gamma(n+1) = n\Gamma(n)$$

Since

$$\Gamma(n = 1) = \int_0^\infty e^{-x}\, dx = 1$$

it follows that $\Gamma(2) = 1!$, $\Gamma(3) = 2!$, ..., or, in general, that $\Gamma(n+1) = n!$ In light of Eq. (B3), we thus conclude that

$$\mathcal{L}\left[t^n\right] = \frac{n!}{s^{n+1}}$$

If $f(t)$ is the unit sinusoid, that is, if

$$f(t) = \sin \omega t = \frac{e^{j\omega t} - e^{-j\omega t}}{2j}$$

then

$$\mathcal{L}[\sin \omega t)] = \int_0^\infty \frac{e^{j\omega t} - e^{-j\omega t}}{2j} e^{-st}\, dt$$

$$= \frac{1}{2j} \int_0^\infty \left[e^{-(s-j\omega)t} - e^{-(s+j\omega)t}\right] dt$$

$$= \frac{1}{2j} \left[\frac{1}{s - j\omega} - \frac{1}{s + j\omega} \right] = \frac{\omega}{s^2 + \omega^2}$$

provided that $\sigma = Re(s) > 0$.

In light of Eq. (B2), the Laplace transform of the time derivative of $f(t)$ is given by

$$\mathcal{L}\left[Df(t) = \frac{df(t)}{dt} \right] = \int_0^\infty \frac{df(t)}{dt} e^{-st}\, dt \tag{B4}$$

Integration by parts now implies that the right side of Eq. (B4) is given by

$$f(t)e^{-st}\Big|_0^\infty + s \int_0^\infty f(t)e^{-st}\, dt = -f(0) + sf(s)$$

so that when $f(0) = 0$, which includes most of the time-varying functions considered in this text,

$$\mathcal{L}[Df(t)] = sf(s)$$

For higher derivatives of t,

$$\mathcal{L}\left[D^k f(t) \right] = s^k f(s) - \lim_{t \to 0} \left[s^{k-1} f(t) + s^{k-2} Df(t) + \cdots + D^{k-1} f(t) \right]$$

$$= s^k f(s) - s^{k-1} f(0) - s^{k-2} f^{(1)}(0) - \cdots - f^{(k-1)}(0)$$

Therefore, if $f(t)$ and its first $k - 1$ derivatives are equal to 0 at $t = 0$,

$$\mathcal{L}[D^k f(t)] = s^k f(s)$$

It often is useful to determine the initial (when $t = 0$) and the final (when $t \to \infty$) value of a function of time $f(t)$ from its Laplace transform $f(s)$. The following two theorems, whose proof can be found in Reference [4], provide such information.

For any given function $f(t)$ of time, the **initial value theorem** states that if $f(s) = \mathcal{L}[f(t)]$, then the initial value of $f(t)$,

$$f(t = 0) = \lim_{s \to \infty} sf(s)$$

In an analogous manner, the **final value theorem** states that if $f(s) = \mathcal{L}[f(t)]$, and if $sf(s)$ is analytic in the half-plane $Re(s) \geq 0$, then the final value of $f(t)$, namely

$$\lim_{t \to \infty} f(t) = \lim_{s \to 0} sf(s)$$

In certain situations, such as physical systems characterized by transportation lag, a time function $f(t)$ can be delayed by a finite amount of time τ. When this occurs, the Laplace transform of the delayed signal is equal to $f(s) = \mathcal{L}[f(t)]$ multiplied by $e^{-s\tau}$; that is, if $r_s(t - \tau)$ is the unit step function delayed by time τ, then

$$\mathcal{L}[f(t - \tau)r_s(t - \tau)] \stackrel{\text{def}}{=} \mathcal{L}[f(t - \tau)] = e^{-s\tau} f(s)$$

TABLE B.1 Some Common Laplace Transform Pairs

Name	Time Function $f(t)$	Laplace Transform $f(s)$
Unit Impulse	$\delta(t)$	1
Unit Step	$r_s(t) = 1$	$\dfrac{1}{s}$
Unit Ramp	$r_r(t) = t$	$\dfrac{1}{s^2}$
nth Order Ramp	t^n	$\dfrac{n!}{s^{n+1}}$
Exponential Decay	e^{-at}	$\dfrac{1}{s+a}$
Sine Wave	$\sin \omega t$	$\dfrac{\omega}{s^2 + \omega^2}$
Cosine Wave	$\cos \omega t$	$\dfrac{s}{s^2 + \omega^2}$
Damped Sine Wave	$e^{-at} \sin \omega t$	$\dfrac{\omega}{(s+a)^2 + \omega^2}$
Damped Cosine Wave	$e^{-at} \cos \omega t$	$\dfrac{s+a}{(s+a)^2 + \omega^2}$
Damped nth Order Ramp	$e^{-at} t^n$	$\dfrac{n!}{(s+a)^{n+1}}$
Scalar Multiplication	$kf(t)$	$kf(s)$
Addition	$f_1(t) \pm f_2(t)$	$f_1(s) \pm f_2(s)$
First Order Differentiation	$\dfrac{df(t)}{dt}$	$sf(s) - f(0)$
nth Order Differentiation (with zero initial conditions)	$\dfrac{d^n f(t)}{dt^n}$	$s^n f(s)$
Integration	$\int_0^t f(t)dt$	$\dfrac{f(s)}{s}$
Transportation Lag	$f(t - \tau)$	$e^{-s\tau} f(s)$

Table B1 lists some of the more common Laplace transform pairs.

Polynomials in the Laplace operator s occur frequently in the analysis and design of linear control systems. For example, a strictly proper, rational transfer function

$$G(s) = \frac{c(s)}{a(s)}$$

of a linear, time-invariant SISO system is the ratio of two polynomials in s, namely

$$c(s) = c_0 + c_1 s + \cdots + c_{n-1} s^{n-1}$$

and

$$a(s) = a_0 + a_1 s + \cdots + a_{n-1}s^{n-1} + s^n$$

The polynomials $c(s)$ and $a(s)$ are said to be **relatively prime**, or **coprime**, if they have no common roots. As shown in References [38] and [62], the $(2n \times 2n)$ **Sylvester matrix**

$$\mathcal{S} \stackrel{\text{def}}{=} \begin{bmatrix} a_0 & 0 & \cdots & 0 & c_0 & 0 & \cdots & 0 \\ a_1 & a_0 & \cdots & 0 & c_1 & c_0 & \cdots & 0 \\ \vdots & \vdots & \ddots & \vdots & \vdots & \vdots & \ddots & \vdots \\ a_{n-1} & a_{n-2} & \cdots & a_0 & c_{n-1} & c_{n-2} & \cdots & c_0 \\ 1 & a_{n-1} & \cdots & a_1 & 0 & c_{n-1} & \cdots & c_1 \\ 0 & 1 & \cdots & a_2 & 0 & 0 & \cdots & c_2 \\ \vdots & \vdots & \ddots & \vdots & \vdots & \vdots & \ddots & \vdots \\ 0 & 0 & \cdots & 1 & 0 & 0 & \cdots & 0 \end{bmatrix}$$

consisting of the "shifted" coefficients of both polynomials, will be non-singular if and only if $c(s)$ and $a(s)$ are coprime.

In light of this definition for $\mathcal{S}$, note that

$$\underbrace{[1 \, s \, s^2 \cdots \, s^{2n-1}]}_{\stackrel{\text{def}}{=} U(s)} \mathcal{S} = \underbrace{[a(s) \, sa(s) \, \ldots \, s^{n-1}a(s) \, c(s) \, sc(s) \, \ldots \, s^{n-1}c(s)]}_{\stackrel{\text{def}}{=} V(s)}$$

Therefore, if $\mathcal{S}$ is nonsingular, that is, if $c(s)$ and $a(s)$ are coprime, it follows that

$$V(s)\mathcal{S}^{-1} = U(s) \qquad (B5)$$

Note further that if

$$k(s) \stackrel{\text{def}}{=} k_0 + k_1 s + \cdots + k_{n-2}s^{n-2} + s^{n-1}$$

and

$$h(s) \stackrel{\text{def}}{=} h_0 + h_1 s + \cdots + h_{n-1}s^{n-1}$$

then

$$V(s) \underbrace{\begin{bmatrix} k_0 \\ k_1 \\ \vdots \\ k_{n-1} = 1 \\ h_0 \\ h_1 \\ \vdots \\ h_{n-1} \end{bmatrix}}_{\stackrel{\text{def}}{=} W} = a(s)k(s) + c(s)h(s)$$

Therefore, if

$$\delta(s) = \delta_0 + \delta_1 s + \cdots + \delta_{2n-2} s^{2n-2} + s^{2n-1} = U(s) \underbrace{\begin{bmatrix} \delta_0 \\ \delta_1 \\ \vdots \\ \delta_{2n-2} \\ \delta_{2n-1} = 1 \end{bmatrix}}_{\stackrel{\text{def}}{=} \bar{\delta}} \tag{B6}$$

is any arbitrary monic polynomial of degree $2n - 1$, a choice of [3]

$$W = \mathcal{S}^{-1} \bar{\delta} \tag{B7}$$

will imply that

$$V(s)W = a(s)k(s) + c(s)h(s) = V(s)\mathcal{S}^{-1}\bar{\delta} = U(s)\bar{\delta} = \delta(s) \tag{B8}$$

using Eqs. (B5) and (B6), or that the coefficients of polynomials $k(s)$ and $h(s)$ can be chosen using Eq. (B7) to solve the so-called **Diophantine Eq. (B8)**; that is, to obtain any arbitrary polynomial $\delta(s)$ of degree $2n - 1$. This observation will be termed the **pole placement algorithm**, since $k(s)$ and $h(s)$ can represent compensator polynomials which place the poles of a closed-loop system at those arbitrary s-plane locations defined by the roots of $\delta(s)$.

---

[3]This is equivalent to solving the matrix/vector equation $\mathcal{S}W = \bar{\delta}$ for W.

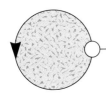

REFERENCES

[1] Ackermann, J. 1972. Der Entwurf Linearer Regelungssysteme im Zustandsraum. *Regelungstechnik und Prozessdatenverarbeitung.* 7:297–300.

[2] AIEE Committee Report. 1951. Proposed symbols and terms for feedback control systems. *Electrical Engineering.* 70:909.

[3] Airy, G. B. 1840. On the regulator of the clockwork for effecting uniform movement of the equatoreals. *Memoirs, Royal Astronomical Society.* 11:249–267.

[4] Aseltine, John A. 1958. *Transform Method in Linear System Analysis.* New York: McGraw-Hill.

[5] Astrom, Karl J., and Bjorn Wittenmark. 1984. *Computer Controlled Systems Theory and Design.* Englewood Cliffs, NJ: Prentice-Hall.

[6] Athans, Michael, and Peter L. Falb. 1966. *Optimal Control: An Introduction to the Theory and Its Applications.* New York: McGraw-Hill.

[7] Bode, H. W. 1945. *Network Analysis and Feedback Amplifier Design.* New York: Van Nostrand.

[8] Bower, J. L., and P. M. Schultheiss. 1958. *Introduction to the Design of Servomechanisms.* New York: John Wiley and Sons.

[9] Black, H. S. 1934. Stabilized feed-back amplifiers. *Bell System Technical Journal.* 13:1–18.

[10] Blakelock, John H. 1965. *Automatic Control of Aircraft and Missiles.* New York: John Wiley and Sons.

[11] Brockett, Roger W. 1965. Poles, zeros, and feedback: State space interpretation. *IEEE Transactions on Automatic Control.* AC-10 (2):129–135.

[12] Brockett, Roger W. 1970. *Finite Dimensional Linear Systems.* New York: John Wiley and Sons.

[13] Chang, S. S. L. 1961. *Synthesis of Optimum Control Systems.* New York: McGraw-Hill.

[14] Chen, Chi-Tsong. 1993. *Analog and Digital Control System Design: Transfer-Function, State-Space, and Algebraic Methods.* Philadelphia: Saunders College Publishing.

[15] Churchill, R. V. 1960. *Introduction to Complex Variables and Applications.* 2nd ed. New York: McGraw-Hill.

[16] D'Azzo, John J., and Constantine H. Houpis. 1988. *Linear Control Systems Analysis: Conventional and Modern.* 3rd ed. New York: McGraw-Hill.

[17] Doyle, J. C., and G. Stein. 1979. Robustness with observers. *IEEE Transactions on Automatic Control.* AC-24 (4):607–611.

[18] Doyle, J. C., and G. Stein. 1981. Multivariable feedback design: Concepts for a classical/modern synthesis. *IEEE Transactions on Automatic Control.* AC-26 (1):4–16.

[19] Doyle, John C., Bruce A. Francis, and Allen R. Tannenbaum. 1992. *Feedback Control Theory.* New York: Macmillan.

[20] Evans, W. R. 1948. Graphical analysis of control systems. *AIEE Transactions Part II.* 67:547–551.

[21] Francis, Bruce A. 1987. *A Course in* $\mathbf{H}_\infty$ *Control Theory.* Springer-Verlag Lecture Notes in Control and Information Sciences. No. 88. New York: Springer-Verlag.

[22] Francis, Bruce A., and W. M. Wonham. 1976. The internal model principle of control theory. *Automatica.* 12:457–465.

[23] Franklin, Gene F., J. David Powell, and Abbas Emami-Naeini. 1991. *Feedback Control of Dynamic Systems*. 2nd ed. Reading, MA: Addison-Wesley.

[24] Freudenberg, J. S., and D. P. Looze. 1985. Right half plane poles and zeros and design tradeoffs in feedback systems. *IEEE Transactions on Automatic Control*. AC-30 (6):555–565.

[25] Freudenberg, J. S., and D. P. Looze. 1988. *Frequency Domain Properties of Scalar and Multivariable Feedback Systems*. Springer-Verlag Lecture Notes on Control and Information Sciences. No. 104. New York: Springer-Verlag.

[26] Friedland, Bernard. 1986. *Control System Design, An Introduction to State-Space Methods*. New York: McGraw-Hill.

[27] Fuller, A. T. 1976. The early development of control theory. *Transactions ASME, Journal of Dynamic Systems, Measurement and Control*. 98:109–118 and 224–235.

[28] Gantmacher, F. R. 1959. *The Theory of Matrices*, vols. I and II. New York: Chelsea.

[29] Graham, D., and R. C. Lathrop. 1953. The synthesis of optimum response: Criteria and standard forms. *AIEE Transactions Part II*. 72:273-288.

[30] Greenwood, Donald T. 1965. *Principles of Dynamics*. Englewood Cliffs, NJ: Prentice-Hall.

[31] Hazen, H. L. 1934. Theory of servomechanisms. *Journal of the Franklin Institute*. 218:279–331.

[32] Horowitz, I. M. 1959. Fundamental theory of linear feedback control systems. *IRE Transactions on Automatic Control*. AC-4:5–19.

[33] Horowitz, I. M. 1992. *Quantitative Feedback Design: Theory (QFT)*. vol 1. Boulder, CO: QFT Publications.

[34] Hostetter, Gene H., Clement J. Savant, Jr., and Raymond T. Stefani. 1989. *Design of Feedback Control Systems*. 2nd ed. Philadelphia: Saunders College Publishing.

[35] Hurwitz, A. 1895. On the conditions under which an equation has

only roots with negative real parts. *Mathematische Annalen*. 46:273–284.

[36] James, H. M., N. B. Nichols, and R. S. Phillips. 1947. *Theory of Servomechanisms*. MIT Radiation Laboratory Series, vol. 25. New York: McGraw-Hill.

[37] Joseph, P. D., and J. Tou. 1961. On linear control theory. *AIEE Transactions Part II*. 80(11)193–196.

[38] Kailath, Thomas. 1980. *Linear Systems*. Englewood Cliffs, NJ: Prentice-Hall.

[39] Kalman, R. E. 1960. Contributions to the theory of optimal control. *Boletin de la Sociedad Matematica Mexicana*.

[40] Kumar, P. R., and Pravin Varaiya. 1986. *Stochastic Systems: Estimation, Identification and Adaptive Control*. Englewood Cliffs, NJ: Prentice-Hall.

[41] Kuo, Benjamin C. 1987. *Automatic Control Systems*. 5th ed. Englewood Cliffs, NJ: Prentice-Hall.

[42] Kuraoka, Hiroaki, Naoto Ohka, Masahiro Ohba, Shigeyuki Hosoe, and Feifei Zhang. 1990. Application of H-infinity design to automotive fuel control. *IEEE Control Systems Magazine*. 10 (3):102–106.

[43] Luenberger, D. G. 1964. Observing the state of a linear system. *IEEE Transactions of Military Electronics*. MIL-8:74–80.

[44] MacFarlane, A. J. G., ed. 1979. *Frequency Response Methods in Control Systems*. New York: IEEE Press.

[45] Maciejowski, J. M. 1989. *Multivariable Feedback Design*. Reading, MA: Addison-Wesley.

[46] *The MATLAB User's Guide*. South Natick, MA: The Math Works.

[47] Maxwell, J. C. 1868. On Governors. *Philosophical Magazine*. 35:385–398.

[48] Mayr, Otto. 1970. *The Origins of Feedback Control*. Cambridge: M.I.T. Press.

[49] Mihalacopoulos, G., and W. A. Wolovich. 1993. Sensitivity function magnitude bounds for plants with RHP poles and zeros. Brown University Division of Engineering Technical Report LEMS-120.

[50] Morari, Manfred, and Evanghelos Zaririou. 1989. *Robust Process Control*. Englewood Cliffs, NJ: Prentice-Hall.

[51] Nyquist, H. 1932. Regeneration theory. *Bell Systems Technical Journal*. 11:126–147.

[52] Ogata, Katsuhiko. 1990. *Modern Control Engineering*. 2nd ed. Englewood Cliffs, NJ: Prentice-Hall.

[53] Palm, William J. 1983. *Modeling, Analysis and Control of Dynamic Systems*. New York: John Wiley and Sons.

[54] Porter, B. 1987. Comments on "On undershoot and nonminimum phase zeros" by T. Norimatsu and M. Ito. *IEEE Transactions on Automatic Control*. AC-32 (3):271.

[55] Routh, E. J. 1877. *A Treatise on the Stability of a Given State of Motion*. London: Macmillan.

[56] Rynaski, E. J. 1982. Flight control synthesis using robust observers. In *Proceedings of* AIAA *Guidance and Control Conference*, September 1982, San Diego, California.

[57] Smith, O. J. M. 1958. *Feedback Control Systems*. New York: McGraw-Hill.

[58] Stein, G., and M. Athans. 1987. The LQG/LTR procedure for multivariable feedback control design. *IEEE Transactions on Automatic Control*. AC-32 (2):105–114.

[59] Thaler, George G. 1989. *Automatic Control Systems*. St. Paul, MN: West Publishing Company.

[60] Truxal, John G. 1955. *Automatic Feedback Control System Synthesis*. New York: McGraw-Hill.

[61] Wolovich, W. A. 1973. On the synthesis of multivariable systems. *IEEE Transactions on Automatic Control*. AC-18 (1):46–50.

[62] Wolovich, W. A. 1974. *Linear Multivariable Systems*. New York: Springer-Verlag.

[63] Wolovich, W. A. 1987. *Robotics: Basic Analysis and Design*. New York: Holt, Rinehart and Winston.

[64] *Xmath*. Santa Clara, CA: Integrated Systems.

[65] Youla, D. C., H. A. Jabr, and J. J. Bongiorno, Jr. 1976. Modern Wiener-Hopf design of optimal controllers—Part II: The multivariable case. *IEEE Transactions on Automatic Control*. AC-21 (3):319-338.

[66] Zames, G. 1981. Feedback and optimal sensitivity: Model reference transformations, multiplicative seminorms, and approximate inverses. *IEEE Transactions on Automatic Control*. AC-26:301–320.

[67] Ziegler, J. G., and N. B. Nichols. 1942. Optimum settings for automatic controllers. *Transactions ASME*. 64:759–768.

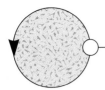

INDEX